This volume contains a set of photomicrographs of technically excellent sections of the brain of the laboratory rat. Each plate consists of matching cell and fiber stains labeled for neuronal groups and axonal tracts, respectively, with abbreviations directly on the appropriate structures and annotations explaining criteria and other problems of nomenclature for structures requiring commentary. Structures can be traced in photomicrographs in sagittal, horizontal, and transverse sections, the latter at 0.3 mm intervals, for both cell and fiber stains. The reasonable page size and sturdy binding render the book suitable for use both at the laboratory bench and as a convenient reference. The text and tabular material provide guidance for the use of this atlas of the rat brain for stereotaxic placement of electrodes or for destruction using cranial landmarks.

PHOTOGRAPHIC ATLAS
OF THE RAT BRAIN

PHOTOGRAPHIC ATLAS OF THE RAT BRAIN

THE CELL AND FIBER ARCHITECTURE ILLUSTRATED IN THREE PLANES WITH STEREOTAXIC COORDINATES

Lawrence Kruger
University of California, Los Angeles

Samuel Saporta
University of South Florida

Larry W. Swanson
University of Southern California

CAMBRIDGE
UNIVERSITY PRESS

Published by the Press Syndicate of the University of Cambridge
The Pitt Building, Trumpington Street, Cambridge CB2 1RP
40 West 20th Street, New York, NY 10011-4211, USA
10 Stamford Road, Oakleigh, Melbourne 3166, Australia

First published 1995

Printed in the United States of America

Library of Congress Cataloging-in-Publication Data
Kruger, Lawrence.
Photographic atlas of the rat brain: the cell and fiber architecture
illustrated in three planes with stereotaxic coordinates / Lawrence
Kruger, Samuel Saporta, Larry W. Swanson.
p. cm.
Includes index.
ISBN 0-521-41342-7 (hc.) – ISBN 0-521-42403-8 (pb)
1. Rats – Nervous system – Atlases. 2. Brain – Atlases. I. Saporta,
Samuel. II. Swanson, Larry W. III. Title.
QL737.R66K78   1995
599.32'33 – dc20   94-11371
                                                                          CIP

A catalog record for this book is available from the British Library.

ISBN 0-521-41342-7 Hardback
ISBN 0-521-42403-8 Paperback

CONTENTS

v

PREFACE

THIS PROJECT ORIGINATED in 1976, instigated by the frustration of successive waves of students struggling with the interpretation of serial sections through the rat brain. The driving force was exerted by an exceptionally industrious and talented undergraduate student, Sanford (Sandy) G. Feldman, and an experienced, fastidious histology technician, Sharon Sampogna. Sandy Feldman, now a practicing ophthalmologist in San Diego County, performed most of the photographic work, including the mounting of the plates; we are most indebted to him for making it possible to pursue this project, albeit many years later. Shortly after labeling of the plates began we became aware of other efforts to complete a thorough stereotaxic atlas of the rat brain, and after a visit from Dr. George Paxinos, who was preparing an atlas based on excellent frozen section material, we suspended our project.

Over the years, various colleagues turned to our photomicrographs, finding that the availability of comparable cell and fiber staining as well as closely spaced sections were indispensable for analyzing many regions of the brain. Eventually, this led us to attempt an analysis of the material in all three planes. As the labeling progressed we succumbed to pressure from our colleagues to complete the task and publish it in an accessible and convenient format. The impetus to complete the project and to add extensive notes defining and explaining the peculiarities of neuroanatomical nomenclature derived from previous collaboration by one of us (L.K.) with Sidney Landau, an expert on the idiosyncrasies of scientific dictionaries. As the editorial director of Cambridge University Press in New York, he offered guidance and encouragement that proved crucial to the final phases of this long journey.

We bear full responsibility for the compromise in the legibility of a few of the labels. This is a problem inherent in the design of such an atlas.

In the course of preparing a work of this size and complexity we have had the help of so many people that it seems impractical to name them all, but certain contributions deserve special mention. We turned to several colleagues for their neuroanatomical expertise and often lengthy arguments about how and what to name specific structures. Among the main contributors were Jeffrey Winer, Harvey Karten, Nicholas Brecha, Paul Micevych, Paul Sawchenko, and Arnold Scheibel. Perhaps a greater influence is attributable to our mentors in guiding us into serious neuroanatomical research; for this we are especially indebted to the late Jerzy E. Rose and to W. Maxwell Cowan.

Typing, preparation of abbreviation lists and bibliography, and numerous secretarial chores were performed with remarkable grace and patience by Anita Roff and David Warren. We were assisted in the labeling of plates by many students but principally by Monalisa Setudehnia and Lisa Chen. Deborah Anderson and Blanca Cervantes provided technical assistance and carried out many laborious checking procedures. Errors in a work of this kind seem inevitable and we were astonished by the enormous number of corrections required in the copyediting stage. W. M. Havighurst saved us from errors ranging from faulty syntax to conflicting abbreviations; he has been the best copy editor we've ever worked with and we are much-indebted admirers. Expert counsel and design ideas in the various efforts of computerized graphics for the covers were contributed by Patrick Gilmore and David Warren. Responsibility for the details of printing and design was assumed with expertise and excellence first by Alan Gold and then by Pauline Ireland. The project was brought successfully to the point of production with the unflagging encouragement and advice of Dr. Robin Smith of Cambridge University Press, who placed us in the able hands of production and editing manager Sophia Prybylski; their cheerfulness, warmth, and intelligence made the final stages a joyous affair.

Last but not least, we acknowledge the patience and encouragement of our wives, Ginny, Cindy, and Neely. This book is dedicated to them with our gratitude.

ABBREVIATIONS

A5	Noradrenergic Cell Group 5
AAA	Anterior Amygdaloid Area
ab	Angular Bundle
$ac_{(o,t)}$	Anterior Commissure (olfactory limb, temporal limb)
Acb	Nucleus Accumbens
AD	Anterodorsal Nucleus (Thalamus)
ADP	Anterodorsal Preoptic Nucleus (Hypothalamus)
AE	Amygdalo-Entorhinal Transition Area
$Ah_{(a,d)}$	Anterior Hypothalamic Nucleus (anterior, dorsal)
AHZ	Amygdalo-Hippocampal Area
AL	Anterior Limbic Area (Cortex)
alv	Alveus
AM	Anteromedial Nucleus (Thalamus)
AMB	Nucleus Ambiguus
AMi	Interanteromedial Nucleus (Thalamus)
$AO_{(e,l,m)}$	Anterior Olfactory Nucleus (external, lateral, medial)
aOB	Accessory Olfactory Bulb
aot	Accessory Optic Tract (basal optic root)
Ap	Anterior Pituitary (adenohypophysis)
ap	Area Postrema
Aq	Cerebral Aqueduct
Arc	Arcuate Nucleus (Hypothalamus)
Ast	Area Striata (Cortex)
Aud	Auditory Area (Cortex)
AV	Anteroventral Nucleus (Thalamus)
BI(s)	Nucleus of the Brachium Inferior Colliculus (subbrachial sector)
bic	Brachium of Inferior Colliculus
$BLA_{(a,p)}$	Basolateral Nucleus of the Amygdala (anterior, posterior)
$BMA_{(p)}$	Basomedial Nucleus of the Amygdala (posterior)
bsc	Brachium of Superior Colliculus
Bst	Bed Nucleus of Stria Terminalis

c	Central Canal
C_1	First Cervical Level of Spinal Cord
C_2	Second Cervical Level of Spinal Cord
CA_{1-3}	Hippocampal Area CA_{1-3} (Ammon's Horn)
Cbl	Cerebellum
cc	Corpus Callosum
CdP	Caudoputamen
CeA	Central Nucleus of the Amygdala
CeC	Central Cervical Nucleus
CG	Central Gray
chp	Choroid Plexus
cic	Commissure of the Inferior Colliculus
Cl	Claustrum
cl	Central Lateral Nucleus (Thalamus)
CM	Centromedial Nucleus (Centrum Medianum) (Thalamus)
cm	Central Medial Nucleus (Thalamus)
cng	Cingulum Bundle
$CO_{(d,p,v)}$	Cochlear Nucleus (dorsal, ventral posterior, ventral anterior)
co	Cochlear Nerve
$CoA_{(a,p)}$	Cortical Nucleus of the Amygdala (anterior, posterior)
cp	Cerebral Peduncle
csc	Commissure of the Superior Colliculus
cst	Corticospinal Tract
CTF	Central Tegmental Field
CU	Cuneate Nucleus
CUe	External Cuneate Nucleus
Cun	Cuneiform Nucleus
D	Nucleus of Darkschewitsch
$DB_{(h,v)}$	Nucleus of the Diagonal Band (horizontal)
db	Diagonal Band of Broca
dc	Dorsal Column
De	Dentate (Lateral) Cerebellar Nucleus
df	Dorsal Fornix
DG	Dentate Gyrus
Dh	Dorsal Horn
dhc	Dorsal Hippocampal Commissure
$DMh_{(a,p,v)}$	Dorsomedial Nucleus (anterior, posterior, ventral) (Hypothalamus)
dMR	Deep Mesencephalic Reticular Nucleus
DR	Dorsal Reticular Nucleus
dsct	Dorsal Spinocerebellar Tract
$DT_{(d,p,v)}$	Dorsal Tegmental Nucleus (dorsal, posterior, ventral)
dtd	Dorsal Tegmental Decussation
ec	External Capsule
eml	External Medullary Lamina
$ENT_{(l,m)}$	Entorhinal Area (lateral, medial)
EP	Endopiriform Nucleus

ep	Ependymal Layer of Ventricle
epl	External Plexiform Layer of Olfactory Bulb
ET	Entopeduncular Nucleus
EW	Edinger-Westphal Nucleus
Fa	Fastigial (Medial) Cerebellar Nucleus
FC	Fasciola Cinerea (gyrus Fasciolaris [NA]-Fasciola cinera cinguli, splenial gyrus)
fc	Cuneate Fascicle
fg	Gracile Fascicle
fi	Fimbria
fl	Foramen of Luschka
floc	Flocculus (Cerebellum)
fo$_{(p,pr)}$	Fornix (postcommissural, precommissural)
fr	Fasciculus Retroflexus
GcR	Gigantocellular Reticular Nucleus
Ge	Nucleus Gelatinosus (Nucleus Submedius) (Thalamus)
gl	Glomerular Layer of Olfactory Bulb
GP	Globus Pallidus (external or lateral)
GR	Gracile Nucleus
H$_1$, H$_2$	H Fields of Forel (H$_1$ and H$_2$)
Hb$_{(l,m)}$	Habenula Nucleus (lateral, medial) (Thalamus)
hc	Habenula Commissure
hf	Hippocampal Fissure
IA	Intercalated Mass of the Amygdala
IC	Inferior Colliculus
Ic	Intercalated Nucleus (Medulla)
ic	Internal Capsule
ICc	Central Nucleus of Inferior Colliculus
icp	Inferior Cerebellar Peduncle
IF	Interfascicular Nucleus
IG	Induseum Griseum
Ig	Intermediate Geniculate Nucleus (leaflet) (Thalamus)
ig	Internal Granular Layer of Olfactory Bulb
IL	Infralimbic Area (Cortex)
iml	Internal Medullary Lamina
In$_{(a,d,g)}$	Insular Area (agranular, dysgranular, granular) (Cortex)
Inc	Interstitial Nucleus of Cajal
Int	Interpositus (Intermediate or Interposed) Cerebellar Nucleus
IO	Inferior Olive
IP	Interpeduncular Nucleus
IPf	Interpeduncular Fossa
is	Infundibular Stalk
itp	Inferior Thalamic Peduncle
KF	Kölliker-Fuse Nucleus
LA	Lateral Nucleus of the Amygdala
LC	Locus Ceruleus

Lce	Lateral Cervical Nucleus
LD	Lateral Dorsal Nucleus (Thalamus)
LDT	Laterodorsal Tegmental Nucleus
LGd	Dorsal of Lateral Geniculate Nucleus (Thalamus)
LGv	Ventral of Lateral Geniculate Nucleus (Thalamus)
LHA	Lateral Hypothalamic Area
LL$_{(d,v)}$	Nucleus Lateral Lemniscus (dorsal, ventral)
ll	Lateral Lemniscus
LM	Lateral Mammillary Nucleus
Lot	Nucleus of Lateral Olfactory Tract
lot	Lateral Olfactory Tract
LP	Lateral Posterior Nucleus (Thalamus)
LPO	Lateral Preoptic Area (Hypothalamus)
LR$_{(l,m,p)}$	Lateral Reticular Nucleus (lateral, medial, posterior)
lT	Terminal Nuclei of the Accessory Optic Root (basal root)
LV	Lateral Ventricle
m5	Motor Trigeminal Nucleus
Ma	Motor Agranular Cortical Area
MaPO	Magnocellular Preoptic Nucleus (Hypothalamus)
mch	Medial Corticohypothalamic Tract
mcp	Middle Cerebellar Peduncle
McR	Magnocellular Reticular Nucleus (Medulla)
MD	Mediodorsal Nucleus (Thalamus)
MDi	Intermediodorsal Nucleus (Thalamus)
ME	Median Eminence (Hypothalamus)
MeA$_{(a,p)}$	Medial Nucleus of the Amygdala (anterior, posterior)
MePO	Median Preoptic Nucleus (Hypothalamus)
mfb	Medial Forebrain Bundle
MG	Medial Geniculate Nucleus (Thalamus)
MGm	Magnocellular Medial Geniculate Nucleus (Thalamus)
MGp	Principal/Parvicellular Medial Geniculate Nucleus (Thalamus)
mi	Mitral Layer of Olfactory Bulb
ml	Medial Lemniscus
mlf	Medial Longitudinal Fascicle
MM	Medial Mammillary Nucleus (Hypothalamus)
MP	Medial Preoptic Nucleus (Hypothalamus)
mp	Mammillary Peduncle
MPO	Medial Preoptic Area (Hypothalamus)
mr	Mammillary Recess
ms5	Mesencephalic Nucleus of the Trigeminal
ms5t	Mesencephalic Tract (root) of the Trigeminal
mT	Medial Terminal Nucleus of Accessory Optic Tract
mt	Mammillothalamic Tract
OB	Olfactory Bulb
Oc	Occipital Area (Cortex)
oc	Optic Chiasm

xii

ocb	Olivocochlear Bundle
Of	Orbitofrontal Area (Cortex)
Olep	External Periolivary Nucleus
Olsp	Superior Paraolivary Nucleus
on	Optic Nerve
Ot	Nucleus of Optic Tract
ot	Optic Tract
ov	Vascular Organ Lamina Terminalis
Pa$_{(a,p)}$	Paraventricular Nucleus (anterior, posterior) (Thalamus)
pa5	Paratrigeminal Nucleus
Pam	Periamygdaloid Cortex
PB$_{(l,m)}$	Parabrachial Nucleus (lateral, medial)
Pbg	Parabigeminal Nucleus
Pc	Paracentral Nucleus (Thalamus)
pc	Posterior Commissure
pcn	Nucleus of Posterior Commissure
PcR	Parvicellular Reticular Nucleus
Pd	Pontine Gray (Deep)
Pe$_{(a,av,p)}$	Periventricular Nucleus (anterior, anteroventral, posterior) (Hypothalamus)
Pf	Parafascicular Nucleus (Thalamus)
PgR	Paragigantocellular Reticular Nucleus
PH	Posterior Hypothalamic Nucleus
Ph	Perihypoglossal Nucleus
Pi	Pineal Gland (Epiphysis)
Pir	Piriform Area (Cortex)
PL	Posterior Limbic Area (Cortex)
PM$_{(d,v)}$	Premammillary Nucleus (dorsal, ventral) (Hypothalamus)
PmR	Paramedian Reticular Nucleus
Po	Posterior Group (Thalamus)
pp	Peripeduncular Nucleus
PpT	Pedunculopontine Tegmental Nucleus
Pr5	Principal Sensory Trigeminal Nucleus
PR$_{(c,o,v)}$	Pontine Reticular Nucleus (caudal, oral, ventral)
prf	Perforant Path
Prl	Prelimbic Area (Cortex)
PS	Parastrial Nucleus
PSCh	Suprachiasmic Preoptic Nucleus
PT	Paratenial Nucleus (Thalamus)
Pta	Anterior Pretectal Nucleus (Thalamus)
Ptm	Medial Pretectal Area (Thalamus)
Pto	Olivary Pretectal Nucleus (Thalamus)
Ptp	Posterior Pretectal Nucleus (Thalamus)
pv	Periventricular Fiber System
PVh	Paraventricular Nucleus (Hypothalamus)
py	Pyramidal Tract
pyx	Decussation of Pyramidal Tract

xiii

R	Reticular Nucleus (Thalamus)
RCh	Retrochiasmatic Area (Hypothalamus)
Rd(l)	Dorsal Raphé Nucleus (lateral extension)
Re	Nucleus Reuniens (Thalamus)
rf	Rhinal Fissure
Rh	Rhomboid Nucleus (Thalamus)
RL	Nucleus Raphé Linearis (rostral)
RLc	Nucleus Raphé Linearis (central)
RM	Nucleus Raphé Magnus
Rm	Median Raphé Nucleus
RO	Nucleus Raphé Obscurus
Rol	Nucleus of Roller
RP	Nucleus Raphé Pontis
rp	Preoptic Recess
RPa	Nucleus Raphé Pallidus
Rpm	Paramedian Raphé Nucleus
Rsp	Retrosplenial Area (Cortex)
Ru(m,p)	Red Nucleus (magnocellular, parvicellular)
S5	Supratrigeminal Nucleus
s5c	Spinal Trigeminal Nucleus (caudal)
s5i	Spinal Trigeminal Nucleus (interpolar)
s5o	Spinal Trigeminal Nucleus (oral)
Sag	Nucleus Sagulum
SC	Superior Colliculus
SCh	Suprachiasmatic Nucleus (Hypothalamus)
sco	Subcommissural Organ
scp	Superior Cerebellar Peduncle
scpx	Decussation of the Superior Cerebellar Peduncle
SF	Septofimbrial Nucleus
sfo	Subfornical Organ
Sg	Suprageniculate Nucleus (Thalamus)
sg	Substantia Gelatinosa
sgd	Deep Gray Layer of Superior Colliculus
Sge	Supragenual Nucleus
sgi	Intermediate Gray Layer of Superior Colliculus
Sgp	Suprageniculate Nucleus of Pretectal Group
sgs	Superficial Gray Layer of Superior Colliculus
SH	Septohippocampal Nucleus
SI	Substantia Innominata
sm	Stria Medullaris
SN(c,l,r)	Substantia Nigra (compact, lateral, reticular)
SO$_{(l,m)}$	Superior Olivary Complex (lateral, medial)
So	Supraoptic Nucleus (Hypothalamus)
so	Optic Layer of Superior Colliculus
soc	Supraoptic Commissure
SomS	Somatosensory Cortex
sPf	Subparafascicular Nucleus (Thalamus)

xiv

SPl$_{(d,i,v)}$	Septal Nucleus (lateral: dorsal, intermediate, ventral)
SPm	Septal Nucleus (medial)
ST(a)	Subthalamic Nucleus (accessory)
st	Stria Terminalis
STF	Striatal Fundus
Sub	Subiculum
SUM$_{(l,m)}$	Supramammillary Nucleus (lateral, medial)
sumx	Supramammillary Decussation
swd	Deep White Layer of Superior Colliculus
swi	Intermediate White Layer of Superior Colliculus
TB	Nucleus of the Trapezoid Body
tb	Trapezoid Body
tc	Thalamic Commissures
ti	Tuberoinfundibular Tract
TM	Tuberomammillary Nucleus (Hypothalamus)
TR	Tegmental Reticular Nucleus
TS$_{(c,l,m)}$	Nucleus of the Solitary Tract (commissural, lateral, medial)
Ts	Triangular Nucleus (Septum)
ts	Solitary Tract
tt	Tenia Tecta
TU	Tuberal Nucleus (Lateral Tuberal Nucleus) (Hypothalamus)
Tu	Olfactory Tubercle
V3	Third Ventricle
V4	Fourth Ventricle
VA	Ventral Anterior Nucleus (Thalamus)
VB$_{(l,m)}$	Ventrobasal Nuclear Complex (lateral, medial) (Ventral Posterior Nucleus) (Thalamus)
Vh	Ventral Horn
vhc	Ventral Hippocampal Commissure
vi	Vellum Interpositum
VL	Ventrolateral Nucleus (Thalamus)
Vl	Lateral Vestibular (Dieter's) Nucleus
VM$_{(p)}$	Ventromedial Nucleus (posterior) (Thalamus)
Vm	Medial Vestibular Nucleus
VMh	Ventromedial Nucleus (Hypothalamus)
Vn	Vomeronasal Nerve
VR	Ventral Reticular Nucleus
Vs	Superior Vestibular Nucleus
vsct	Ventral Spinocerebellar Tract
Vsp	Spinal Vestibular Nucleus
VT	Ventral Tegmental Nucleus
VTA	Ventral Tegmental Area (of Tsai)
vtd	Ventral Tegmental Decussation
X	Nucleus X
x	Needle Track
Y	Nucleus Y

Z	Nucleus Z
zb	Zuckerkandl Bundle
ZI	Zona Incerta
zo	Zonal Lamina of Superior Colliculus
3	Oculomotor Nucleus
3n	Oculomotor Nerve
4	Trochlear Nucleus
4n	Trochlear Nerve
5m	Trigeminal Nerve Root (motor or minor division)
5n	Trigeminal Nerve
5r	Trigeminal Root
5s	Trigeminal Nerve Root (sensory or major division)
5t	Trigeminal Tract
6	Abducens Nucleus
6n	Abducens Nerve
7	Facial Nucleus
7g	Genu of Facial Nerve
7n	Facial Nerve
7r	Facial Nerve Root
8n	Vestibular-Cochlear (Acoustic) Nerve
9n	Glossopharyngeal nerve
10	Dorsal Motor Nucleus of Vagus
10n	Vagus Nerve
11n	Spinal Accessory Nerve
12	Hypoglossal Nucleus
12n	Hypoglossal Nerve
12r	Hypoglossal Nerve Root

INTRODUCTION

T HERE ARE SEVERAL published stereotaxic atlases of the rat brain available, each with its own set of limitations and compromises, the present volume included. The three brains upon which this atlas is based were prepared for celloidin embedment in 1975 with high hopes of filling a void by creating an atlas capable of overcoming deficiencies that we had encountered, principally a lack of detail due to the wide spacing between sections. Contemplating a venture of this sort enabled us to evaluate a set of criteria for producing an "ideal" atlas and to determine whether one could realistically hope to achieve a result worthy of the substantial effort required.

CRITERIA FOR AN "IDEAL" ATLAS

The theoretical ideal atlas would be a complete set of perfect sections stained suitably for both cell bodies and axons with no distortions and no sections lost. Assuming such technical perfection was achievable, this would involve thousands of photomicrographs. For a variety of other reasons, it would be impossible to prepare such material. Thus we are obliged to evaluate the variety of practical alternatives currently available and to choose among them, recognizing that we are essentially yielding to a set of compromises that should be stated explicitly.

With the appearance of the excellent atlas by Paxinos and Watson in 1982 and a subsequent edition in 1986, many of our concerns were met; indeed, because of their use of frozen sections, their stereotaxic coordinates were likely to be more accurate than those of any brain subjected to the distortions consequent to embedment. Recognizing that the "ideal" atlas was much like an asymptote that can be approached but never reached, we decided to proceed for a variety of reasons that we shall attempt to outline briefly for the purpose of enabling users to evaluate the limitations inherent in their referral to this form of morphological presentation.

1

We jokingly call this work "The Ratlas" for practical reasons derived from computer file contractions and because it seems crisp and euphonious. However, Cambridge University Press felt that the slightly prosaic title, *Photographic Atlas of the Rat Brain,* more accurately informs the potential buyer. *Photographc Atlas of the Rat Brain* is intended to fill a gap for both reference and daily lab bench use by providing closely spaced cell- and fiber-stained sections that can serve as a guide for anyone trying to analyze and interpret histological material from the rat brain.

The diversity of neuroanatomical nomenclature and the failure to achieve an international standard account for much of the uncertainty attached to employment of any description or atlas. Most of the original terms derive from early studies of the human brain and presumptive homologies in various mammalian orders were based almost solely on form and relations, only recently to be supplemented by experimental studies of connections, immunoreactivity, and so forth.

This atlas is based upon fiber and cellular architecture despite our awareness that many functional subdivisions can be recognized through experimental studies. The conservative decision reflects the importance of acknowledging only what can be seen readily in "normal" material without contamination of inferential, and often multiple, criteria for naming structures. It is apparent that many tracts in the brain stem and spinal cord have been identified through degeneration studies, but unless they can be seen as distinct, encapsulated fibrous entities, we have refrained from indicating tracts of known location when their margins cannot be identified in "normal" stained material.

Cellular architecture is especially complex, and thus often controversial, but it constitutes the main guide to the naming of structures in this atlas. Cortical fields and nuclei of the brain are rarely homogeneous and, with the possible exception of the cerebellar cortex, they are usually susceptible to subdivision. Such splitting can often be justified or substantiated on functional or connectional grounds, but for purposes of consistency, adherence to the rule of relying solely upon the readily seen architectural features enables avoidance of some of the abuses and discrepancies in neuroanatomical nomenclature. There are some easily recognized subdivisions that are not indicated in this atlas for the practical reason that higher magnification would be required both to recognize and to label many subtle features. For example, it would seem pointless, as well as cluttering, to label the separate leaflets of the inferior olivary nucleus. Fine cytoarchitectonic distinctions that can be discerned at higher magnification but are not recognizable in our photomicrographs were deliberately avoided in order to protect the user from inferring misleading criteria. In essence, the rule has been: what you see is what you get.

One consequence of such a rigid position is the omission of uncontroversial distinctions. For example, a distinct component of the locus ceruleus involved in cardiovascular regulation readily separated by labeling with a specific antibody (CRF) has been recognized as "Barrington's nucleus," but examination of Nissl-stained sections does not easily reveal such distinctions. Reference to mapping principles (e.g., Swanson, '92) or to detailed comprehensive reviews such as those in the *Handbook of Chemical Neuroanatomy* provide the supportive material for a functional understanding that is not easily derived from examining Nissl-stained sections.

ATLAS AND MAPS

The deliberate decision to provide an atlas of labeled photomicrographs without accompanying outline drawings was based on some philosophical considerations and our understanding of the differences between maps, atlases, and guides. Many users require a map for plotting experimental findings in an anatomical context. It would seem simple and eminently desirable to have a universally accepted chart of the various complex entities constituting the brain, but alas, the subjectivity from which maps derive inevitably reflects the techniques, biases, and intent of the mapmaker. Thus, there are serious discrepancies among the various maps of the rat brain. Part of this problem may be due to the use of limited samples based on a fixed, relatively small number of sections. These are likely to differ significantly in different publications because the sections are quite commonly at a slightly different location and in a small brain, less than 0.1 mm can be considerable for many structures. Furthermore, maps, by their very nature, are based on explicit decisions imposed by the mapmaker, and different individuals commonly employ different criteria.

Cartography of the brain has proven elusive for other distinct reasons that will probably continue to deter acceptance of international standards. The more recent atlases generally rely upon cytoarchitectonic criteria for setting boundaries for cartography, but the historical tradition of many divisions was founded in the 19th century with the utilization of fiber architecture, often based upon Weigert-stained preparations of human brains. Many structures are most obviously delineated by fibrous capsules, but are difficult to define on strictly cytoarchitectural grounds. It should also be noted that fiber architecture is a somewhat misleading concept because it is principally based on myeloarchitecture. It is possible to consider axonal architecture by employing silver impregnation methods, but this has proven impractical for most uses. A stain for a specific axon class can be instructive, and indeed, Paxinos and Watson ('86) employ this strategy, staining for acetylcholinesterase because axons containing this enzyme are prevalent in many structures, although it is not specific for cholinergic fibers. In the current era of emphasis on chemical neuroanatomy, scores of other specific biochemical "markers" have been applied and have revealed distinctive architectural features, many of which are not easily related to the historically established features generally referred to as "classical" in contemporary parlance.

3

Although we adopt a conservative approach to nomenclature, there is no compelling need to sustain traditional naming when better alternatives become available. In practice, such changes evolve rarely and slowly. Thus, it seemed best to indicate the location of key structures on closely spaced photomicrographs, with emphasis on the transverse (coronal) axis because this is most widely used. Maps of such presentations can be constructed in the form of a series of outline drawings and this has been provided in the most recent and best atlases now available (Paxinos and Watson, '86; Swanson, '92). However, many of the outlines of nuclei are apparently discrepant with the opinions of other scientists; in some instances, there is a forced, artificial recognition of distinctions that are not recognizable as separate nuclei. This occurs whenever a cell group becomes scattered in development and is interspersed with other functional entities. There are numerous examples of nuclei consisting partly of "straggling" small neuronal aggregates that can be misleadingly outlined. The nucleus ambiguus, the salivatory nuclei, and the spinal vestibular nucleus typify the range of problem nuclei, but other examples abound and they are dealt with individually in the commentary accompanying the plates. When a nucleus cannot be readily identified in a given plane but may be inferred from sections in another plane, we have generally demurred from labeling it. In this manner we hope to avoid intimidating the user by reserving labels for what can be seen and identified with reasonable security. Some nuclei and tracts are visible but unlabeled, conforming to the dictum of "when in doubt, leave it out."

Essentially, we have elected to present an *atlas* with sufficient explanatory text to provide a *guide* from which the user can construct a *map* with outlines, nomenclature, and even abbreviations appropriate to individual need and understanding. The *commentary* is crucial for guidance because, unlike geographic atlases that provide maps at successive stages of magnification, brain atlases generally have been limited to a single low magnification presented in large format as a compromise. For detailed analysis of many structures, more closely spaced sections, higher magnification, and the invocation of multiple criteria for subdivision, the user requires some understanding of how and where to seek further information. To this end, we here provide an atlas and a guide (commentary) to the structure of the rat brain without a map; a set of maps has been presented most recently with schemata for functional generalization in large format by Swanson ('92). The page size and number of plates more closely approximate those of the atlas volume of encyclopedia sets rather than a shelf-resistant oversize-page volume such as the Rand-McNally World Atlas of maps. Each format serves the idiosyncratic needs of the individual user.

There are several obvious defects and deficiencies in this atlas, some deliberate decisions, others unavoidable. On the technical end we were fortunate in obtaining adjacent ("kissing") sections for cell and fiber strains at the same level, but there were a few mishaps with lost or damaged sections, especially with the larger horizontal and sagittal sections. Thus, there are a few lacunae, but the close spacing of sections compared with other atlases should compensate for some of the omissions. We were obliged to illustrate the rostral pole and olfactory bulb from a separate series used by Swanson ('92) for his *Brain Maps*. The small size of these sections in the transverse plane would not seem to justify the added pages and extra cost for this volume, and thus they are presented in a single plate to avoid confusion. In addition, there is little that is controversial or ambiguous in this region, and there are excellent alternative sources available, including the atlas devoted specifically to this region by Slotnick and Hersch ('80) as well as the nicely illustrated material in Paxinos and Watson ('86) and Swanson ('92).

SURGERY AND TISSUE PREPARATION

Three male Sprague-Dawley rats weighing 320, 328, and 387 grams were anesthetized with sodium pentobarbital (35 mg/kg) and placed in a stereotaxic instrument. Stainless steel insect pins were positioned in the brain through holes drilled in the calvarium and cemented in place. The incisor bar was placed 3.0 mm below the interaural line in order to position the dorsal surface of the brain in a horizontal position. With the head held firmly in place by the stereotaxic frame, careful measurements were made of the location of the midline sagittal suture, bregma (the convergence of the coronal sutures and the sagittal suture), and the interaural line indicated by a scribed line on most earbars that bisects the earbar (earbar 0). Reference marks were established in each brain with straight stainless steel insect pins (0.3 mm diameter) positioned through holes drilled in the calvarium. The pins were secured by cementing them to the calvarium with dental cement.

Reference marks for transverse sections were established with pins placed bilaterally at earbar 0, 10 mm anterior to the earbar 0 (approximately 1 mm anterior to bregma) and 11 mm anterior to the earbar 0. The latter set of pins was used to align the cerebral hemispheres in the microtome, establishing the plane of transverse sections. All pins were positioned 2 mm from the midline. To establish landmarks for sagittal sectioning, pins were positioned 2 mm rostral to the interaural line and 4 mm lateral to either side of the midline. An additional pin was inserted on one side 10 mm rostral to the 0 earbar position (and 4 mm lateral to the midline) to establish the plane of section. Landmarks for horizontal sections were established by inserting pins horizontally 8 mm above the interaural line and 2 mm lateral to either side of the midline. Two additional pins were placed in one hemisphere 4 mm from the midline, and 2 mm and 10 mm rostral to earbar 0.

The animals were removed from the stereotaxic apparatus, anesthesia was increased to a very deep level, and the animals were perfused with warm saline followed by 10% buffered formalin. The vertebral column and spinal cord were transected at the C_2–C_3 interspace. The skull and vertebral col-

5

umn, with the brain in situ, were stored in cold fixative for 36 hours. The calvarium was then opened widely without disturbing the pins and the tissue was stored in fixative for an additional two weeks. The cerebral hemispheres, brain stem, and spinal segments were removed from the skull as a unit and blocked along the set of pins specifically placed to determine plane of section, and all pins were removed.

The tissue was dehydrated through an ascending series of ethanol, ethanol-ether, and ether-nitrocellulose, and embedded in nitrocellulose. This procedure took approximately six weeks to ensure complete infiltration of the tissue. The brain was aligned in the microtome with the aid of the pin holes specifically placed to determine the plane of section, and sectioned at 20 μm. Every section was collected. A 1-in-8 series of sections was stained with cresyl violet. The immediately adjacent series of sections was stained with the Loyez fiber stain (Anderson, '29), an optimal method for visualizing thin myelinated axon bundles. When the next appropriate section in a series was damaged or unusable, an adjacent section was taken as a replacement.

Sections were photographed on Kodak 4127 and 6127 4×5 sheet film and printed as contact prints on Kodabromide papers.

The decision whether to embed brains or to cut frozen sections is one of the most serious of compromises. Frozen sections generally undergo less shrinkage and distortion than observed even in the best conditions of embedment, and consequently this yields the greatest accuracy for stereotaxic purposes. Unfortunately, it is technically overwhelming to handle fragile frozen sections without breaking and losing sections or parts of sections, and this becomes even more hazardous when the sections are thin enough for optimal staining. The losses can be compensated for by preparing several brains for each plane of section – the solution resorted to by Paxinos and Watson ('86), who used three brains for a transverse series that still contains large and irregular spatial gaps, for purposes of visualizing some structures. This would be a trivial and acceptable compromise if variability proved minimal, and this is indeed reasonable if some details and achievement of serial reconstruction can be relinquished as essential features for atlas use.

We have chosen the alternative of celloidin embedment to avoid the "mix and match" problems of interleafing sections from different specimens and in the hope that this atlas could provide the basis for an "electronic version" for computer reconstruction. Such specimens are far less fragile, rendering imperfection or loss of sections less likely. It also allowed us to collect adjacent sections for cell and fiber staining so that photomicrographs based on opposite faces of a section could be illustrated on a single page.

There are some distinct advantages to direct, simultaneous comparison of cell and fiber architecture at the same level aside from the obvious elimination of redundancy and economy of space. It allows labeling one side for tracts and the other for neuronal aggregates and results in substantially less clutter, while enabling the user to distinguish whether fiber or cellular architecture is the

6

basis for naming of structures. It also enables the comparison of different criteria for delimiting structures, a feature of profound importance in recognizing some boundaries of nuclear aggregates in the diencephalon. For example, the thalamic pretectal nuclear group and its subdivisions are more easily recognized in fiber-stained material, lending credence to cytoarchitectural distinctions that are often difficult even with optimum high-magnification Nissl-stained micrographs. Attempts to identify nuclei in micrographs of fiber-stained sections have proven cumbersome and confusing in earlier atlases, and the strategy of simultaneous employment of multiple criteria can be helpful in understanding the basis for seemingly obscure distinctions. The space saving also allows presentation of more material at higher density without increasing the number of pages or cost.

PLANE OF SECTION

One of the basic decisions in the development of any atlas is determining the plane in which sections of the brain will be cut. In aligning the brain for sagittal sections, the plane is established by the midline and is unambiguous. However, establishing a plane for transverse and horizontal sections is more difficult and, to date, a standard stereotaxic plane for transverse or horizontal sections of the rat brain has not been established. While all rat stereotaxic instruments have used fixations of the head at each external auditory meatus as two of the points necessary to establish a plane for orienting the brain, the third point, usually determined by the height of a tooth bar placed behind the upper incisors, has varied from atlas to atlas. Indeed, most stereotaxic instruments allow placement of the tooth bar over a range of at least ±10 mm from horizontal.

The first instrument designed specifically for rat stereotaxic surgery arbitrarily established a plane of orientation with the incisor bar 2.5 mm higher than the interaural line (Krieg, '46). This plane tilted the frontal pole of the brain above the occipital pole. Other atlases have varied this plane for a number of reasons. At one extreme, the atlases of de Groot ('59) and Pellegrino et al. ('79) have used a plane defined by placing the incisor bar as high as 5.0 mm above the interaural line, which tilts the head sharply upward and places the anterior commissure at the same horizontal level as the posterior commissure. At the other extreme is the atlas by Paxinos and Watson ('82), which places the incisor bar 3.3 mm below the interaural line, a "flat skull" position that gives a slight downward tilt to the frontal pole of the brain. Examples of atlases with coordinate systems between these two extremes can also be found (König and Klippel, '63; Albe-Fessard et al., '71).

The plane of section chosen for this atlas reflects our wish to establish a plane that can be accurately, consistently, and quickly reproduced when the brain is sectioned for histological analysis, as well as one that is easily used during stereotaxic surgery. The plane most easily and consistently established for transverse histological sections is one that is perpendicular to the dorsal surface of the cerebral hemisphere; or one that is parallel to this surface, for horizontal sections. This plane is approximated by setting the incisor bar 3.0 mm below horizontal 0 (i.e., 3.0 mm below the interaural line).

DISTORTION

The goal of any brain atlas is to present high-quality photographs of well-stained brain sections that have as little tissue distortion as possible. However, this goal is never realized because tissue preparation requires varying degrees of tissue manipulation. Therefore, it is important to indicate clearly the type and amount of distortion that has occurred in tissue, and how this distortion has been dealt with in the figures of the atlas.

Regardless of the manner of tissue processing, some degree of distortion invariably occurs when tissue is processed and sections are cut and mounted. Tissue preservation inevitably introduces some swelling or shrinkage of tissue regardless of the choice of preservative or the manner in which it is introduced. Embedding in wax, nitrocellulose, or plastic requires that the tissue be dehydrated through a series of solvents, which causes tissue shrinkage. Further, the tissue may not shrink equally in all dimensions, causing additional distortion. The introduction of steel needles for establishing stereotaxic coordinates provides an unmeasured increment of distortion, and although this is small and probably can be ignored, it has a deleterious influence on the staining of sections in the vicinity, especially with the Loyez method.

An unexpected source of distortion, detected only in the sagittal plane, is due to slight buckling in the long axis. This is readily evident by observing a near midline parasagittal section passing once through the hypothalamic third ventricle and again through the central canal in the upper cervical spinal cord (see Plate 100). Thus, the section encounters the midline but extends from ~0.1–0.3 mm laterally in the same sections in the entire series. For a variety of technical reasons, it is exceedingly difficult to obtain a true median sagittal section, and any deviation from flatness in cutting the midline extends to other parallel sections. This provides a source of error quite independent from that due to the flexure angle of the brain stem.

1. Fixation. Fixation will introduce some distortion of tissue regardless of the manner of fixation or fixative used. While attempting to minimize this by fixing the tissue *in situ*, some artifact is still apparent from compression of the tissue against the calvarium. The degree of distortion of various surface features caused by this compression varies from animal to animal and structure to structure. For example, distortion of the cerebral hemispheres, cerebellar hemispheres, or inferior surface of the diencephalon was minimal, while distortion of the superior and inferior colliculi was pronounced.

2. Shrinkage. Based on measurements of the known distance between various sets of pinholes, the average degree of shrinkage due to dehydration

and tissue processing for the Nissl material was 29% ± 3%. It was apparent during these calculations that the degree of shrinkage in the anterior-to-posterior (28.6%), medial-to-lateral (28.4%), and dorsal-to-ventral (31.2%) planes was not equal. We have adjusted the scales in each of these dimensions to reflect accurately this inequality. The degree of shrinkage in the cresyl violet series was slightly less than that in the Loyez series. Separate sets of scales reflecting these differences are provided for each Nissl- and Loyez-stained photomicrograph. Dorsoventral shrinkage was estimated by measuring the depth from the surface of the brain to the base at the two anterior (alignment) pins and then measuring the same position on slide-mounted sections through these pinholes. This measurement implies a symmetrical shrinkage in all directions determined by the nature of the embedment and its response to aqueous solvents used in tissue processing. The effect of immersion on tissue section dimensions is generally ignored, but it is readily evident in the adjacent pairs of fiber- and cell-stained sections of this atlas, all of which display consistent but different shrinkage ratios. In addition, there is some distortion due to knife compression that is most pronounced in thin sections when they are uncurled and mounted on slides. In order to avoid dorso-ventral compression, the transverse series was cut from side to side.

ACCURACY OF THE STEREOTAXIC COORDINATES

The accuracy with which stereotaxic scales can be placed on coronal, sagittal, and horizontal sections from different animals is subject to a great deal of variation. In addition to such obvious differences as the degree of tissue shrinkage from brain to brain as well as between different axes of the same brain noted above, and the normal variation that may occur in the brains of animals of the same species, distortion of the brain during the placement of pins and slight variation in the placement of the animal's head in the stereotaxic apparatus will introduce error in the location of a nucleus or fiber tract deep within the brain. Some variation in stereotaxic coordinates is, therefore, to be expected in comparing the location of a structure in three planes of section. The most accurate alignment of structures occurs at the level of interaural 0, and 10 mm rostral to interaural 0, where pins were always placed to determine these levels unambiguously. The degree of error in the anterior-to-posterior direction increases with the distance from these points.

One of the greatest difficulties in establishing accurate stereotaxic coordinates lies in precisely and consistently defining a starting point from which to measure in an anterior-to-posterior (A–P) or a medial-to-lateral (M–L) direction in each animal. The sagittal suture is usually tortuous in its course and some degree of judgment must be used in establishing midline 0 for M–L measurements from this suture. Likewise, the junction of the two coronal sutures and the sagittal suture – bregma – may not always be a single well-defined point. It is not uncommon for the coronal sutures to meet at two

9

different points on the sagittal suture or for small sutural (wormian) bones to be present at the junction of the coronal and sagittal sutures, requiring estimation of the position of bregma. Therefore, while the most convenient and widely used reference point of measurement for rat stereotaxic surgery is bregma because it establishes both A–P zero and the midline with one measurement on the dorsal surface of the skull, it is often an ambiguous stereotaxic landmark. If care is taken in estimating bregma, it may be a very accurate reference point for structures within the telencephalon (Slotnick and Brown, '80). However, the most consistent point from which to establish a stereotaxic reference in the A–P plane is the theoretical line passing between the midline of the external auditory meatus of each side, the interaural line. This line, usually scribed on the earbar, is referred to in this atlas as earbar 0 because it is an approximation of the true interaural line.

This point was used as the A–P zero reference in all the brains used in this atlas, though it is the most inconvenient stereotaxic reference point to establish in practice. There was excellent agreement, however, in the distance of bregma from earbar 0. The average value of bregma for the animals used in this atlas was found to be 9.0 ± 0.1 mm rostral to earbar 0 (see also Paxinos et al., '85). The midline was always estimated from the sagittal suture. Additional differences, such as the degree of tissue shrinkage from brain to brain, the normal variation that may occur in the brains of animals of the same species, distortion of the brain during the placement of pins, and slight variation in the placement of the animal's head in the stereotaxic apparatus due to inaccuracies introduced when inserting the earbars or variations in the formation of the external auditory meatus (see Slotnick and Brown, '80), will introduce additional error in the location of a nucleus or fiber tract deep within the brain. Thus, some variation in stereotaxic coordinates is to be expected in comparing the location of a structure in different animals or planes of section.

In assessing the accuracy of our stereotaxic coordinate system, we compared the stereotaxic location of 40 structures (nuclei and fiber tracts) in the diencephalon and brain stem in the three sets of sections presented in this atlas. Differences in the position of nearly all structures in the three planes were less than 0.5 mm. The position of some nuclei could be appreciated more easily in one particular plane of section, making accurate comparison of its location in all three planes of section difficult. In these instances, there was satisfactory agreement in stereotaxic coordinates in two of the three planes of section.

METHODOLOGY

Standardization of a coordinate system and measurement of error and variance is difficult with a sample limited to one specimen for each of the three principal planes. As noted, consistency for several easily measurable landmarks such as the boundaries of encapsulated tracts (e.g., anterior and poste-rior commissures, corpus callosum, etc.) indicates that the accuracy of local-ization is usually significantly better than 0.5 mm. However, comparison of a given nucleus in all three planes may yield larger discrepancies. For exam-ple, the rostral limit of the parvicellular sector of the red nucleus is relatively easy to identify in horizontal and sagittal sections but quite difficult to detect in transverse sections. We have deliberately resisted the temptation to adjust our criteria for recognition of difficult boundaries by *post hoc* inferences derived from sections in the other axes in the belief that conservative esti-mates would be more useful than forcing artificial and uncertain recognition factors.

We were interested not only in the accuracy of our own stereotaxic coordi-nates, but in how they compared with other published coordinate systems. To date, the most accurate set of stereotaxic coordinates for adult rats has been published by Paxinos and Watson ('86) based on material from Wistar rats. In comparing our coordinates with those of Paxinos and Watson, it became apparent that a discrepancy of nearly 1 mm in the A–P location of structures appeared rather quickly in the midbrain and that this A–P shift persisted throughout the remainder of the brain stem.

Because our material is based on Sprague-Dawley rats, we wondered if a strain difference existed in these two types of rat. A series of 12 Wistar rats was compared with 23 Sprague-Dawley rats to determine if differences ex-isted in the basic stereotaxic measurements of these animals. Animals that were to be used for other experiments were anesthetized with sodium pento-barbital and placed in a stereotaxic instrument. The skull was exposed and cleaned, and the foramen magnum enlarged to expose clearly the obex. An insect pin was placed in the electrode holder of the stereotaxic instrument and used to determine the position of bregma, earbar 0, and the obex. Figure 1 is a graph plotting the distance from bregma to the obex in Sprague-Dawley and Wistar rats of various weights. The rate at which the central nervous system increases in length in these two strains of rat appears comparable, though the Sprague-Dawley rat consistently has a central nervous system that is ~0.5 mm longer in this A–P axis than the Wistar rat for animals of the same weight. When the distance from bregma to earbar 0 is examined (Figure 2), there is essentially no significant difference between the two strains of rat. Therefore, the difference in the length of the brain seen in Figure 1 must be due to a difference in the distance from earbar 0 to the obex in the two strains of rat. That is, the brain stem of the Sprague-Dawley rat must be longer than that of the Wistar. Figure 3 confirms that the distance from earbar 0 to the obex is indeed ~0.5 mm greater for the Sprague-Dawley than for the Wistar rat. This difference was not found in earlier craniometric measurements of a variety of rats (Paxinos et al., '85), though the magnitude of deviation from the mean for the distance from earbar 0 to the nucleus of the seventh cranial nerve fell within this degree of brain stem lengthening for the Sprague-

11

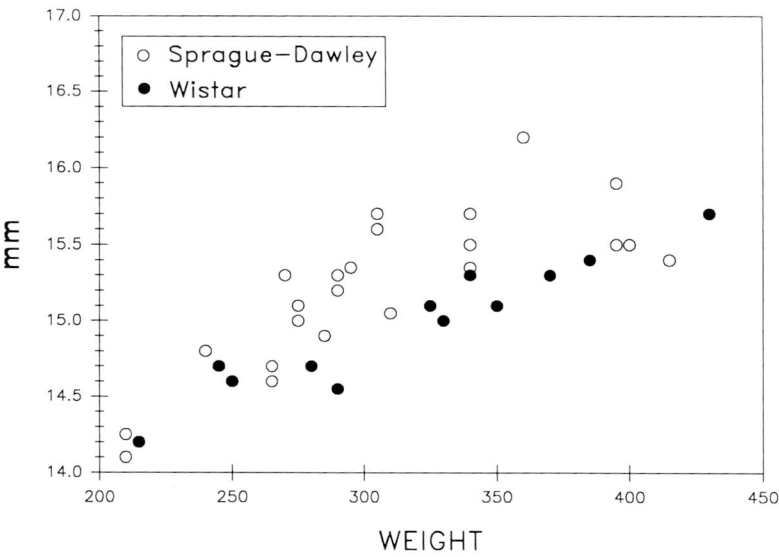

Fig. 1. The rate at which the central nervous system increases in length in adolescent and adult Sprague-Dawley and Wistar rats in relation to a skull stereotaxic landmark, bregma, is comparable. However, the Sprague-Dawley rat's central nervous system is consistently about 0.5 mm longer than that of the Wistar rat for animals of the same weight (in grams).

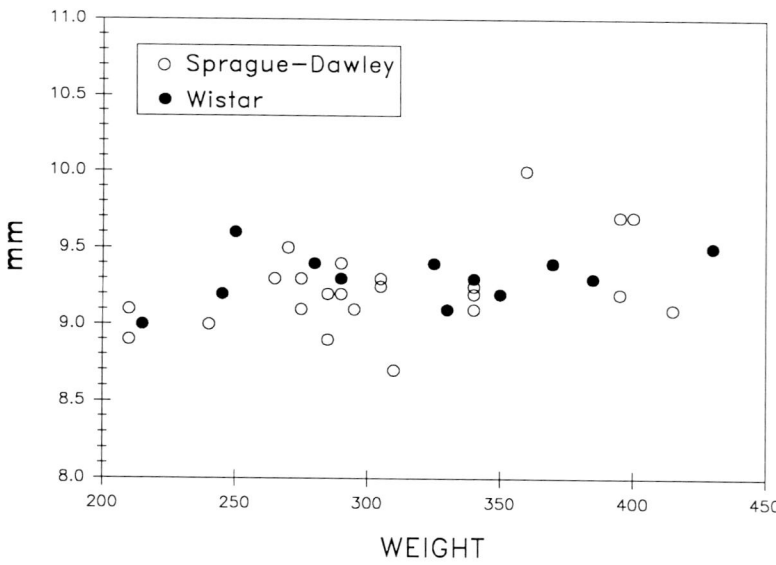

Fig. 2. The distance from bregma to earbar 0 is similar in the Sprague-Dawley and Wistar rat for animals of the same weight (in grams).

12

DISTANCE FROM EARBAR 0 TO OBEX

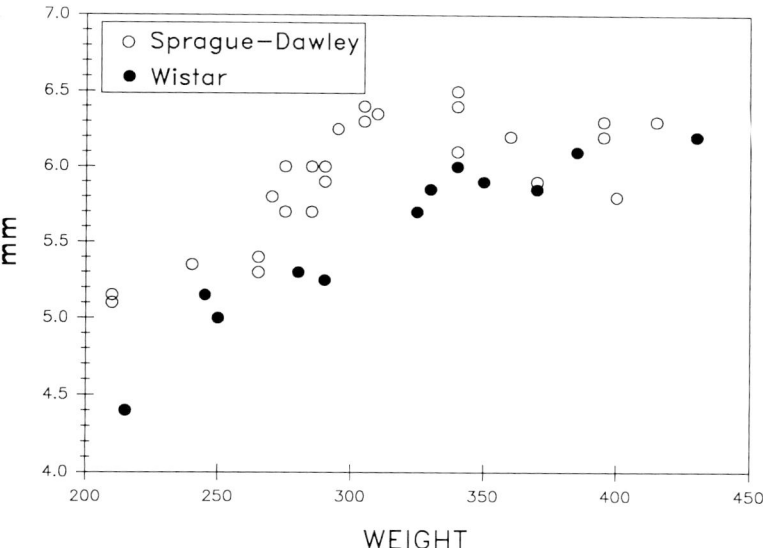

Fig. 3. The distance from earbar 0 to the obex is greater for the Sprague-Dawley rat than for the Wistar rat for animals of the same weight (in grams). This increase in the length of the Sprague-Dawley brain stem is approximately 0.5 mm.

Dawley as compared to the Wistar rat. Based on measurements of the distance between nuclei in various regions of the brain stem, the increase in length of the brain stem of Sprague-Dawley rats occurs primarily between −5 mm and −9 mm caudal to bregma, within the midbrain.

However, this strain variation does not account for all the difference seen in the A–P brain stem coordinates of the Sprague-Dawley as compared to the Wistar rat. The difference in the location of some structures within the ventral midbrain appears to be greater than structures located in the dorsal midbrain. As one compares the degree of cephalic flexure (i.e., the angle formed by the axis of the brain stem and the diencephalon) of sagittal sections taken from these two strains of animal, the brains used in this atlas appear to have a less obtuse angle (lie flatter) than those used by Paxinos and Watson. The different degree of cephalic flexure between Wistar (137°) and Sprague-Dawley (141°) rats may be an additional difference between these two strains. Alternatively, one may speculate that this difference in cephalic flexure may be not entirely a strain difference, but may also reflect some degree of tissue distortion. The weight of the brain resting on its ventral surface in a series of solutions for eight weeks during fixation and embedding may be sufficient to flatten the cephalic flexure, or at least contribute to the degree to which the cephalic flexure is made less acute. One can imagine a pivot point for this deformation somewhere in the dorsal aspect of the rostral midbrain, with deformation

13

spreading both rostrally and caudally. The brain's bulk rostral to this point, toward the diencephalon, could allow a slight degree of deformation, but significant deformation of this parenchyma would be unlikely. Caudally, however, the brain stem would be sufficiently pliant to render more significant deformation possible. This would predict that a greater alteration in A–P coordinates would occur within the midbrain and rostral pons, rather than rostral toward the diencephalon. Whatever the reason for this decreased flexure, however, the results would be the same. As would be expected from deformation around a dorsally located pivot, the location of structures deep within the midbrain of the Sprague-Dawley rat would be expected to shift caudally to a greater degree than more dorsally located structures. Measurements in the A–P plane made 4 mm below the surface of the midbrain may be altered by as much as 0.6 mm caudally merely due to the flattening of the cephalic flexure, while measurements made 1 mm below its surface will vary by 0.2 mm.

STEREOTAXIC SCALES

Two sets of anterior-to-posterior scales have been included on each photomicrograph for the Nissl-stained sections. One scale is based on measurements from the interaural (earbar) line. The second is based on measurements from bregma, which is 9.0 mm rostral to interaural 0. Since the most widely used stereotaxic reference is bregma, all A–P scales in the three planes of section are based on this reference. However, the location of earbar 0, the A–P reference point originally used in this material, is also indicated on each plate. Measurements from bregma are given in the lower right corner; the equivalent measurement from earbar 0 is given in the lower left corner of each transverse plate for the convenience of the user. Similarly, a line labeled earbar 0 is extended on the A–P scales of sagittal and horizontal sections indicating its location. The medial-to-lateral scale uses the midline as the 0 reference. The scale for depth on transverse and sagittal plates has been measured from the tip of the earbars (the interaural line) to the surface of the cortex, because the sloping contour of the cerebral cortex makes it difficult to present a scale starting at the cortical surface.

An additional scale, labeled P&W, is provided on each plate. This scale is an attempt to correlate the differences in stereotaxic coordinates inherent when using the Wistar rat, as presented in the atlas of Paxinos and Watson ('86), with those of the Sprague-Dawley rat used in this atlas, as discussed above. On transverse sections, from +1.8 mm to −5.1 mm from bregma, one figure is presented at the top left of the plate that refers to an approximately equivalent plate in the Paxinos and Watson atlas. Because of the uneven spacing of their sections, there are some instances when we estimated the closest section as the best approximate fit. However, beyond this point, the difference in the cephalic flexure of the two strains of rat and the

14

difference in the length of the midbrain resulted in a sufficiently different angle of cut through the midbrain and pons in our material from −5.4 mm to −12.3 mm that it was not possible to specify a single equivalent plate. Instead, we chose to designate two planes on these sections. The top number refers to the equivalent position of the tectum or the dorsal area of the tegmentum in the Paxinos and Watson atlas, while the bottom number refers to the equivalent position of the ventral tegmentum. The remainder of the sections through the medulla can again be referenced to one plate in the Paxinos and Watson atlas since the orientation of sections is again approximated. Paxinos and Watson do not present sections posterior to −15.9 mm from bregma. Additionally, their use of several brains for the frontal planes compounds the inaccuracy of scaling.

In sagittal sections, a P&W scale has been placed beneath the brain section. This scale begins at the midbrain-diencephalic junction and extends caudally past the obex. However, rather than a linear scale that reflects the slight differences in head position used in the two atlases, the differences in the cephalic flexure and the length of the rostral midbrain in these two strains of rat made it necessary to construct a nonlinear scale to account for differences in the A–P coordinates of specific nuclei. Corrections to this scale due to the change in angle of the cephalic flexure were calculated for a dorsal–ventral position 2 mm above earbar 0, a depth passing through a maximum amount of brain stem. The pivot point of the cephalic flexure appears to be near the midbrain-diencephalic junction. The lengthening of the midbrain also appears to occur in this general vicinity and was calculated from the caudal shift of the motor and principal sensory trigeminal nuclei, the superior olivary nucleus and the facial nucleus in the Sprague-Dawley rat. The greatest A–P distortion of the scale, therefore, appears at its beginning segment with a decreasing A–P alteration caudally through the midbrain. Caudal to the midbrain the scale is linear, since once past the midbrain the distance between nuclei in the remainder of the brain stem remains remarkably consistent in both strains of rat. For example, the distance between the end of the motor nucleus of 5 and the beginning of the facial nucleus in transverse sections is 0.6 mm in both the Sprague-Dawley and Wistar rat, and the distance from the end of the facial nucleus to the beginning of the inferior olivary complex, as judged from transverse sections, varies by less than 0.1 mm in the two strains of rat. The very slight change in the dorsal–ventral position of nuclei that would be expected from the more acute cephalic flexure of the Wistar rat has been ignored.

The P&W scale on horizontal sections is similar to that for sagittal sections. We have assumed that the degree of midbrain lengthening is consistent at all D–V levels. However, the apparent difference in the cephalic flexure of the two strains of animal would not distort the midbrain to the same degree at each horizontal level. Sections near the dorsal surface of the midbrain would be expected to have less distortion than deeper sections. Therefore, the magnitude of distortion, because of differences in the cephalic flexure, has been calculated for each horizontal plane. Plate 90 corresponds to the scale dimensions presented in sagittal sections. Sections progressively dorsal to Plate 90 require decreasing amounts of correction for the change in cephalic flexure, while sections ventral to Plate 90 require increasing correction of the initial segments of the scale.

NOMENCLATURE

Steadfast rules of biological taxonomy have been violated periodically in the history of neuroanatomy, sufficiently to provoke a suspicion that basic principles of nomenclature are lacking. The history of unsuccessful attempts to achieve accord through international panels of experts or modernization of the *Nomina Anatomica* illustrates the difficulty and magnitude of the task. This is partly a circumstance determined by building on names derived initially from early gross dissections of the human brain, upon which terms drawn from characteristics of stained histological sections have been superimposed. Our aim in applying terms in this atlas has been to portray some sense of common usage, and when decisions could not be applied comfortably, we have attempted to confess our uncertainty and indecisiveness. This approach imposes the necessity of explaining the nature of alternatives in a manner that cannot be found in a list of approximately synonymous terms. In so doing, the user can often render his or her own judgment of the wisdom or foolishness of our decisions as well as those of our predecessors. Some names maintain only a tenuous hold on their position, or even existence, on many of the plates, and we invite each reader to challenge even those names that we did not recognize as deserving of tentative status. Clearly, we cannot claim comprehensive scholarship, and deliberate omissions will not be distinguishable from inadvertent ones. Nor have we attempted to explain the rationale for rejecting some terms and descriptions.

Examples abound throughout; we hope that they provide more than simply economical use of available blank space on the pages listing abbreviations and definitions. If it were indeed possible to apply terms without explanation, we could simply follow the lead of precedent atlases, most notably the excellent, detailed and stereotaxically accurate atlas for the same species by Paxinos and Watson. As might be expected, there is little disagreement concerning the naming of most structures between our material and that of others. At times we have deliberately resisted the temptation to use a term we prefer, so as not to add complexity and cause confusion. We have assumed that trespasses, especially of current usage, require an explanation and defense.

We draw on one complicated example here because it requires more than the space available on an abbreviation page, and also because it enables us to provide an explanation that should help the reader to choose among alternatives that may differ from the labels applied to the photographs in this book. The ventral tier of thalamic nuclei is probably the region most frequently bearing inconsistent nomenclature. The entire aggregate constitutes the ventral nuclear *group*, within which one can easily discern collections of neurons of similar architecture called *nuclei*. The largest of these *nuclei* in the rat thalamus is recipient of the largest somatosensory ascending tract, the medial lemniscus, and electrophysiological mapping has established that this *nucleus* constitutes the principal tactile region of the thalamus. Such studies have also revealed the topography of the cutaneous projection onto this nucleus in representative species of several mammalian Orders. Although the same basic *pattern* prevails in each species examined, discrete subdivision is clearly evident in some species. Thus, in some advanced primates, the arm, leg, and face representations are separated by distinct fibrous laminae, thereby splitting a single functional entity into separate components. Such components have variously been given distinct names based on positional relations or been awarded the status of *subnuclei* within a nuclear *complex*.

For most primate brains, the tradition has been to designate medial and lateral sectors of the ventral posterior (VP) "nucleus" as separate nuclei, ventral posterior lateral (VPL) and ventral posterior medial (VPM). In rodent and lagomorph brains, the fibrous lamina separation is less pronounced in the functionally homologous region, which had been designated the *ventrobasal* (VB) "complex" in the late 19th century. It has been argued that the VB usage has historical priority as well as general mammalian applicability (e.g., Rose and Mountcastle, '52), but the VP nomenclature is probably the more popular contemporary term. However, this adds further burdensome consequences because it is obvious that the medial "nucleus" (VPM) consists of two architecturally distinct entities: the lateral portion, containing larger neurons, and the medial portion, containing smaller, densely packed neurons. The small-celled division, generally called the parvicellular portion (VPMpc), was recognized as a separate entity on the basis of its distinctive fiber architecture and in earlier descriptions was usually called the *arcuate nucleus* (because of its shape), principally in primate brains.

The two strikingly different components of what is generally called the ventral posterior medial nucleus (VPM) probably constitute one of the best examples of nomenclature gone awry, because there is little doubt that VPMpc deserves separate functional status and it is universally agreed that the lateral (or external) portion of VPM resembles and belongs with VPL as a functional nuclear complex. For practical purposes, the tactile thalamic nucleus recipient of the medial lemniscus input is easily recognizable and only its rostral boundary is variously described in different accounts. Whether its subdivisions are placed into VP or VB contexts is of little consequence when these terms are precisely synonymous.

The parvicellular (pc) portion of the medial component enjoys a more complicated nuclear status. An extensive scholarly account outlining the problematic issues is beyond the scope or pretensions of this atlas, but the

17

reasons for failure to achieve a generally accepted nomenclature for this region may provide valuable insight into the process and limitations of morphological methods.

Standard textbook surveys usually assign this medial pc nucleus of the ventral group to the gustatory system, and early designations of "arcuate" based on myeloarchitecture or VPMpc based on cytoarchitecture are easily recognized. The medial, dorsal, and ventral limits of the pc nucleus remain problematic, and there are numerous seemingly discrepant descriptions in the mammalian thalamus generally and one problem specific to the rodent thalamus. The ventral portion of "VPMpc" in rat thalamus has been described as possessing a medial and a lateral sector ("VPLpc") containing a "viscerotopic" organization (Cechetto and Saper, '87) distinct from the mediodorsal "gustatory" sector, which is apparently quite small and has not been precisely delimited by experimental methods (see Kruger and Mantyh, '89, for review). The dorsal limit of "VPMpc" has not been consistently separated from the several nuclei of the "intralaminar wing" within the internal medullary lamina, a portion of which is known to be recipient of "spinothalamic" tract input in several mammals (but is relatively small in the rat thalamus), for which the designation *nucleus submedius* has become current usage (Craig and Burton, '81). The rat thalamus contains a distinctive medial nucleus generally called *nucleus gelatinosus*, which some authors have suggested is the homologue of nucleus submedius. It is more likely that only the dorsal portion is recipient of spinothalamic fibers; thus the status of this nucleus, in terms of homology with other species, remains uncertain.

Nomenclature based on topographic descriptors dependent upon other structures has been avoided to the extent that this would not exacerbate uncertainty. For example, such terms as *retrorubral nucleus* and *subincertal nucleus* are not assigned so that these nuclear aggregates are not subsumed under the red nucleus or zona incerta because such a relation has not been established. Similarly, the accessory nuclei of the cranial nerve nuclei (e.g., the accessory hypoglossal and retrofacial nuclei) generally are not specifically identified because this can only be achieved properly by examining closely spaced serial sections and at higher magnification than in these illustrations (see Székely and Matesz, '82). Other examples include the cell groups identified by positional descriptors in relation to the gigantocellular reticular nucleus called "paragigantocellular . . . ," the retrotrapezoid nucleus and the parapyramidal nucleus. Our failure to indicate these aggregates does not imply that their existence is in doubt, rather it reflects frugal assignment of names on positional grounds when functional or connectional criteria for separate designations may be lacking or deficient.

There are some general terms in current use that are a source of confusion to the nonspecialist; they require only brief comment for orientation purposes:

Corpus striatum and basal ganglia. Those portions of the telencephalon that do not migrate outward to form a cortex (or pallium) in development are commonly subsumed under the general term *basal ganglia*, which, in turn, is often used synonymously with *corpus striatum* or simply *striatum*. In the strict sense, the striped or "striated" appearance of the fiber bundles piercing the human caudate, putamen, and globus pallidus is the source of the striatal usage, but the striations are less pronounced in some basal nuclei and seem far-fetched as a descriptor in the amygdala, for example. Thus, all basal nuclei generically called *basal ganglia* should not be subsumed under the umbrella of *striatum*, although the common practice of using these terms interchangeably (e.g., Cajal, '09) is deeply embedded in the neuroanatomical literature.

In modern usage, the term *corpus striatum* might generally refer to the striatum and pallidum only (Graybiel and Ragsdale, '79), with dorsal and ventral parts of the corpus striatum. Broadly speaking, the dorsal corpus striatum is related preferentially to isocortical (neocortical, hemispheric) areas, whereas the ventral corpus striatum is related preferentially to allocortical (archi- and paleocortical, limbic) areas. Major parts of the former include the globus pallidus and caudoputamen (dorsal pallidum and striatum, respectively); while major parts of the latter are usually said to include the substantia innominata and nucleus accumbens (ventral pallidum and ventral striatum, respectively). The organizational principles underlying various schemes and nomenclature as well as bibliographical sources can be found in Swanson ('92). In the rat it is difficult to recognize the marked striations and the separation of putamen and caudate easily evident in the human brain.

Cerebral cortex. This is especially difficult to present in a consistent and useful manner for a variety of reasons, ranging from the nature of criteria employed in cyto- and myeloarchitectonic descriptions and maps to the technical requirements of staining for cortex differing from that of deeper structures for optimal analysis. There are several extensive, detailed, and well-illustrated atlases, reconstructions, and descriptions of the rat cerebral cortex. These range from the widely used illustrated account by Krieg ('46a and b), based essentially on the Brodmann nomenclature system, to modern detailed presentations by Zilles and Wree ('85), Schober ('86), and Kolb and Tees ('90). The photomicrographs of this atlas are unsuited for justifiable application of any of these accounts, and we were uncomfortable with attempting to force ourselves into a framework that is subject to extensive debate and criticism. We have compromised by labeling main functional areas that are consistent with electrophysiological maps and experimental studies of connections without succumbing to indicating limits of specific cortical fields.

Raphé nuclei. The definition of raphé nuclei is problematic. From a strictly anatomical point of view, the brain-stem *raphé* refers to the midline region of the reticular formation where the rapidly growing ventromedial walls of the neural tube fuse (ventral to the aqueduct and fourth ventricle) during development, in a region equivalent to the ventral median fissure of the spinal cord. On the other hand, the Falck-Hillarp method revealed a generally unappreciated series of midline nuclei with many serotonergic neurons (Dahlström and Fuxe, '64) often referred to as the raphé nuclei (for review, see Steinbusch and Nieuwenhuys, '83). The problem is complicated by the observation that some parts of the raphé (e.g., in the midbrain) contain few if any serotonergic neurons, and that many serotonergic neurons are scattered through adjacent parts of the brain-stem reticular formation and are thus not associated with any clearly recognizable nuclei, a feature that often leads to indistinct boundaries for most of the raphé nuclei. A modern, systematic, cytoarchitectonic analysis of the raphé nuclei was carried out by Meessen and Olszewski ('49) in the rabbit, and this has been largely adopted in human (Olszewski and Baxter, '54), cat (Taber et al., '60), and rat (Valverde, '62; Wünscher et al., '65).

Eponyms. Avoidance of eponyms has been the rule for a variety of reasons, principally because most are derived from human neuroanatomy, but also to avoid the problems attached to establishing priority or precedent. In some cases, when the name is used with reasonable frequency — for example, *Deiter's nucleus* for the lateral vestibular nucleus — we list the eponym, but it would seem misleading to apply the names of Schwalbe for the medial and Bechterew for the superior vestibular nuclei to the rat brain where the separation is far less obvious than in Weigert-stained sections of the human brain stem.

Abbreviations. The selection of readily identifiable, brief, and uncumbersome abbreviations is an ideal that is difficult to achieve but crucially important. Precedent is an important factor, and long-established common usage should prevail whenever possible, provided it does not conflict irreconcilably with sound taxonomic principles. Nuclei are designated by group and an uppercase first letter, and tracts by an initial lowercase letter, each followed by subgroups and modifiers. This requires some flexibility for practical purposes in dealing with multiple, and often conflicting, variables.

For nuclei, the general rule is to begin with the group and then the subdivision. Thus the anterior thalamic group is all A followed by its three subnuclei: AD, AM, and AV. Regional modifiers are avoided because the location would indicate it is thalamic and not easily confused with another similar designation — for example, the anteroventral thalamic nucleus and the anteroventral subnucleus of the cochlear nuclei. The ventromedial nucleus of the thalamus, hypothalamus, and vestibular complex required arbitrary solution in choice of uppercase and lowercase and acquiescence to a regional descriptor, a practice irregularly employed in some atlases. The medial and lateral geniculate bodies are designated MG and LG and their subdivisions by lowercase modifiers. Thus, the dorsal lateral geniculate nucleus becomes LGd, and similarly, the midline component of the antero-medial nucleus (interanteromedial originally) is AMi.

When confronted with choosing between English and Latin usage, we have attempted to use English, except where this conflicts with long-established practice. Thus, the fasciculus retroflexus is fr because the angli-cized version is rarely employed. The eponym of Meynert frequently at-tached to this tract is avoided, but in those cases where eponyms alone are in common use, we have employed them – for example, the nucleus of Edinger-Westphal. Subnuclei and subsectors of nuclei are generally lower-case modifiers and subscripts, respectively, unless a different common us-age prevails.

An example follows: the raphé nuclei as a group are designated by an R, and each subdivision by a modifier (e.g., dorsal and thus Rd); but if separate subsectors of the dorsal raphé nucleus must be noted, subscripts may be applied. Extra letters are sometimes included to facilitate identification and memorization (e.g., RM for raphé magnus). Unfortunately, violations seemed unavoidable: the amygdaloid nuclei were not abbreviated first by group but are indicated by an A preceded by commonly used preceding modifiers; thus the central amygdaloid nucleus is CeA. Similarly, the anterodorsal preoptic nucleus is abbreviated ADP, thereby conforming to commonly accepted usage for hypothalamic nuclei (e.g., Swanson, '92), especially in such re-gions where there are several conflicting systems of nomenclature.

When a regional modifier is required, as when a nucleus of the hypothala-mus might be confounded with one with the same name elsewhere, the nucleus is abbreviated with a lowercase h, or if it is an area of the hypothala-mus, with an uppercase H.

Spelling adheres to common contemporary practice in the United States and is largely consistent with British usage, except that e is preferred to ae in such terms as tenia (rather than taenia) and anesthesia (rather than anaesthesia).

SIZE, COST, AND SPACE UTILIZATION

Several atlases provide photomicrographs of both sides of the brain. This has the advantage of depicting some of the variability relating to the exact level of a section, especially when the two sides are not identical. On the other hand, when symmetry is maintained accurately, as in the Paxinos and Watson atlas, the redundancy doubles the page size without adding information. The large page format results in a cumbersome volume for use, shelving, and transport, as well as contributing to increased cost. The binding on oversize volumes rarely survives heavy use, and page tears are difficult to avoid.

For purposes of efficient use of space, we have combined photomicro-graphs of adjacent sections stained for cells and fibers rather than alternating them. This also allows less cluttered lettering, with nuclear groups labeled on one side and fiber tracts on the other. This strategy has enabled a substantial

21

reduction in the size and number of pages while increasing their informational content.

MAGNIFICATION AND DENSITY OF SAMPLING

The major deficiency of most atlases is the small number of sections through a given structure and the occasional omission of some features because they are not encountered within the sections selected for illustration. We therefore decided to provide closer spacing in the transverse (coronal) plane by photographing an adjacent cell and fiber-stained section at approximately 300 μm intervals; that is, exceeding six photomicrographs for three levels per mm. The principle of employing adjacent cell and fiber-stained sections is maintained for the three principal axes but with greater spacing for the horizontal and sagittal planes.

The choice of magnification involves some difficult decisions as well as some obvious ones. Maximum utilization of a full standard page size was the first aim. A photographic negative at a magnification suitable for contact prints provides the highest quality images. The sections in this atlas were photographed to fill the largest area of the negative and we were able to print the transverse and horizontal sections as contact prints. We have compromised for slight magnification in the photo offset process in the sagittal sections without recognizing very significant image degradation due to "empty" magnification. This was largely based on accommodation to page and linear scale dimensions. It should be noted that sections in each plane are presented at slightly different magnifications.

The disadvantage in the size of these illustrations is that the details of many structures are lost or are too closely spaced for labeling. On the other hand, even a two- or threefold increase would also often be inadequate. For detailed high-magnification analysis of many structures, more closely spaced sections as well as greater magnification would be necessary for satisfactory depiction. Recognizing this need, we decided to provide relevant reference information and commentary for those structures that cannot be analyzed suitably with a uniform-size presentation. It is our intention to collect a set of closely spaced higher magnification micrographs of selected structures to be placed on the unused space of the facing page containing the abbreviation list in a future edition. For stereotaxic purposes, the inherent inaccuracy of the coordinates precludes the usefulness of most detailed subdivisions. We believe that we can rectify this shortcoming in the future without compromising the convenience of the size and format used here.

Amputation of the rostral pole of the hemispheres in order to establish the plane of section in the microtome for the transverse series eliminated the olfactory bulb. In order to compensate for this deficiency, we have included in Plate 1 a series of eight Nissl-stained transverse sections obtained from another specimen that was prepared for and used in Swanson's ('92) *Brain*

Maps. The stereotaxic coordinates derive from that source. Although fiber-stained matching sections are lacking for this region, the only distinctive tract at this level, the lateral olfactory tract (lot), can be distinguished readily from the underlying molecular (zonal) layer of the anterior olfactory nucleus and piriform cortex by the density of glia, and this is easily traced in the ensuing fiber-stained sections by its heavily myelinated composition.

BOUNDARIES

The limits of many structures are often striking at low magnification, but there are numerous circumstances where high magnification cannot resolve the problems inherent in attempting to impose boundaries in regions of gradual gradients of change or, in the case of fiber architecture, where there is no way of recognizing functionally distinct contiguous tracts. The criteria for drawing lines between neural structures often differ for individual investigators, and in current practice, boundaries have been refined on the basis of cytoarchitecture, fiber architecture, connections, and chemoarchitecture. One can find numerous examples where the same authors have vacillated or even used different names for the same structures on the same photomicrographs. There are dozens of examples of such imprecision in studies of the rat brain published by distinguished neuroscientists.

The final product in each description thus is determined by the flexibility, and even correctness, of criteria employed. When these are not explicitly stated, the reader loses a sense of how to evaluate a wide range of opinions often lacking congruence. In these circumstances, drawing boundaries within gradients can be unwarranted and when forced distinctions are made, this can be a disservice. We have deliberately rejected the strategy of outlining structures in a set of drawings in the belief that the major separations should be evident in the photomicrographs. At higher magnifications and with employment of more numerous sections as well as other criteria, many other distinctions can be identified with reasonable certainty, but that must be postponed for a future, more extensive edition. We await the development of algorithms for three-dimensional reconstruction of major brain structures in a manner outlined by Toga et al. ('89). Feedback from colleagues concerning the designations of this photographic presentation should provide guidance for the outline drawings necessary for a useful electronic version based on the brain sections of this atlas. For practical purposes, the resolution achieved with digitized images is not yet sufficient to match even photo offset reproduction of photomicrographs.

Plate 1

AO$_e$ Anterior Olfactory Nucleus (external)
AO$_l$ Anterior Olfactory Nucleus (lateral)
AO$_m$ Anterior Olfactory Nucleus (medial)
aOB Accessory Olfactory Bulb
EP Endopiriform Nucleus
epl External Plexiform Layer of Olfactory Bulb
gl Glomerular Layer of Olfactory Bulb
ig Internal Granular Layer of Olfactory Bulb
IL Infralimbic Area (Cortex)
lot Lateral Olfactory Tract
Ma Motor Agranular Cortical Area
mi Mitral Layer of Olfactory Bulb
OB Olfactory Bulb
Of Orbitofrontal Area (Cortex)
on Optic Nerve
Pir Piriform Area (Cortex)
tt Tenia Tecta
Tu Olfactory Tubercle
Vn Vomeronasal Nerve

The scale coordinates on the photographic plates are based on referents of the midline, bregma, and earbar location (see pages 9–16 for explanation). The P & W scale refers to corresponding coordinates of the Paxinos and Watson ('86) atlas, which is based on brains fixed with the cephalic flexure displaced downward compared to this atlas for which the brains were removed and allowed to harden with the brain stem elevated due to the base of the brain lying flat in the fixative solution vessel.

Note. As noted in the introductory comments, the rostral pole of the hemispheres was amputated in order to establish the plane of the frontal pole and the microtome knife; this resulted in the loss of the plane of olfactory bulb from the series. Another series of sections prepared for Swanson's *Brain Maps* ('92) proved suitable for a representation of this portion of the brain, although it is tilted about 10° from the plane of the remainder of the transverse sections and it lacks comparable fiber-stained sections. Establishing stereotaxic coordinates is especially difficult because the angle of each atlas differs, but fortunately a detailed stereotaxic analysis of this region has been provided by Slotnick and Hersch ('80), who recognized the special requirements and usefulness of establishing a coordinate system for the olfactory bulb and related frontal region.

AO$_{e,l,m}$: Anterior Olfactory Nucleus. This "nucleus" is an olfactory cortical area embedded in the caudal portion of the main olfactory bulb in the olfactory peduncle. It displays a molecular layer and a cellular (pyramidal) layer (Price, '87). On topographic grounds the cellular layers can be divided into external, lateral, and medial components (see Haberly and Price, '78). Some argument endures concerning whether the deeper parallel leaflets of neurons constitute a nucleus or deserve cortical status. The medial part of the nucleus ends caudally beneath the tenia tecta.

IL: Infralimbic Area (Cortex). See AL: Anterior Limbic Area (Plate 12).

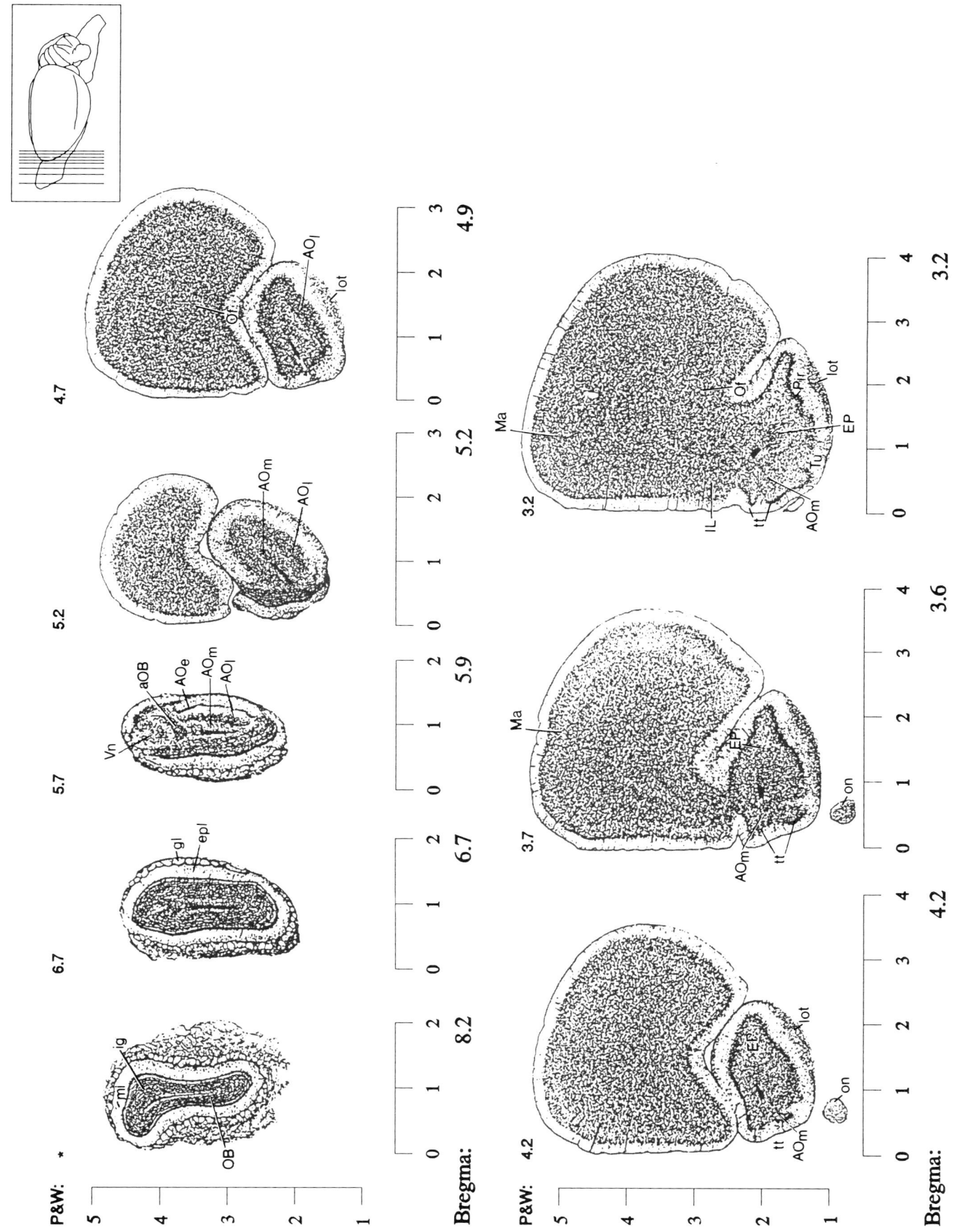

Plate 2

ac Anterior Commissure
Acb Nucleus Accumbens
AL Anterior Limbic Area (Cortex)
CdP Caudoputamen
Cl Claustrum
ec External Capsule
EP Endopiriform Nucleus
ep Ependymal Layer of Ventricle
IL Infralimbic Area (Cortex)
In$_a$ Insular Area (agranular) (Cortex)
In$_d$ Insular Area (dysgranular) (Cortex)
In$_g$ Insular Area (granular) (Cortex)
lot Lateral Olfactory Tract
Ma Motor Agranular Cortical Area
mfb Medial Forebrain Bundle
on Optic Nerve
Pir Piriform Area (Cortex)
Prl Prelimbic Area (Cortex)
rf Rhinal Fissure
SH Septohippocampal Nucleus
SomS Somatosensory Cortex
STF Striatal Fundus
tt Tenia Tecta
Tu Olfactory Tubercle

Ina: Agranular Insula. This cortical field constitutes the isocortical boundary with the rhinal fissure. The term is misleading in that it does not totally lack a granular layer, which dissipates ventrally, forming "dysgranular" and "agranular" sectors progressively (Cechetto and Saper, '87). The term *insula*, meaning "island," refers to its buried position within the depths of the Sylvian fissure in the human brain, but in all mammals this cortical field can be identified by its location overlying the claustrum. The agranular and dysgranular portions contain the gustatory projection (Kosar et al., '86), and the granular sector constitutes the main viscerosensory projection (see review of Kruger and Mantyh, '89).

tt: Tenia (or Taenia) Tecta. Accounts of the tenia tecta and induseum griseum have a long, complex history. Usage is still inconsistent, and both terms have been employed for the thin dorsal and anterior extensions of the hippocampus sometimes called the *hippocampal rudiment*. The nomenclature adopted here is based on the observation that the induseum griseum (of modern accounts) extends rostrally around the genu of the corpus callosum, along the rostral border of the lateral septal nucleus, whereas the tenia tecta displays different morphology, lies more ventral, and is intimately associated with the anterior olfactory nucleus (see Davis et al., '78; Wyss and Sripanidkulchai, '83; Swanson, '92). The part of the induseum griseum extending around the genu was called the *dorsal part of the taenia tecta* by Haberly and Price ('78) and labeled *taenia tecta* by Paxinos and Watson ('86). The taenia tecta of the present account was called *ventral taenia tecta* by Haberly and Price ('78), who divided it into superior (dorsal) and inferior (ventral) parts.

P&W 1.7

Bregma
1.8

Earbar 0
10.8

Plate 3

ac Anterior Commissure
Acb Nucleus Accumbens
AL Anterior Limbic Area (Cortex)
cc Corpus Callosum
CdP Caudoputamen
Cl Claustrum
cng Cingulum Bundle
DB Nucleus of the Diagonal Band
ec External Capsule
EP Endopiriform Nucleus
ep Ependymal Layer of Ventricle
IG Induseum Griseum
In$_a$ Insular Area (agranular) (Cortex)
In$_d$ Insular Area (dysgranular) (Cortex)
In$_g$ Insular Area (granular) (Cortex)
lot Lateral Olfactory Tract
Ma Motor Agranular Cortical Area
mfb Medial Forebrain Bundle
on Optic Nerve
Pir Piriform Area (Cortex)
rf Rhinal Fissure
SH Septohippocampal Nucleus
SomS Somatosensory Cortex
SPl Septal Nucleus (lateral)
STF Striatal Fundus
tt Tenia Tecta
Tu Olfactory Tubercle

Cl: Claustrum. This structure underlying the rhinal fissure generally defines the overlying insular cortex in mammals and is sometimes thought to be part of the basal ganglia, an interpretation supported by its broad projection upon the isocortex (Narkiewicz, '65). M. Rose ('35) gave it a separate status, on ontogenetic grounds of its later migration, from the overlying insular field and considered the region a "bicortex," a view shared by some modern authors who suggest that it may be the homologue of cortical layer 6b (Divac et al., '87; Valverde et al., '89; Swanson, '92).

Tu: Olfactory Tubercle (also called Tuberculum Olfactorium). This unusual olfactory cortical area (see Price, '73) corresponds to the anterior perforated space of human anatomy. It is generally regarded as a three-layered cortex with a thick molecular layer, a thin highly convoluted layer of modified pyramidal cells, and a thin polymorphic layer that lies adjacent to the ventral extension of the substantia innominata (see Millhouse and Heimer, '84; Swanson, '92). The cellular layer of this cortex contains dense neuronal clusters called the *islands of Calleja*.

P&W 1.6

Bregma
1.5

Earbar 0
10.5

Plate 4

ac   Anterior Commissure
Acb   Nucleus Accumbens
AL   Anterior Limbic Area (Cortex)
cc   Corpus Callosum
CdP   Caudoputamen
Cl   Claustrum
cng   Cingulum Bundle
DB   Nucleus of the Diagonal Band
ec   External Capsule
EP   Endopiriform Nucleus
ep   Ependymal Layer of Ventricle
IG   Induseum Griseum
In   Insular Area (Cortex)
lot   Lateral Olfactory Tract
LV   Lateral Ventricle
Ma   Motor Agranular Cortical Area
mfb   Medial Forebrain Bundle
on   Optic Nerve
Pir   Piriform Area (Cortex)
rf   Rhinal Fissure
SH   Septohippocampal Nucleus
SomS Somatosensory Cortex
SPl   Septal Nucleus (lateral)
STF   Striatal Fundus
tt   Tenia Tecta
Tu   Olfactory Tubercle
x   Needle Track
zb   Zuckerkandl Bundle

cng: Cingulum. An anteroposteriorly oriented bundle of myelinated axons in the white matter of the cingulate gyrus of the human brain following the medial curve of the corpus callosum was considered a belt or cincture, hence the name *cingulum*. The homologous bundle in the rat brain is not easily distinguished from the other tracts in the white matter underlying the medial cortex, so that the entire zone of supracallosal white matter is called *cingulum* by some authors (Swanson, '92), although it undoubtedly contains the superior longitudinal (arcuate) fascicle.

STF: Striatal Fundus. The ventrolateral region of the striatum just dorsal or deep to the olfactory tubercle and accompanying the substantia innominata is considerably more heterogeneous cytoarchitectonically than the caudoputamen, or even the nucleus accumbens. This general region has been referred to sporadically in the recent literature as the *fundus of the striatum (fundus striati* of Heimer, '72), and seems to correspond to the *substriatal gray of* Crosby and Humphrey ('41). Although its exact boundaries and connections have not been established, it has occasionally been regarded as part of the nucleus accumbens.

P&W 1.2

Bregma
1.2

Earbar 0
10.2

Plate 5

ac Anterior Commissure
Acb Nucleus Accumbens
AL Anterior Limbic Area (Cortex)
cc Corpus Callosum
CdP Caudoputamen
Cl Claustrum
cng Cingulum Bundle
DB Nucleus of the Diagonal Band
db Diagonal Band of Broca
ec External Capsule
ep Ependymal Layer of Ventricle
IG Induseum Griseum
In Insular Area (Cortex)
lot Lateral Olfactory Tract
Ma Motor Agranular Cortical Area
mfb Medial Forebrain Bundle
on Optic Nerve
Pir Piriform Area (Cortex)
rf Rhinal Fissure
SH Septohippocampal Nucleus
SomS Somatosensory Cortex
SPl_d Septal Nucleus (lateral: dorsal)
SPl_i Septal Nucleus (lateral: intermediate)
SPl_v Septal Nucleus (lateral: ventral)
SPm Septal Nucleus (medial)
STF Striatal Fundus
Tu Olfactory Tubercle
x Needle Track
zb Zuckerkandl Bundle

Acb: Nucleus Accumbens. The rostromedial head of the basal ganglia contiguous with the caudoputamen is difficult to separate on cytoarchitectural grounds and is commonly considered a single complex (see Heimer et al., '85) or a part of the "ventral striatum." The sector medial to the lateral ventricle is often designated Acb in many species, avoiding indication of a sharp ventrolateral boundary (e.g., Swanson, '92), but others (e.g., Paxinos and Watson, '86) recognize a densely packed "core" surrounding the base of the lateral ventricle and a radiating surround. Such distinction probably must await the imposition of other criteria (see Chronister et al., '81).

SH: Septohippocampal Nucleus. While the connections of this long, thin cell group are unknown, it has long been recognized (Johnston, '13) as stretching between the hippocampal formation caudally and the tenia tecta rostrally; Fox ('40) even suggested that it might correspond to a subcallosal equivalent of the induseum griseum (that is, a component of the primordial hippocampus).

P&W 1.0

Bregma
0.9

Earbar 0
9.9

Plate 6

Pir: Piriform Area (Cortex) (also called Prepyriform or Praepiriform Cortex). This term has been applied in various fashions and derives from the gross anatomy of the human brain, in which the rostral pole of the hippocampal (or, more commonly now, parahippocampal) gyrus (or lobe) was called "pear-shaped," referring to the bulbous expansion of the amygdala. The overlying cortex is distinctive in possessing three layers and in containing the lateral olfactory tract in the superficial layer. It is tempting to follow Cajal's reasoning; he called this area *sphenoidal olfactory cortex*, but early 20th-century neuroanatomists followed Brodmann ('09) in calling the portion overlying the amygdala the *periamygdaloid cortex*, and the contiguous, architecturally similar cortex rostral to the amygdala, the *prepyriform cortex*. The British school substituted the more logical spelling of *piriform* (from Latin *pirium*) and some authors have used the term *periamygdaloid* cortices. Arguing this issue in the rat brain, where there is nothing even remotely resembling a pear, seems a fruitless exercise. We offer no defense in acquiescing to the traditionally accepted homology. *Piriform* in this atlas corresponds to the continental European *prepyriform cortex*, and we refer to that portion of piriform cortex overlying the amygdala as *periamygdaloid* (Pam) as a gesture of conformity. The term *olfactory cortex* for both might be a felicitous substitute but for olfactory tract termination in other zones such as the distinctive olfactory tubercle (Tu) and anterior olfactory nucleus. A modern account of this region (see Haberly and Price, '78; Price et al., '87) should be consulted for useful photomicrographs and additional references.

ac	Anterior Commissure
Acb	Nucleus Accumbens
AL	Anterior Limbic Area (Cortex)
Bst	Bed Nucleus of Stria Terminalis
cc	Corpus Callosum
CdP	Caudoputamen
Cl	Claustrum
cng	Cingulum Bundle
DB	Nucleus of the Diagonal Band
db	Diagonal Band of Broca
ec	External Capsule
EP	Endopiriform Nucleus
ep	Ependymal Layer of Ventricle
IG	Induseum Griseum
In	Insular Area (Cortex)
lot	Lateral Olfactory Tract
LV	Lateral Ventricle
Ma	Motor Agranular Cortical Area
mfb	Medial Forebrain Bundle
oc	Optic Chiasm
Pir	Piriform Area (Cortex)
rf	Rhinal Fissure
SH	Septohippocampal Nucleus
SI	Substantia Innominata
SomS	Somatosensory Cortex
SPl$_d$	Septal Nucleus (lateral: dorsal)
SPl$_i$	Septal Nucleus (lateral: intermediate)
SPl$_v$	Septal Nucleus (lateral: ventral)
SPm	Septal Nucleus (medial)
STF	Striatal Fundus
Tu	Olfactory Tubercle
zb	Zuckerkandl Bundle

P&W 0.7

Bregma
0.6

Earbar 0
9.6

9
8
7
6
5
4
3
2
1
0

6
5
4
3
2
1
0

Plate 7

ac_o Anterior Commissure (olfactory limb)
AL Anterior Limbic Area (Cortex)
Bst Bed Nucleus of Stria Terminalis
cc Corpus Callosum
CdP Caudoputamen
Cl Claustrum
cng Cingulum Bundle
DB Nucleus of the Diagonal Band
db Diagonal Band of Broca
ec External Capsule
EP Endopiriform Nucleus
ep Ependymal Layer of Ventricle
IG Induseum Griseum
In Insular Area (Cortex)
lot Lateral Olfactory Tract
Ma Motor Agranular Cortical Area
mfb Medial Forebrain Bundle
oc Optic Chiasm
Pir Piriform Area (Cortex)
rf Rhinal Fissure
SH Septohippocampal Nucleus
SI Substantia Innominata
SomS Somatosensory Cortex
SPl_d Septal Nucleus (lateral: dorsal)
SPl_i Septal Nucleus (lateral: intermediate)
SPl_v Septal Nucleus (lateral: ventral)
SPm Septal Nucleus (medial)
STF Striatal Fundus
Tu Olfactory Tubercle
zb Zuckerkandl Bundle

SPm: Medial Septal Nucleus and (DB) Nucleus of the Diagonal Band. There is no clear boundary between these two separately named nuclei, which together provide a roughly topographic cholinergic and GABAergic innervation of the hippocampal formation (see Gritti et. al., '93), and in a broad sense are thus part of the larger basal forebrain region providing such innervation to the entire cortical mantle (this region also includes the substantia innominata, magnocellular preoptic nucleus, and lateral preoptic area). The nucleus of the diagonal band has been divided for descriptive purposes into vertical and horizontal limbs (Raisman, '66). An arbitrary boundary between the medial septal and diagonal band nuclei is usually drawn at the level of a lateral expansion of the complex that lies at about the level of the anterior commissure. Unfortunately, Price and Powell ('70) referred to the well-established hypothalamic magnocellular preoptic nucleus (MaPO) as the horizontal limb of the nucleus of the diagonal band, although it does not project to the hippocampal formation (see MaPO: Magnocellular Preoptic Nucleus at Plate 10).

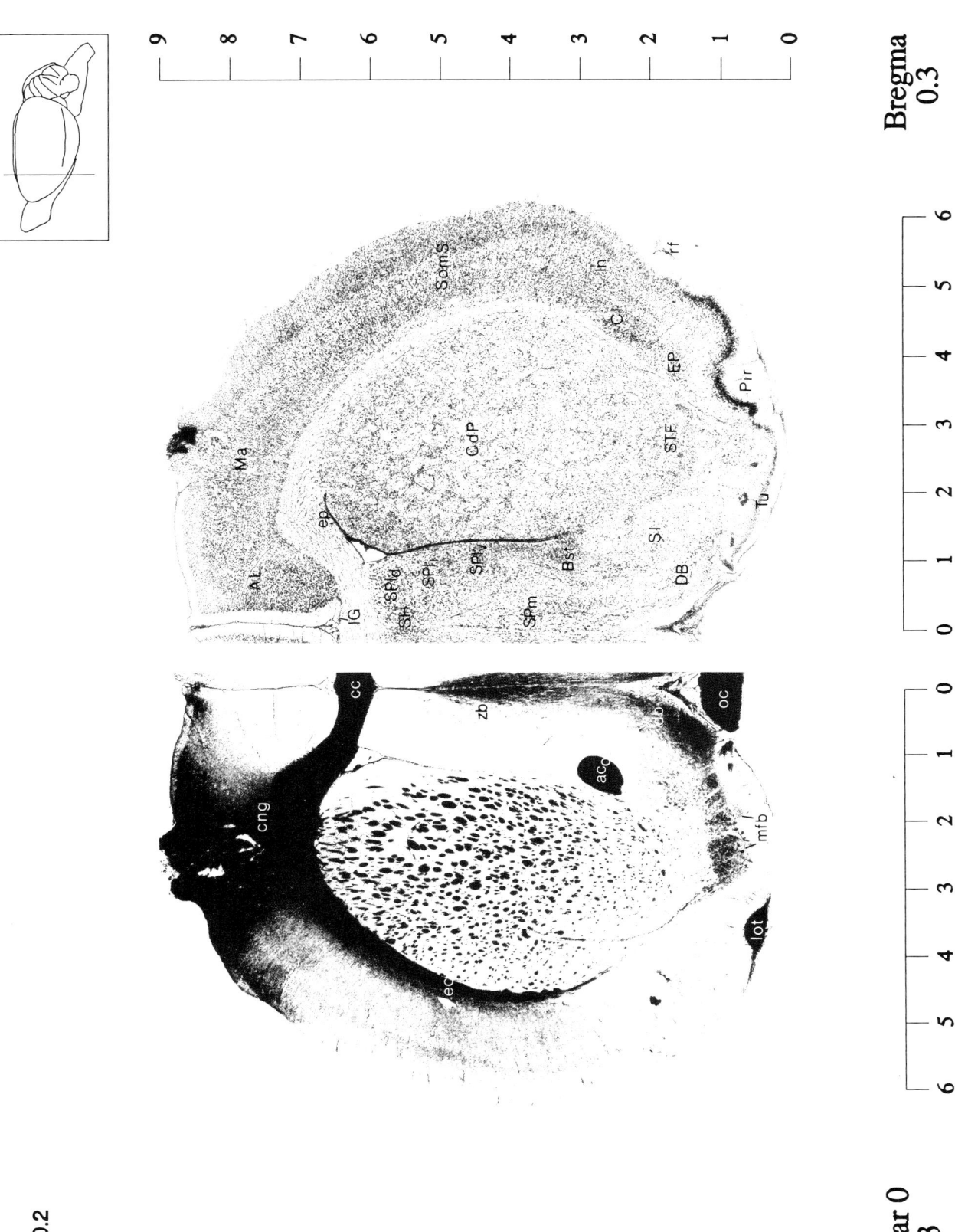

P&W 0.2

Bregma
0.3

Earbar 0
9.3

SomS

In

Cl

EP

Pir

STF

Tu

Ma

CdP

SI

ep

DB

AL

SPi

SPd

SPv

Bst

IG

StH

SPm

cc

zb

cng

aco

mfb

lot

oc

Plate 8

ov: Vascular Organ of the Lamina Terminalis (also called the Organum Vasculosum Vasculosum Laminae Terminalis). A thin, midline, circumventricular organ lacking a blood–brain barrier, resembling the area postrema in being highly vascularized – a feature that renders it most easily recognizable (see illustrations by Yamaguchi et al., '93). It is difficult to recognize in routinely stained rat brain because it is very small in this species, but by careful examination it can be seen just dorsal to the rostral tip of the optic chiasm.

SF: Septofimbrial Nucleus. This nucleus resembles a caudal extension of the lateral septal nucleus embedded in the precommissural fornix, but connections of the two cell groups apparently differ. Swanson and Cowan ('79) group the SF and the triangular nucleus of the septum (Ts) as the posterior septal nuclei.

SPl$_{d,i,v}$: Lateral Septal Nucleus (dorsal, intermediate, and ventral). This nucleus can be subdivided into dorsal and ventral sectors adjacent to the ventricles and a large, loosely packed medial intermediate sector (Swanson and Cowan, '79), but boundaries are not evident in either fiber or cell architecture. The ventral sector and the anterodorsal preoptic nucleus (ADP) are designated the *septohypothalamic nucleus* (SHy) in Paxinos and Watson ('86); we do not employ this usage here because of the prominent cytoarchitectonic difference between these nuclei in our material.

zb and db: Zuckerkandl's Bundle and Diagonal Band of Broca. These fibers are concentrated in medial parts of the septal region. Zuckerkandl's original description dealt with a vast system of fibers in what are now referred to as the *medial septal nucleus and nucleus of the diagonal band*, whereas Broca's description emphasized ventral parts of the system in what is now known as the *nucleus of the diagonal band (of Broca)*. Fibers in the vicinity of the medial septal nucleus are referred to here as *Zuckerkandl's bundle*, and those in the vicinity of the nucleus of the diagonal band as the *diagonal band of Broca*.

ac$_o$	Anterior Commissure (olfactory limb)
AL	Anterior Limbic Area (Cortex)
Bst	Bed Nucleus of Stria Terminalis
cc	Corpus Callosum
CdP	Caudoputamen
Cl	Claustrum
cng	Cingulum Bundle
DB	Nucleus of the Diagonal Band
db	Diagonal Band of Broca
ec	External Capsule
EP	Endopiriform Nucleus
ep	Ependymal Layer of Ventricle
fo$_{pr}$	Fornix (precommissural)
IG	Induseum Griseum
In	Insular Area (Cortex)
lot	Lateral Olfactory Tract
LPO	Lateral Preoptic Area (Hypothalamus)
LV	Lateral Ventricle
Ma	Motor Agranular Cortical Area
MePO	Median Preoptic Nucleus (Hypothalamus)
mfb	Medial Forebrain Bundle
oc	Optic Chiasm
ov	Vascular Organ Lamina Terminalis
Pe$_{av}$	Periventricular Nucleus (anteroventral) (Hypothalamus)
Pir	Piriform Area (Cortex)
rf	Rhinal Fissure
rp	Preoptic Recess
SF	Septofimbrial Nucleus
SH	Septohippocampal Nucleus
SI	Substantia Innominata
SomS	Somatosensory Cortex
SPl$_d$	Septal Nucleus (lateral: dorsal)
SPl$_i$	Septal Nucleus (lateral: intermediate)
SPl$_v$	Septal Nucleus (lateral: ventral)
SPm	Septal Nucleus (medial)
STF	Striatal Fundus
Tu	Olfactory Tubercle
zb	Zuckerkandl Bundle

P&W 0.2

Bregma
0.0

Earbar 0
9.0

9 8 7 6 5 4 3 2 1 0

6 5 4 3 2 1 0   0 1 2 3 4 5 6

9 8 7 6 5 4 3 2 1 0

SomS
fn
Cl
EP
Pir
STF
SI
Tu
DB
Bst
LPO
Peav
MePO
rp
ov
Ma
ep
LV
SHd
SH
SHi
SF
SPm
SPiv
SPi
IG
AL
cc
opr
zb
aco
mfb
oc
db
mfb
cng
ec
lot
CdP

Plate 9

ac Anterior Commissure
ADP Anterodorsal Preoptic Nucleus (Hypothalamus)
AL Anterior Limbic Area (Cortex)
Bst Bed Nucleus of Stria Terminalis
cc Corpus Callosum
CdP Caudoputamen
Cl Claustrum
DB Nucleus of the Diagonal Band
ec External Capsule
EP Endopiriform Nucleus
ep Ependymal Layer of Ventricle
fo$_p$ Fornix (postcommissural)
fo$_{pr}$ Fornix (precommissural)
ic Internal Capsule
IG Induseum Griseum
In Insular Area (Cortex)
lot Lateral Olfactory Tract
LPO Lateral Preoptic Area (Hypothalamus)
Ma Motor Agranular Cortical Area
MaPO Magnocellular Preoptic Nucleus (Hypothalamus)
MePO Median Preoptic Nucleus (Hypothalamus)
mfb Medial Forebrain Bundle
MPO Medial Preoptic Area (Hypothalamus)
oc Optic Chiasm
Pe$_{av}$ Periventricular Nucleus (anteroventral) (Hypothalamus)
Pir Piriform Area (Cortex)
PS Parastrial Nucleus
PSCh Suprachiasmatic Preoptic Nucleus
rf Rhinal Fissure
SF Septofimbrial Nucleus
SH Septohippocampal Nucleus
SI Substantia Innominata
So Supraoptic Nucleus (Hypothalamus)
SomS Somatosensory Cortex
SPI$_d$ Septal Nucleus (lateral: dorsal)
SPI$_i$ Septal Nucleus (lateral: intermediate)
SPI$_v$ Septal Nucleus (lateral: ventral)
STF Striatal Fundus
Tu Olfactory Tubercle

MPO: Medial Preoptic Area (Hypothalamus). This regional designation is the source of great confusion consequent to diverse schemes of nomenclature. The MPO lies between the nucleus of the diagonal band and the anterior hypothalamic area, is dominated by the large medial preoptic nucleus (see MP: Medial Preoptic Nucleus at Plate 10), and is recipient via the stria terminalis of vomeronasal organ input through the medial amygdala and portions of the bed nucleus of the stria terminalis. It has been widely studied for sexual dimorphism (see review by Segovia and Guillamon, '93). Neurons in this region concentrate gonadal steroids. The larger nuclear volume in the MPO and anterior hypothalamic area of female rats have been known since Dorner and Staudt ('68) (see also PS: Parastrial Nucleus at Plate 10). Gorski et al. ('78) then noted a small dense subgroup in the MPO that was larger in males and called it the *sexually dimorphic nucleus* (SDN) of the preoptic area. It was then recognized by Simerly et al. ('84) that the central part of the medial preoptic nucleus, which they call MPNc (and is the MPOC of Paxinos and Watson, '86), is significantly larger in males. Although they follow Gurdjian's ('27) description, they expand upon it by recognizing anterodorsal, anteroventral and posterior ventral preoptic nuclei as well. The issue became more complicated with a further revision by Bloch and Gorski ('88), who argued that sexual dimorphism is most significant in the anteroventral region of the medial preoptic nucleus, which is "virtually absent in females." Bleier et al. ('82) provide still another schema for both nomenclature and sexual dimorphism. Under the circumstances, we are reluctant to recommend any nomenclature and employ a conservative compromise that should be regarded as tentative. A final note of warning to the uninitiated is the unfortunate common practice of recognizing a medial preoptic nucleus (MP), a median preoptic nucleus (MePO), and a magnocellular preoptic nucleus (MaPO). These are functionally distinct entities.

PSCh: Suprachiasmatic Preoptic Nucleus. This thin band of cells lies along the optic chiasm between the anteroventral periventricular and suprachiasmatic nuclei. It was included together with the anteroventral periventricular nucleus by Bleier et al. ('79) in the medial preoptic nucleus, whereas the name used here was proposed by Simerly et al. ('84) to distinguish it from the traditional medial preoptic nucleus (and anteroventral periventricular nucleus).

P&W -0.3

9
8
7
6
5
4
3
2
1
0

Bregma
-0.3

6
5
4
3
2
1
0

SomS
In
C
EP
STF
Pir
MaPO
Tu
Ma
AL
GP
CdP
Bst
SI
So DB
ADP PS
LPO
MPO
PSCh
MePO
Pe av
IG
SH SPld
SPt
SPl
SF

cc
fopr
fop
ac
C
mfb
oc
lot
ec

Earbar 0
8.7

0
1
2
3
4
5
6

9
8
7
6
5
4
3
2
1
0

Plate 10

MaPO: Magnocellular Preoptic Nucleus (Hypothalamus). This distinct group of large neurons was named by Loo ('31) and is a part of the magnocellular basal cell group that supplies cholinergic fibers specifically to the olfactory bulb. It is cytoarchitecturally and connectionally distinct from the medially adjacent lateral tip of the nucleus of the diagonal band, which is another component of the magnocellular basal group that sends cholinergic fibers to the hippocampal formation. Some confusion has resulted from Price and Powell ('70), who designated the horizontal limb of the diagonal band, the MaPO.

MP: Medial Preoptic Nucleus (Hypothalamus). This nucleus is rather prominent in mammals like rats with a large vomeronasal system. It was named by Gurdjian ('27), and this designation has been used by most workers since, although Bleier and her colleagues ('79) referred to it as the *anterior hypothalamic nucleus*. The MP as a whole may be divided into three major parts (medial, lateral, and central), all of which are sexually dimorphic (Simerly et al., '84), with the central part corresponding to the sexually dimorphic nucleus of the preoptic area (Gorski et al., '78). The borders of the MP are relatively clear in Nissl preparations, except dorsolaterally where it merges almost imperceptibly with the bed nuclei of the stria terminalis.

PS: Parastrial Nucleus. In 1973, Raisman and Field showed that the distribution of synapses in a triangular region of low cell density in what they called the strial part of the preoptic area is sexually dimorphic. Between this area and the bed nuclei of the stria terminalis they also described a round nucleus, which Simerly et al. ('84) combined with a caudally adjacent lens-shaped group of fusiform cells and named the *parastrial nucleus*.

Ts: Triangular Nucleus (Septum). This nucleus at the caudal end of the septum was identified by Cajal ('01). It is occasionally confused with the much smaller dorsal tip of the median preoptic nucleus (e.g., König and Klippel, '63), which lies dorsal to the anterior commissure in the midline. The triangular nucleus has been thought of as a bed nucleus of the ventral hippocampal commissure (e.g., Gurdjian, '25) and is grouped with the septofimbrial nucleus (SF) as posterior septal nuclei by Swanson and Cowan ('79).

ac   Anterior Commissure
ac_t   Anterior Commissure (temporal limb)
Al   Anterior Limbic Area (Cortex)
Bst   Bed Nucleus of Stria Terminalis
cc   Corpus Callosum
CdP   Caudoputamen
Cl   Claustrum
cng   Cingulum Bundle
DB   Nucleus of the Diagonal Band
db   Diagonal Band of Broca
df   Dorsal Fornix
ec   External Capsule
EP   Endopiriform Nucleus
fo_p   Fornix (postcommisural)
fo_pr   Fornix (precommisural)
GP   Globus Pallidus (external or lateral)
ic   Internal Capsule
IG   Induseum Griseum
In   Insular Area (Cortex)
lot   Lateral Olfactory Tract
LPO   Lateral Preoptic Area (Hypothalamus)
LV   Lateral Ventricle
Ma   Motor Agranular Cortical Area
MaPO   Magnocellular Preoptic Nucleus (Hypothalamus)
MePO   Median Preoptic Nucleus (Hypothalamus)
MP   Medial Preoptic Nucleus (Hypothalamus)
oc   Optic Chiasm
Pir   Piriform Area (Cortex)
PS   Parastrial Nucleus
rf   Rhinal Fissure
SF   Septofimbrial Nucleus
SI   Substantia Innominata
SomS   Somatosensory Cortex
SPl_d   Septal Nucleus (lateral: dorsal)
SPl_i   Septal Nucleus (lateral: intermediate)
SPl_v   Septal Nucleus (lateral: ventral)
st   Stria Terminalis
STF   Striatal Fundus
Ts   Triangular Nucleus (Septum)
Tu   Olfactory Tubercle

P&W −0.4

9 8 7 6 5 4 3 2 1 0

Bregma
−0.6

7 6 5 4 3 2 1 0

Earbar 0
8.4

0 1 2 3 4 5 6 7

Plate 11

AAA Anterior Amygdaloid Area
Ah Anterior Hypothalamic Nucleus
AL Anterior Limbic Area (Cortex)
Bst Bed Nucleus of Stria Terminalis
cc Corpus Callosum
CdP Caudoputamen
Cl Claustrum
cng Cingulum Bundle
CoA$_a$ Cortical Nucleus of the Amygdala (anterior)
df Dorsal Fornix
ec External Capsule
EP Endopiriform Nucleus
fi Fimbria
fo Fornix
GP Globus Pallidus (external or lateral)
IA Intercalated Mass of the Amygdala
ic Internal Capsule
IG Induseum Griseum
In Insular Area (Cortex)
Lot Nucleus of Lateral Olfactory Tract
lot Lateral Olfactory Tract
LPO Lateral Preoptic Area (Hypothalamus)
Ma Motor Agranular Cortical Area
MaPO Magnocellular Preoptic Nucleus (Hypothalamus)
mch Medial Corticohypothalamic Tract
mfb Medial Forebrain Bundle
MP Medial Preoptic Nucleus (Hypothalamus)
oc Optic Chiasm
Pa$_a$ Paraventricular Nucleus (anterior) (Thalamus)
Pir Piriform Area (Cortex)
PT Paratenial Nucleus (Thalamus)
PVh Paraventricular Nucleus (Hypothalamus)
rf Rhinal Fissure
SCh Suprachiasmatic Nucleus (Hypothalamus)
SF Septofimbrial Nucleus
sfo Subfornical Organ
SI Substantia Innominata
sm Stria Medullaris
So Supraoptic Nucleus (Hypothalamus)
SomS Somatosensory Cortex
st Stria Terminalis
STF Striatal Fundus
ti Tuberoinfundibular Tract
Ts Triangular Nucleus (Septum)
vhc Ventral Hippocampal Commissure

Nucleus Basalis. This region of the ventromedial pallidus containing scattered cholinergic neurons that project to isocortex is sometimes called the basal nucleus of Meynert (Saper, '84). It is a prominent structure in the primate brain but consists of poorly segregated "large" neurons in the rat. The region is probably sufficiently ambiguous to have inspired use of alternative terms such as *substantia innominata* and the *magnocellular preoptic nucleus*, although this probably deserves separate status in the rat as a cholinergic nucleus projecting upon the olfactory bulb (Swanson, '76a and '92). The matter is confounded by application of the neologism *nucleus of the horizontal limb of the diagonal band* (Price and Powell, '70), and other authorities on this region (Paxinos and Butcher, '85) indicate that the "horizontal limb of the diagonal band is otherwise known as the magnocellular preoptic nucleus." None of these alternatives seem satisfactory, and we seriously doubt that the nucleus basalis designated by Paxinos and Watson ('86) constitutes a recognizable "nucleus" in the rat brain. Forcing such homologies has been the source of much confusion in neuroanatomical nomenclature. The alternative of relying on a specific labeling method for cholinergic neurons (Sofroniew et al., '82; Houser et al., '83) enables a useful functional identification of this purported "nucleus" that we cannot recognize, and therefore do not label, in Nissl-stained sections. These neurons can be distinguished from surrounding noncholinergic neurons by the intensity of their hybridization signal for choline acetyltransferase in mRNA (Lauterborn et al., '93). In the context of "chemical markers" for defining a "nucleus," it is of some interest to note that the magnocellular neurons of the n. basalis express the peptide galanin in most mammals but not in anthropoids (Chan-Palay, '88).

ti: Tuberoinfundibular Tract. This fiber tract, at the base of the hypothalamus underlying the third ventricle in the tuber cinereum (ashen protuberance), appears to be continuous with the infundibular stalk, but its connections have not been established.

P&W -0.92

Bregma
-0.9

Earbar 0
8.1

Plate 12

AAA Anterior Amygdaloid Area
Ah_a Anterior Hypothalamic Nucleus (anterior)
AL Anterior Limbic Area (Cortex)
AV Anteroventral Nucleus (Thalamus)
Bst Bed Nucleus of Stria Terminalis
cc Corpus Callosum
CdP Caudoputamen
Cl Claustrum
cng Cingulum Bundle
CoA_a Cortical Nucleus of the Amygdala (anterior)
df Dorsal Fornix
ec External Capsule
EP Endopiriform Nucleus
fi Fimbria
fo Fornix
GP Globus Pallidus (external or lateral)
IA Intercalated Mass of the Amygdala
ic Internal Capsule
IG Induseum Griseum
In Insular Area (Cortex)
LHA Lateral Hypothalamic Area
Lot Nucleus of Lateral Olfactory Tract
lot Lateral Olfactory Tract

LV Lateral Ventricle
Ma Motor Agranular Cortical Area
mch Medial Corticohypothalamic Tract
mfb Medial Forebrain Bundle
MP Medial Preoptic Nucleus (Hypothalamus)
oc Optic Chiasm
ot Optic Tract
Pa_a Paraventricular Nucleus (anterior) (Thalamus)
Pir Piriform Area (Cortex)
PT Paratenial Nucleus (Thalamus)
PVh Paraventricular Nucleus (Hypothalamus)
R Reticular Nucleus (Thalamus)
rf Rhinal Fissure
SCh Suprachiasmatic Nucleus (Hypothalamus)
SF Septofimbrial Nucleus
sfo Subfornical Organ
SI Substantia Innominata
sm Stria Medullaris
So Supraoptic Nucleus
SomS Somatosensory Cortex
st Stria Terminalis
STF Striatal Fundus
Ts Triangular Nucleus (Septum)
vhc Ventral Hippocampal Commissure

AL: Anterior Limbic Area (Cortex). This definition is difficult and the large anterior and medial cortical field surrounding the corpus callosum is also called *anterior cingulate* and *area 24 of Brodmann*. The "agranular" characteristic of this field is sufficient for separation from the posterior limbic (or posterior cingulate) field (Rose and Woolsey, '48) containing a heavy granular layer, but this is of little value in distinguishing the rostroventral adjoining fields variously called infralimbic and prelimbic. Rose and Woolsey and later Krettek and Price ('77a) provided support for recognizing this cortical field as a functional entity on the basis of its thalamic and amygdaloid inputs, but lacking a sound rationale for an assessment of popular usage, we demur from entering into further obfuscation and refer the reader to original papers and the review of Vogt and Peters ('81).

fo: Fornix. As the fimbria enters the caudal end of the septal region, it divides into two major components, the pre- and postcommissural fornix. The latter is a condensed bundle that curves ventrally and then passes just caudal to the anterior commissure to enter the hypothalamus and extend through the lateral hypothalamic area to the mammillary body. It has one major branch, the medial corticodiencephalic tract (see mch: Medial Corticohypothalamic Tract at Plate 12). The precommissural fornix passes through the septo-fimbrial nucleus to end in the lateral septal nucleus. Another component, the dorsal fornix (fornix longus), extends along the midline just ventral to the corpus callosum and is particularly associated with field CA₁ of Ammon's horn (Powell and Cowan, '55).

mch: Medial Corticohypothalamic Tract. As the postcommissural fornix descends caudal and ventral to the anterior commissure, it gives rise to the medial corticodiencephalic tract (Swanson, '87). The dorsal component of this tract (which is indistinct in fiber preparations) ends in the anterior group of the thalamus, while the ventral component, the medial corticohypothalamic tract, extends throughout the medial zone of the hypothalamus (Canteras and Swanson, '92).

P&W -1.3

Bregma
-1.2

Earbar 0
7.8

Plate 13

AAA Anterior Amygdaloid Area
AD Anterodorsal Nucleus (Thalamus)
Ah Anterior Hypothalamic Nucleus
AL Anterior Limbic Area (Cortex)
AM Anteromedial Nucleus (Thalamus)
AV Anteroventral Nucleus (Thalamus)
Bst Bed Nucleus of Stria Terminalis
cc Corpus Callosum
CdP Caudoputamen
Cl Claustrum
cng Cingulum Bundle
CoA_a Cortical Nucleus of the Amygdala (anterior)
df Dorsal Fornix
ec External Capsule
EP Endopiriform Nucleus
fi Fimbria
fo Fornix
GP Globus Pallidus (external or lateral)
IA Intercalated Mass of the Amygdala
ic Internal Capsule
IG Induseum Griseum
iml Internal Medullary Lamina
In Insular Area (Cortex)

itp Inferior Thalamic Peduncle
LHA Lateral Hypothalamic Area
Lot Nucleus of Lateral Olfactory Tract
lot Lateral Olfactory Tract
Ma Motor Agranular Cortical Area
mfb Medial Forebrain Bundle
oc Optic Chiasm
ot Optic Tract
Pa_a Paraventricular Nucleus (anterior) (Thalamus)
Pir Piriform Area (Cortex)
PT Paratenial Nucleus (Thalamus)
PVh Paraventricular Nucleus (Hypothalamus)
R Reticular Nucleus (Thalamus)
Re Nucleus Reuniens (Thalamus)
rf Rhinal Fissure
SF Septofimbrial Nucleus
SH Septohippocampal Nucleus
SI Substantia Innominata
sm Stria Medullaris
So Supraoptic Nucleus
SomS Somatosensory Cortex
st Stria Terminalis
STF Striatal Fundus
vhc Ventral Hippocampal Commissure

CoA: Cortical Nucleus of the Amygdala. This is often included in the corticomedial division (divided into cortical and medial "nuclei") or in earlier usage subsumed under *periamygdaloid cortex*, Pam. The argument for calling this region *cortex* is its similarity to the contiguous piriform cortex (*prepyriform* in earlier European usage) in possessing a zona lamina (molecular or plexiform layer) where axons from mitral cells of the olfactory bulb synapse on the distal apical dendrites of underlying modified pyramidal cells (see Price, '73). Thus, there is some merit to designating this sector as primary olfactory cortex (i.e., recipient of olfactory tract input), with recognition of a distinct medial nucleus (see MeA: Medial Nucleus of the Amygdala at Plate 91) that deserves separate status as the main recipient of axons from the *accessory olfactory bulb*. The cortical nucleus is readily divisible into anterior and posterior portions (see De Olmos et al., '85, and Canteras et al., '92a). The posterior position is further subdivided into lateral and medial sectors (Swanson, '92), with the lateral sector receiving axons from the main olfactory bulb and the medial sector receiving accessory olfactory bulb input (see Scalia and Winans, '75, who also recognize a *bed nucleus of the accessory olfactory bulb*, which we have not identified). A separate, circumscribed region of this "olfactory cortex" is generally recognized as the *nucleus of the lateral olfactory tract* (Lot).

soc: Supraoptic Commissures or Decussations. These massive fiber systems cross the midline at the level of and caudal to the optic chiasm (Tsang, '40). They are traditionally divided into anterior, dorsal, and ventral components, although they are rarely identified in the rat (see Nauta and Haymaker, '69; Swanson, '87). There are five recognized commissures in the human brain, each bearing an eponym.

P&W -1.4

Earbar 0
7.5

Plate 14

AD Anterodorsal Nucleus (Thalamus)
Ah Anterior Hypothalamic Nucleus
AM Anteromedial Nucleus (Thalamus)
AMi Interanteromedial Nucleus (Thalamus)
Arc Arcuate Nucleus (Hypothalamus)
AV Anteroventral Nucleus (Thalamus)
BLA Basolateral Nucleus of the Amygdala
CA$_3$ Hippocampal Area CA$_3$ (Ammon's Horn)
cc Corpus Callosum
CdP Caudoputamen
CeA Central Nucleus of the Amygdala
Cl Claustrum
cm Central Medial Nucleus (Thalamus)
cng Cingulum Bundle
CoA$_a$ Cortical Nucleus of the Amygdala (anterior)
df Dorsal Fornix
DG Dentate Gyrus
ec External Capsule
eml External Medullary Lamina
EP Endopiriform Nucleus
ET Entopeduncular Nucleus
fi Fimbria
fo Fornix
GP Globus Pallidus (external or lateral)
IA Intercalated Mass of the Amygdala
ic Internal Capsule
IG Induseum Griseum

In Insular Area (Cortex)
LHA Lateral Hypothalamic Area
LV Lateral Ventricle
Ma Motor Agranular Cortical Area
MeA$_a$ Medial Nucleus of the Amygdala (anterior)
mfb Medial Forebrain Bundle
ot Optic Tract
Pa$_a$ Paraventricular Nucleus (anterior) (Thalamus)
Pam Periamygdaloid Cortex
Pe$_a$ Periventricular Nucleus (anterior) (Hypothalamus)
Pir Piriform Area (Cortex)
PT Paratenial Nucleus (Thalamus)
PVh Paraventricular Nucleus (Hypothalamus)
R Reticular Nucleus (Thalamus)
RCh Retrochiasmatic Area (Hypothalamus)
Re Nucleus Reuniens (Thalamus)
rf Rhinal Fissure
Rh Rhomboid Nucleus (Thalamus)
SI Substantia Innominata
sm Stria Medullaris
So Supraoptic Nucleus
soc Supraoptic Commissure
SomS Somatosensory Cortex
st Stria Terminalis
VA Ventral Anterior Nucleus (Thalamus)
vi Vellum Interpositum
ZI Zona Incerta

AD: Anterodorsal Nucleus (Thalamus). This nucleus caps the anterior nuclear group and is easily delimited from the underlying anteroventral nucleus in Nissl and fiber stains. It can be divided into dorsal and ventral components in rabbit (M. Rose, '35), but this is less evident in rat. An extension to the midline to form a central interanterodorsal nucleus (ADi) is endorsed by several authors, although this can be ambiguous because of difficulty in separating the wing of the paracentral nucleus (e.g., Paxinos and Watson, '86; Swanson, '92).

Ah: Anterior Hypothalamic Nucleus. The Ah was named by Gurdjian ('27), a designation that has generally been followed since, although Bleier et al. ('79) divided this region into anterior and dorsal hypothalamic areas, a lateral anterior hypothalamic nucleus, and an area of the tuber cinereum. On the other hand, Saper et al. ('78) divided the more traditional Ah into anterior, central, and posterior parts, and Swanson ('92) later referred to the tiny *dorsal tuberal nucleus* of Bleier et al. ('79) or *stigmoid nucleus* of Paxinos and Watson ('86) as the dorsal part of the Ah because it is embedded within the posterior part of the nucleus. In Paxinos and Watson ('86) the *lateroanterior hypothalamic nucleus* may correspond to the anterior part of the Ah of Saper et al. ('78), and their anterior part of the Ah corresponds to anterior regions of the central part of Saper et al. ('78). The Ah is surrounded by a less cell-dense region with indeterminate borders known as the *anterior hypothalamic area*. The caudal end of the Ah is relatively distinct and easy to distinguish from the poorly differentiated anterior part of the dorsomedial nucleus and is best observed in sagittal Nissl preparations.

PVh: Paraventricular Nucleus (Hypothalamus). A heterogeneous nuclear complex that includes the densely packed dorsal hypothalamic covering of the third ventricle. Most authors recognize numerous subdivisions, some far removed from the ventricle and architecturally quite distinct; these are illustrated nicely and in some detail by Kiss et al. ('91) and Swanson ('92), but we have avoided imposing subnuclear status upon a functional designation that probably deserves tentative acceptance except for a posterior sector caudal to the arcuate nucleus (see Plate 100), a region that is called the *posterior periventricular nucleus* by Swanson ('92), and here labeled Pe$_p$ to separate it from the PVh system.

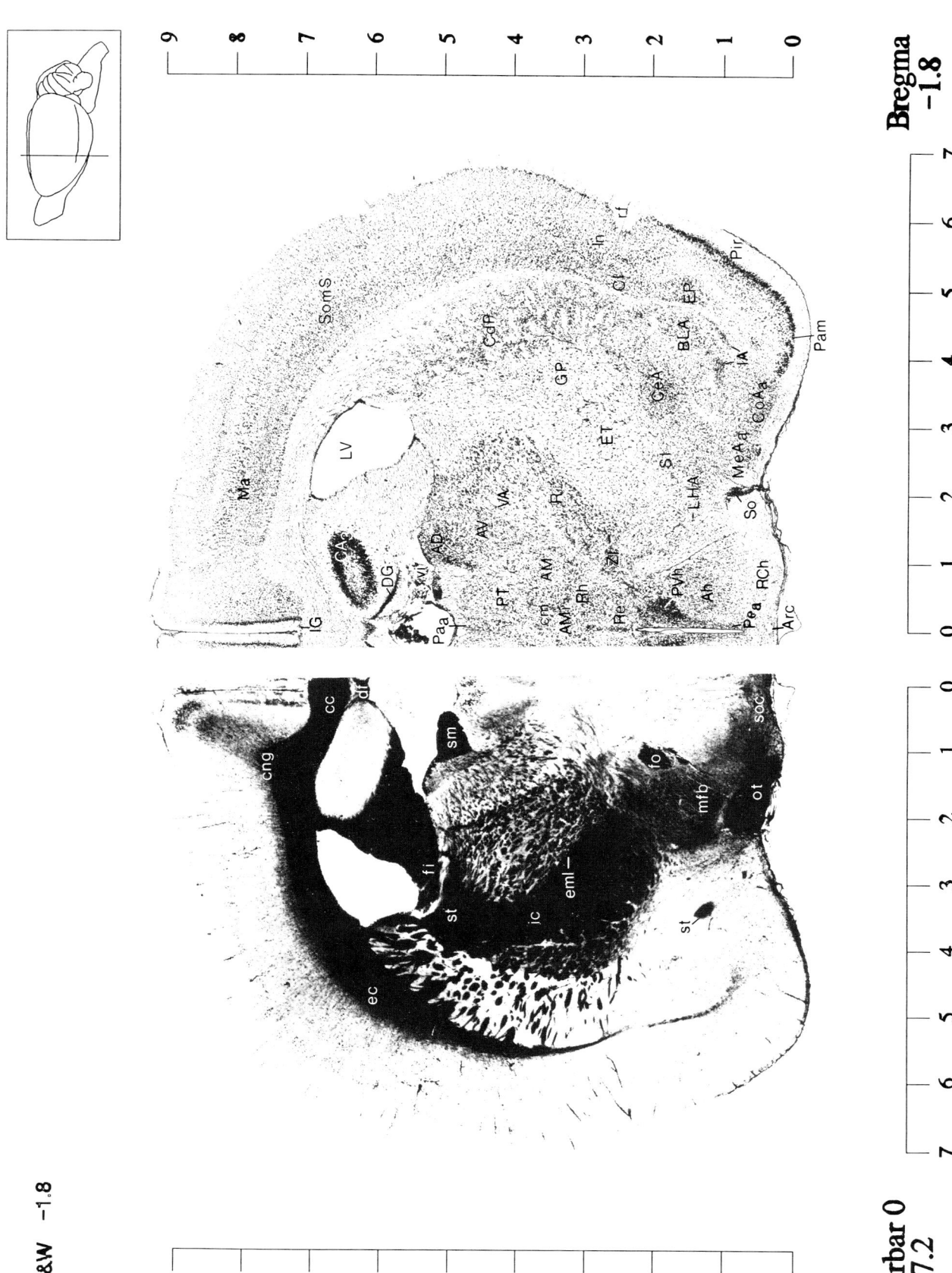

R&W -1.8

Bregma
-1.8

Earbar 0
7.2

Plate 15

AD Anterodorsal Nucleus (Thalamus)
Ah Anterior Hypothalamic Nucleus
Ah$_d$ Anterior Hypothalamic Nucleus (dorsal)
alv Alveus
AM Anteromedial Nucleus (Thalamus)
AMi Interanteromedial Nucleus (Thalamus)
AV Anteroventral Nucleus (Thalamus)
BLA Basolateral Nucleus of the Amygdala
CA$_3$ Hippocampal Area CA$_3$ (Ammon's Horn)
cc Corpus Callosum
CdP Caudoputamen
CeA Central Nucleus of the Amygdala
Cl Claustrum
cm Central Medial Nucleus (Thalamus)
cng Cingulum Bundle
CoA$_a$ Cortical Nucleus of the Amygdala (anterior)
DG Dentate Gyrus
ec External Capsule
eml External Medullary Lamina
EP Endopiriform Nucleus
ET Entopeduncular Nucleus
fi Fimbria
fo Fornix
GP Globus Pallidus (external or lateral)
IA Intercalated Mass of the Amygdala
ic Internal Capsule
IG Induseum Griseum
iml Internal Medullary Lamina

In Insular Area (Cortex)
LA Lateral Nucleus of the Amygdala
LD Lateral Dorsal Nucleus (Thalamus)
LHA Lateral Hypothalamic Area
Ma Motor Agranular Cortical Area
MD Mediodorsal Nucleus (Thalamus)
MeA$_a$ Medial Nucleus of the Amygdala (anterior)
mfb Medial Forebrain Bundle
mt Mammillothalamic Tract
ot Optic Tract
Pa Paraventricular Nucleus (Thalamus)
Pam Periamygdaloid Cortex
Pc Paracentral Nucleus (Thalamus)
Pir Piriform Area (Cortex)
PT Paratenial Nucleus (Thalamus)
PVh Paraventricular Nucleus (Hypothalamus)
R Reticular Nucleus (Thalamus)
RCh Retrochiasmatic Area (Hypothalamus)
Re Nucleus Reuniens (Thalamus)
rf Rhinal Fissure
Rh Rhomboid Nucleus (Thalamus)
SI Substantia Innominata
sm Stria Medullaris
soc Supraoptic Commissure
SomS Somatosensory Cortex
st Stria Terminalis
VA Ventral Anterior Nucleus (Thalamus)
VL Ventrolateral Nucleus (Thalamus)
VM Ventromedial Nucleus (Thalamus)
ZI Zona Incerta

LHA and LPO: Lateral Hypothalamic Area and Lateral Preoptic Area. The lateral zone of the hypothalamus is exceedingly heterogeneous, complex, and poorly understood. More than 50 components of the medial forebrain bundle are believed to arise within, pass through, and/or end within this rostral continuation of the brain stem reticular formation (see Nieuwenhuys et al., '82; Swanson, '87). Thus, while it has been subdivided in a number of ways, none of these schemes is entirely satisfactory. For descriptive purposes reference to the fiber architecture of preoptic, anterior, tuberal and mammillary levels of the lateral zone can be convenient. There is no suitable cytoarchitectonic criterion for distinguishing between the lateral preoptic and lateral hypothalamic areas, and it is hard to distinguish a border with the substantia innominata, which lies rostral, lateral, and dorsal. It has been designated variously by different authors on topographic grounds. Thus, the region below the zona incerta can be called the *subincertal nucleus* (e.g., Paxinos and Watson, '86), which in functional and developmental terms implies a "subthalamic" origin rather than belonging to the hypothalamus. In the absence of explicit persuasive criteria such nomenclatural conflicts seem inevitable. Our usage should be considered tentative.

RCh: Retrochiasmatic Area (Hypothalamus). This relatively neuron-sparse hypothalamic zone serves merely as a commonly recognized landmark and includes caudally directed axons from the supraoptic commissures.

VA: Ventral Anterior Nucleus (Thalamus) (also called Nucleus Ventralis Anterior). The rostral pole of the ventral nuclear group consisting of the transitional zone between the ventrolateral nucleus (VL) and the thalamic reticular nucleus (R) is small in the rat and usually difficult to recognize as an entity, except in the sagittal plane (see Plate 109). The existence of VA in the rat is not always recognized and it can be argued that this is a somewhat forced distinction based upon attempts to recognize a large VA nucleus in human and monkey thalamus.

P&W -2.12

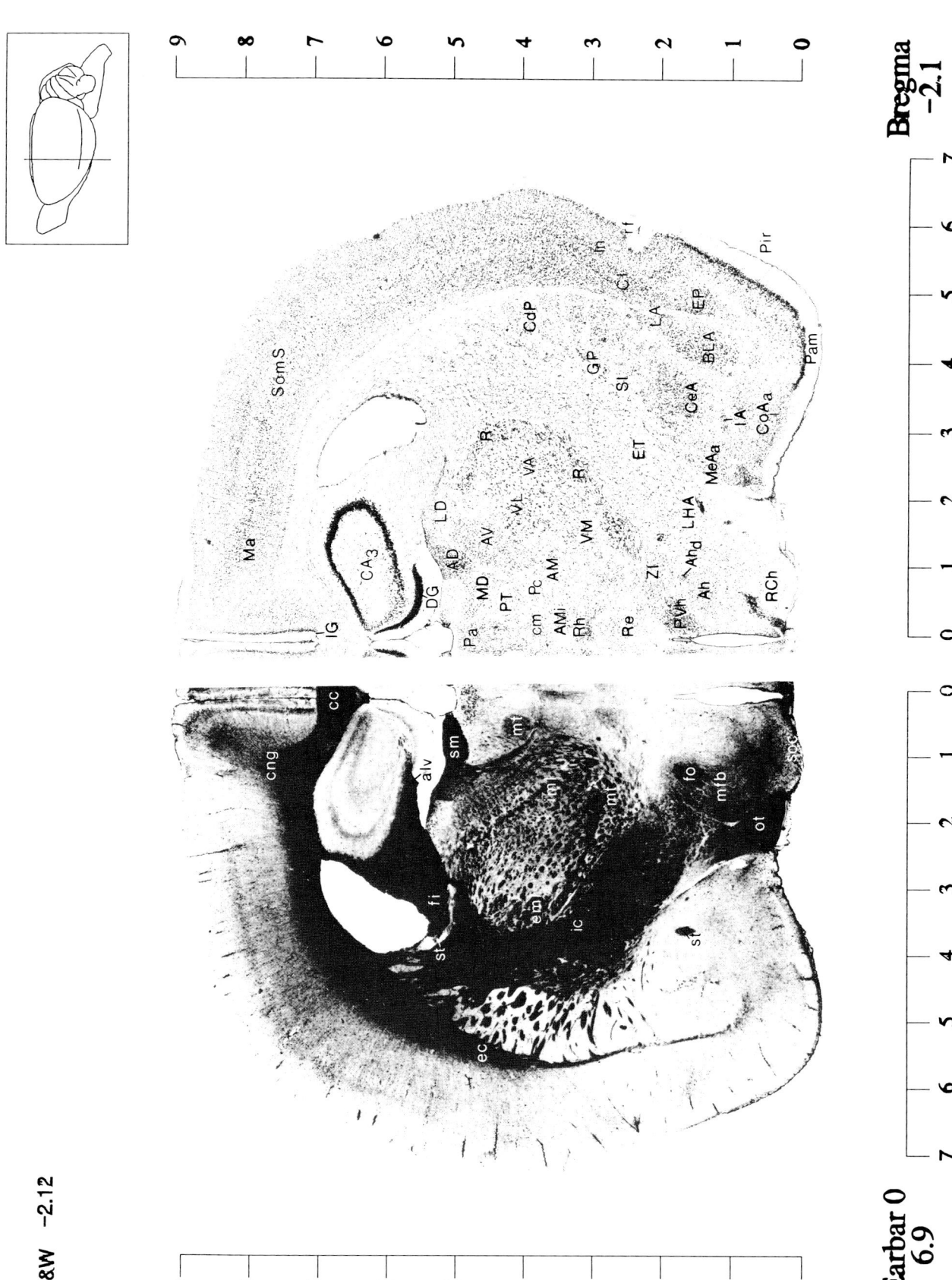

Bregma
-2.1

Earbar 0
6.9

Plate 16

AD Anterodorsal Nucleus (Thalamus)
Ah Anterior Hypothalamic Nucleus
alv Alveus
AM Anteromedial Nucleus (Thalamus)
AMi Interanteromedial Nucleus (Thalamus)
Arc Arcuate Nucleus (Hypothalamus)
AV Anteroventral Nucleus (Thalamus)
BLA Basolateral Nucleus of the Amygdala
BMA Basomedial Nucleus of the Amygdala
CA₁ Hippocampal Area CA₁ (Ammon's Horn)
CA₃ Hippocampal Area CA₃ (Ammon's Horn)
cc Corpus Callosum
CdP Caudoputamen
CeA Central Nucleus of the Amygdala
Cl Claustrum
cm Central Medial Nucleus (Thalamus)
cng Cingulum Bundle
CoA_a Cortical Nucleus of the Amygdala (anterior)
DG Dentate Gyrus
ec External Capsule
eml External Medullary Lamina
EP Endopiriform Nucleus
ET Entopeduncular Nucleus
fi Fimbria
fo Fornix
Ge Nucleus Gelatinosus (Nucleus Submedius) (Thalamus)
GP Globus Pallidus (external or lateral)
ic Internal Capsule
IG Induseum Griseum

iml Internal Medullary Lamina
In Insular Area (Cortex)
LA Lateral Nucleus of the Amygdala
LD Lateral Dorsal Nucleus (Thalamus)
LHA Lateral Hypothalamic Area
LV Lateral Ventricle
Ma Motor Agranular Cortical Area
MD Mediodorsal Nucleus (Thalamus)
MDi Intermediodorsal Nucleus (Thalamus)
ME Median Eminence (Hypothalamus)
MeA_a Medial Nucleus of the Amygdala (anterior)
mfb Medial Forebrain Bundle
mt Mammillothalamic Tract
ot Optic Tract
Pa Paraventricular Nucleus (Thalamus)
Pam Periamygdaloid Cortex
Pc Paracentral Nucleus (Thalamus)
Pir Piriform Area (Cortex)
PL Posterior Limbic Area (Cortex)
PVh Paraventricular Nucleus (Hypothalamus)
R Reticular Nucleus (Thalamus)
Re Nucleus Reuniens (Thalamus)
rf Rhinal Fissure
Rh Rhomboid Nucleus (Thalamus)
SI Substantia Innominata
sm Stria Medullaris
SomS Somatosensory Cortex
st Stria Terminalis
V3 Third Ventricle
VL Ventrolateral Nucleus (Thalamus)
VM Ventromedial Nucleus (Thalamus)
VMh Ventromedial Nucleus (Hypothalamus)
ZI Zona Incerta

AM: Anteromedial Nucleus (Thalamus). This medial component of the anterior group is distinctly encapsulated and easier to delimit by its fiber architecture. It can be divided into dorsal and ventral sectors (Swanson, '92), supported by experimental findings (Canteras and Swanson, '92), but some authors (e.g., Paxinos and Watson, '86) recognize part of the ventral sector as the rhomboid intralaminar wing and caudally the ventral portion can be difficult to separate from the ventromedial nucleus. A midline interanteromedial nucleus (AMi) with lateral extensions is recognized in most descriptions, although there are some differences in each atlas. This is probably confounded by a complex cytoarchitecture readily amenable to even further subdivision.

AV: Anteroventral Nucleus (Thalamus). This largest nucleus of the anterior nuclear group in rat is easily recognized and possesses distinct boundaries in cellular and fiber architecture. Subdivision into upper and lower portions in the rabbit by M. Rose ('35) has been adopted by Paxinos and Watson ('86) as dorsomedial and ventrolateral sectors or subnuclei, but this is rarely employed by most authors despite experimental evidence supporting functional subdivision.

In: Insular Area (Cortex). The insular cortical fields in the rodent brain surround the rhinal fissure but do not form an "island" as in the human brain, from which the name is derived. It is unique among cortical areas in that it is usually defined by the limits of the underlying claustrum (Cl) and has been variously subdivided on the basis of the granular layer, most recently (Cechetto and Saper, '87) into granular, agranular, and dysgranular fields. It is generally believed to contain the principal visceral and gustatory projections, the borders of which remain controversial (see review by Kruger and Mantyh, '89).

P&W -2.3

Bregma
-2.4

Earbar 0
6.6

9
8
7
6
5
4
3
2
1
0

7
6
5
4
3
2
1
0

9
8
7
6
5
4
3
2
1
0

7
6
5
4
3
2
1
0

Som S.

Ma

PL

IG

CA1

DG

CA3

LV

AD

LD

MD

MDl

Pa

Pc

AM

AM

Rh

Re

V3

cm

R

VL

VM

Ge

ZI

LHA

Ah

PVh

VMh

ME

Arc

ET

SI

GP

CdP

CeA

MeA

BMA

CoAa

LA

BLA

EP

Pam

Pir

Cl

ln

f

f

cc

cng

alv

sm

fi

LV

st

ec

ic

alv

mt

ml

sm

aml

st

mt

fo

ot

mfb

Plate 17

AM Anteromedial Nucleus (Thalamus)
AMi Interanteromedial Nucleus (Thalamus)
Arc Arcuate Nucleus (Hypothalamus)
BLA Basolateral Nucleus of the Amygdala
BMA Basomedial Nucleus of the Amygdala
CA$_1$ Hippocampal Area CA$_1$ (Ammon's Horn)
CA$_3$ Hippocampal Area CA$_3$ (Ammon's Horn)
cc Corpus Callosum
CdP Caudoputamen
CeA Central Nucleus of the Amygdala
cl Central Lateral Nucleus (Thalamus)
cm Central Medial Nucleus (Thalamus)
cng Cingulum Bundle
DG Dentate Gyrus
DMh$_a$ Dorsomedial Nucleus (anterior) (Hypothalamus)
ec External Capsule
EP Endopiriform Nucleus
ET Endopeduncular Nucleus
fi Fimbria
fo Fornix
Ge Nucleus Gelatinosus (Nucleus Submedius) (Thalamus)
GP Globus Pallidus (external or lateral)
Hb$_l$ Habenula Nucleus (lateral) (Thalamus)
Hb$_m$ Habenula Nucleus (medial) (Thalamus)
hf Hippocampal Fissure
IA Intercalated Mass of the Amygdala
ic Internal Capsule

IG Induseum Griseum
iml Internal Medullary Lamina
In Insular Area (Cortex)
LA Lateral Nucleus of the Amygdala
LD Lateral Dorsal Nucleus (Thalamus)
LHA Lateral Hypothalamic Area
Ma Motor Agranular Cortical Area
MD Mediodorsal Nucleus (Thalamus)
MDi Intermediodorsal Nucleus (Thalamus)
ME Median Eminence (Hypothalamus)
MeA$_p$ Medial Nucleus of the Amygdala (posterior)
mfb Medial Forebrain Bundle
mt Mammillothalamic Tract
ot Optic Tract
Pa Paraventricular Nucleus (Thalamus)
Pam Periamygdaloid Cortex
Pc Paracentral Nucleus (Thalamus)
Pir Piriform Area (Cortex)
PL Posterior Limbic Area (Cortex)
R Reticular Nucleus (Thalamus)
Re Nucleus Reuniens (Thalamus)
rf Rhinal Fissure
Rh Rhomboid Nucleus (Thalamus)
SI Substantia Innominata
sm Stria Medullaris
SomS Somatosensory Cortex
st Stria Terminalis
VB$_l$ Ventrobasal Nuclear Complex (lateral) (Thalamus)
VL Ventrolateral Nucleus (Thalamus)
VM Ventromedial Nucleus (Thalamus)
VMh Ventromedial Nucleus (Hypothalamus)
ZI Zona Incerta

Arc: Arcuate Nucleus (Hypothalamus). This ventral, wing-shaped differentiation of the periventricular nucleus is essentially coextensive with, and lies just dorsal to, the median eminence (see Everitt et al., '86). Its lateral border is quite distinct and is formed by the cell-sparse capsule surrounding the ventromedial nucleus; separation of the dorsal border of the arcuate nucleus from the remainder of the periventricular nucleus is arbitrary. It is commonly referred to as the *infundibular nucleus* in human anatomy.

Rh: Rhomboid Nucleus (Thalamus) (also called Nucleus Rhomboidalis). This rhomboid-shaped, darkly stained midline nucleus contains large, tightly packed neurons overlying the lightly stained *n. reuniens* (also called *n. medialis ventralis*; e.g., Jones, '83 and '85). The dorsal boundary presents a problem in various descriptions because a pale region intervenes between the rhomboid nucleus and the condensed portion of the central medial nucleus, part of which may constitute the interanteromedial nucleus, although this extends caudal to the anteromedial nucleus in Paxinos and Watson ('86). Jones ('85) places the rhomboid nucleus above cm in the region that other authors have designated *interanteromedial* (AMi) and *intermediodorsal* (MDi), the convention used here to avoid confusion and to be more consistent with the earlier account of these sections by Jones ('83). The ventrolateral boundary of the caudal portion of the rhomboid nucleus constitutes the distinctly outlined *n. gelatinosus* (also called *n. submedius*) according to Paxinos and Watson ('86) and Swanson ('92), but this region is the central medial nucleus in the Jones ('83 and '85) atlases, which we suggest may be inaccurate. This nucleus is nicely illustrated in several accounts (Gurdjian, '27; Krieg, '44; Paxinos and Watson, '86; Swanson, '92).

VMh: Ventromedial Nucleus (Hypothalamus). This is one of the most easily defined nuclei in the rat hypothalamus because it is surrounded by a cell-sparse capsule that essentially corresponds to the undifferentiated capsule (or zone) surrounding the medial preoptic and anterior hypothalamic nuclei (the medial preoptic and anterior hypothalamic areas, respectively). Because of these similarities, the capsule around the VMh has been called the *tuberal area* (Swanson, '92). The nucleus can be divided into dorsomedial, central, ventrolateral, and anterior parts (Gurdjian, '27; Van Houten and Brawer, '78).

P&W -2.8

Bregma
-2.7

Earbar 0
6.3

9 8 7 6 5 4 3 2 1 0

7 6 5 4 3 2 1 0

7 6 5 4 3 2 1 0

0 1 2 3 4 5 6 7

SomS

PL

Ma

CA1

CA3

DG

IG

Hbm

Hbl

LD

cl

CdP

R

VB

VL

VM

GP

SI

LA

EP

CeA

BLA

BMA

IA

MeAp

Pam

Pir

In

rf

ET

ZI

LHA

DMha

VMh

Arc

ME

Re

Rh

Ge

AM

AMi

cm

Pc

MDi

MD

Pa

cc

sm

cng

hf

fi

st

ec

iml

mt

fo

mfb

ot

ic

Plate 18

Arc Arcuate Nucleus (Hypothalamus)
BLA Basolateral Nucleus of the Amygdala
BMA Basomedial Nucleus of the Amygdala
CA₁ Hippocampal Area CA₁ (Ammon's Horn)
CA₃ Hippocampal Area CA₃ (Ammon's Horn)
cc Corpus Callosum
CdP Caudoputamen
CeA Central Nucleus of the Amygdala
cl Central Lateral Nucleus (Thalamus)
cm Central Medial Nucleus (Thalamus)
cng Cingulum Bundle
DG Dentate Gyrus
DMh$_a$ Dorsomedial Nucleus (anterior) (Hypothalamus)
ec External Capsule
eml External Medullary Lamina
EP Endopiriform Nucleus
FC Fasciola Cinerea
fi Fimbria
fo Fornix
Ge Nucleus Gelatinosus (Nucleus Submedius) (Thalamus)
GP Globus Pallidus (external or lateral)
Hb$_l$ Habenula Nucleus (lateral) (Thalamus)
Hb$_m$ Habenula Nucleus (medial) (Thalamus)
hf Hippocampal Fissure
IA Intercalated Mass of the Amygdala
ic Internal Capsule
IG Induseum Griseum

iml Internal Medullary Lamina
In Insular Area (Cortex)
LA Lateral Nucleus of the Amygdala
LD Lateral Dorsal Nucleus (Thalamus)
LHA Lateral Hypothalamic Area
LV Lateral Ventricle
MD Mediodorsal Nucleus (Thalamus)
MDi Intermediodorsal Nucleus (Thalamus)
ME Median Eminence (Hypothalamus)
MeA$_p$ Medial Nucleus of the Amygdala (posterior)
mfb Medial Forebrain Bundle
mt Mammillothalamic Tract
ot Optic Tract
Pa Paraventricular Nucleus (Thalamus)
Pam Periamygdaloid Cortex
Pc Paracentral Nucleus (Thalamus)
Pir Piriform Area (Cortex)
Po Posterior Group (Thalamus)
R Reticular Nucleus (Thalamus)
Re Nucleus Reuniens (Thalamus)
rf Rhinal Fissure
Rh Rhomboid Nucleus (Thalamus)
sm Stria Medullaris
SomS Somatosensory Cortex
st Stria Terminalis
V3 Third Ventricle
VB Ventrobasal Nuclear Complex (Thalamus)
VL Ventrolateral Nucleus (Thalamus)
VM Ventromedial Nucleus (Thalamus)
VMh Ventromedial Nucleus (Hypothalamus)
ZI Zona Incerta

Ge: Nucleus Gelatinosus (Thalamus) (also called Nucleus Submedius). This thalamic nucleus, lying ventral to the internal medullary lamina, is easily outlined with both cell and fiber stains. The nucleus submedius, described in carnivores by Rioch ('29) as a ventral component of the intralaminar wing nuclei, was shown by Craig and Burton ('81) to receive spinothalamic and medullary input, and their finding of a comparable input to the dorsal part of *n. gelatinosus* led to applying *n. submedius* to this region in rat (Burton and Craig, '83). The ventral portion of this nucleus was designated the olfactory portion of *n. submedius* by Price and Slotnick ('83), but some electrophysiological studies claim that the entire nucleus belongs to the "spinothalamic" input system (Dostrovsky and Guilbaud, '88; Miletic and Coffield, '89). The homology of Rioch's *n. submedius* in dog and cat was apparently introduced by Krieg ('44) for the rat *n. gelatinosus*, although several authors argue that there is only partial homology (e.g., Burton and Craig, '83; Paxinos and Watson, '86). The Washington University group (Craig and Burton, '81; Price and Slotnick, '83; Jones, '85; Swanson, '92) have substituted *n. submedius* for *n. gelatinosus*, but in consideration of the various disclaimers of exact homology and functional subdivision, the traditional taxonomic approach of employing the precedent term (*gelatinosus*) adopted earlier (e.g., König and Klippel, '70; Slotnick and Leonard, '75; Paxinos and Watson, '86) is followed here.

ME: Median Eminence (Hypothalamus). A circumventricular organ that lacks a traditional blood–brain barrier. It may be divided into an internal lamina adjacent to the third ventricle that contains axons descending to the posterior pituitary, and an external lamina that contains the proximal end of the hypophysial portal system.

VL: Ventrolateral Nucleus (Thalamus) (also called Nucleus Ventralis Lateralis). Capping the rostral limits of the ventrobasal complex (VB) is the region designated VL in mammals; the boundary is reasonably distinct, especially in horizontal and sagittal sections. The sectors recipient of cerebellar and basal ganglia inputs are recognizable as separate zones in many larger brains, but we are unable to impose such subdivision in the rat brain.

P&W −3.14

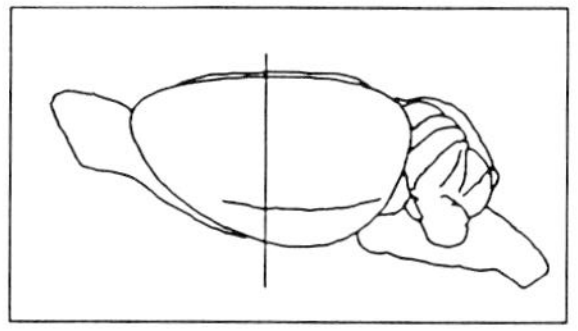

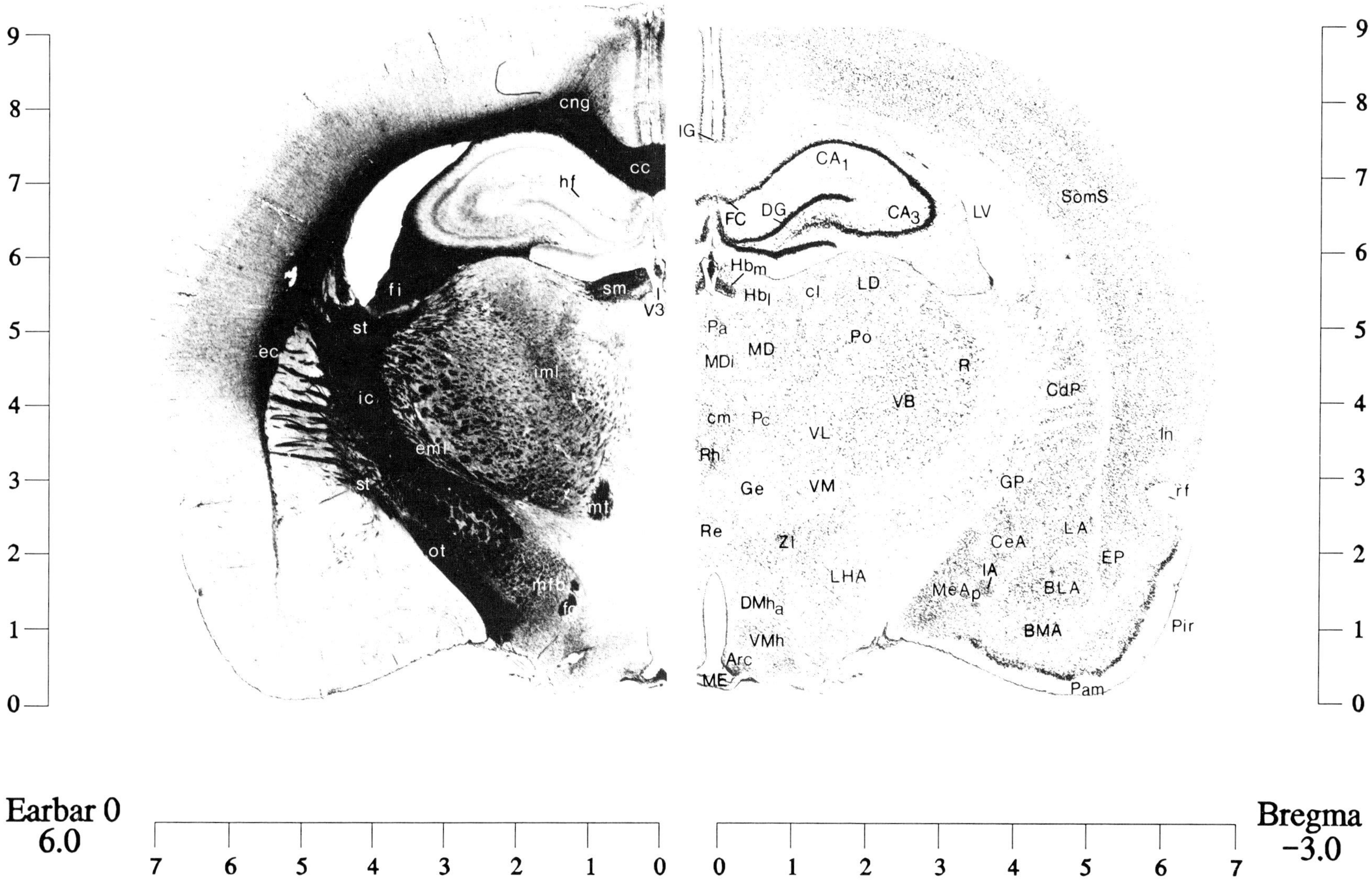

cng
cc
hf
fi
sm
V3
st
ec
ic
iml
eml
st
mt
ot
mfb
fc
IG
CA1
DG
FC
CA3
LV
SomS
Hbm
Hbl
cl
LD
Pa
Po
MD
MDi
R
CdP
VB
cm
Pc
VL
In
Rh
Ge
VM
GP
rf
Re
ZI
CeA
LA
EP
IA
LHA
MeAp
BLA
DMha
BMA
Pir
VMh
Arc
ME
Pam
Earbar 0
6.0
Bregma
−3.0

Plate 19

alv Alveus
Arc Arcuate Nucleus (Hypothalamus)
BLA Basolateral Nucleus of the Amygdala
BMA Basomedial Nucleus of the Amygdala
CA₁ Hippocampal Area CA₁ (Ammon's Horn)
CA₃ Hippocampal Area CA₃ (Ammon's Horn)
cc Corpus Callosum
CdP Caudoputamen
CeA Central Nucleus of the Amygdala
Cl Claustrum
cl Central Lateral Nucleus (Thalamus)
cm Central Medial Nucleus (Thalamus)
cng Cingulum Bundle
DG Dentate Gyrus
DMh_p Dorsomedial Nucleus (posterior) (Hypothalamus)
ec External Capsule
eml External Medullary Lamina
EP Endopiriform Nucleus
FC Fasciola Cinerea
fr Fasciculus Retroflexus
Ge Nucleus Gelatinosus (Nucleus Submedius) (Thalamus)
Hb₁ Habenula Nucleus (lateral) (Thalamus)
Hb_m Habenula Nucleus (medial) (Thalamus)
hf Hippocampal Fissure
ic Internal Capsule
IG Induseum Griseum
iml Internal Medullary Lamina
In Insular Area (Cortex)
LA Lateral Nucleus of the Amygdala
LD Lateral Dorsal Nucleus (Thalamus)
LHA Lateral Hypothalamic Area
Ma Motor Agranular Cortical Area
MD Mediodorsal Nucleus (Thalamus)
MeA_p Medial Nucleus of the Amygdala (posterior)
mfb Medial Forebrain Bundle
mt Mammillothalamic Tract
ot Optic Tract
Pa Paraventricular Nucleus (Thalamus)
Pam Periamygdaloid Cortex
Pc Paracentral Nucleus (Thalamus)
pc Posterior Commissure
PH Posterior Hypothalamic Nucleus
Pir Piriform Area (Cortex)
PL Posterior Limbic Area (Cortex)
Po Posterior Group (Thalamus)
R Reticular Nucleus (Thalamus)
Re Nucleus Reuniens (Thalamus)
rf Rhinal Fissure
sm Stria Medullaris
SomS Somatosensory Cortex
st Stria Terminalis
V3 Third Ventricle
VB Ventrobasal Nuclear Complex (Thalamus)
VL Ventrolateral Nucleus (Thalamus)
VM Ventromedial Nucleus (Thalamus)
VMh Ventromedial Nucleus (Hypothalamus)
ZI Zona Incerta

DMh: Dorsomedial Nucleus (Hypothalamus). The heterogeneous architecture of this nuclear complex lends itself to subdivision with only a distinct ventral border overlying the ventromedial nucleus. The dorsal portion is described by Krieg ('32) as the "most poorly defined and inconspicuous cell group of the hypothalamus." We follow the nomenclature employed by Swanson ('92) dividing the DMh into an anterior portion, more loosely packed and ill-defined, and a dense, distinct posterior portion; these were called *dorsal* and *ventral*, respectively, by Gurdjian ('27). Paxinos and Watson ('86) recognize a central "dense" division surrounded by a "diffuse" division, and Swanson ('92) supports separating the more scattered neurons in the ventral region as a distinct subdivision.

VM_(p): Ventromedial Nucleus (posterior) (Thalamus) (also called Nucleus Ventralis Medialis). The medial edge of the VB complex is formed by a nucleus or nuclear complex containing smaller and generally more compactly spaced neurons in mammals. The gustatory nucleus is found in this region in those species in which it has been mapped, lying adjacent to the tactile tongue representation at the medial limit of VB. The gustatory region of thalamus and cortex in the rat appears to be rather small (see Kruger and Mantyh, '89, for review); but a gustatory thalamic nucleus is recognized in this region in several descriptions (e.g., Paxinos and Watson, '86) and this is approximately correct despite possible dispute over boundaries and further subdivision of VM to conform to the more elaborate differentiation recognized in larger mammals.

A substantial sector of the VM contains small cells and is often called VPMpc (for parvicellular) and probably includes a visceral input as well as a gustatory component. This sector constitutes the arcuate nucleus of primates, a name derived from its form in fiber architecture. In the rat thalamus the parvicellular component extends laterally below the ventrobasal (VB) complex in a viscerotopic arrangement (see Cechetto and Saper, '87), and where it can be recognized we indicate this as VM_p to distinguish it from the larger-celled main VM nucleus, more prominently related to motor pathways. The portion of the VM adjacent to, or sometimes included within, the intralaminar wing also presents a problem in nomenclature and is sometimes called *nucleus submedius* (see Ge: Nucleus Gelatinosus at Plate 18). It is likely that a new nomenclature for this "nucleus" will emerge, guided by experimental studies. In a strict sense, this is a nuclear group and we have resisted the temptation to offer more than crude tentative labeling until the key issues of conflict are resolved.

P&W -3.3

9
8
7
6
5
4
3
2
1
0

Bregma
-3.3

7 6 5 4 3 2 1 0

SomS
In
CdP Cl
CeA LA
MeAp BLA
BMA
Pam
Pir
EP

Ma
CA1
CA3
DG
FC
IG
PL
LD
Po
R
VB
VM
VL
Pc
cl
MD
pc
Re
cm
Ge
Zi
LHA
DMh
PH
VMh
Arc
Pa
Hbm
V3
fr

cng
cc
hf
alv
alvv
st
eml
ec
ic
ot
mfb
mt
sm
st

Earbar 0
5.7

0 1 2 3 4 5 6 7

9
8
7
6
5
4
3
2
1
0

Plate 20

alv Alveus
Arc Arcuate Nucleus (Hypothalamus)
Aud Auditory Area (Cortex)
BLA Basolateral Nucleus of the Amygdala
BMA Basomedial Nucleus of the Amygdala
CA₁ Hippocampal Area CA₁ (Ammon's Horn)
CA₃ Hippocampal Area CA₃ (Ammon's Horn)
cc Corpus Callosum
CdP Caudoputamen
cl Central Lateral Nucleus (Thalamus)
cm Central Medial Nucleus (Thalamus)
cng Cingulum Bundle
CoA_p Cortical Nucleus of the Amygdala (posterior)
cp Cerebral Peduncle
DG Dentate Gyrus
DMh_p Dorsomedial Nucleus (posterior) (Hypothalamus)
ec External Capsule
eml External Medullary Lamina
EP Endopiriform Nucleus
FC Fasciola Cinerea
fi Fimbria
fo Fornix
fr Fasciculus Retroflexus
Hb_l Habenula Nucleus (lateral) (Thalamus)
Hb_m Habenula Nucleus (medial) (Thalamus)
ic Internal Capsule

IG Induseum Griseum
iml Internal Medullary Lamina
In Insular Area (Cortex)
LA Lateral Nucleus of the Amygdala
LD Lateral Dorsal Nucleus (Thalamus)
LHA Lateral Hypothalamic Area
LV Lateral Ventricle
Ma Motor Agranular Cortical Area
MD Mediodorsal Nucleus (Thalamus)
MeA_p Medial Nucleus of the Amygdala (posterior)
ml Medial Lemniscus
mt Mammillothalamic Tract
ot Optic Tract
Pa Paraventricular Nucleus (Thalamus)
Pam Periamygdaloid Cortex
PH Posterior Hypothalamic Nucleus
Pir Piriform Area (Cortex)
PL Posterior Limbic Area (Cortex)
Po Posterior Group (Thalamus)
R Reticular Nucleus (Thalamus)
Re Nucleus Reuniens (Thalamus)
rf Rhinal Fissure
sm Stria Medullaris
SomS Somatosensory Cortex
st Stria Terminalis
VB Ventrobasal Nuclear Complex (Thalamus)
VL Ventrolateral Nucleus (Thalamus)
VM Ventromedial Nucleus (Thalamus)
VMh Ventromedial Nucleus (Hypothalamus)
ZI Zona Incerta

FC: Fasciola Cinerea. The FC is thought of as a thin band of cortex that courses around the splenium of the corpus callosum (see Hjorth-Simonsen, '72) and continues rostrally as the induseum griseum (see Wyss and Sripanidkulchai, '83). Thus, the fasciola cinerea and induseum griseum are contiguous infra- and supra-callosal parts of the same formation, constituting the hippocampal rudiment (see tt: Tenia Tecta at Plate 2 and SH: tenia tecta, Septohippocampal Nucleus at Plate 5). The induseum griseum lies at the base of the longitudinal cerebral fissure, immediately dorsal to the corpus callosum, and (like the caudally adjacent dentate gyrus) forms the medial edge of the cortical mantle (see Swanson, '92); it underlies the grossly visible longitudinal striae of Lancisi in the human brain, but these fiber bands are not evident in the rat.

st: Stria Terminalis. The condensed segment of this complex fiber system extends between the amygdala and septal region (bed nuclei of the stria terminalis) along the embryonic sulcus terminalis between the telencephalon and thalamus. Various distinct and not so distinct groups of fibers diverge from either end of the stria terminalis proper, although a clear picture of their organization has not yet emerged (see De Olmos et al., '85). The stria terminalis contains ascending and descending fibers. Many fibers between the amygdala and other parts of the basal forebrain take an alternate route through the ansa peduncularis, which lies within the substantia innominata; these fibers are sometimes referred to as the *ventral amygdalar pathway*, which is more appropriate than the *ventral amygdalofugal pathway*, a misnomer because it is bidirectional (see Nauta and Haymaker, '69).

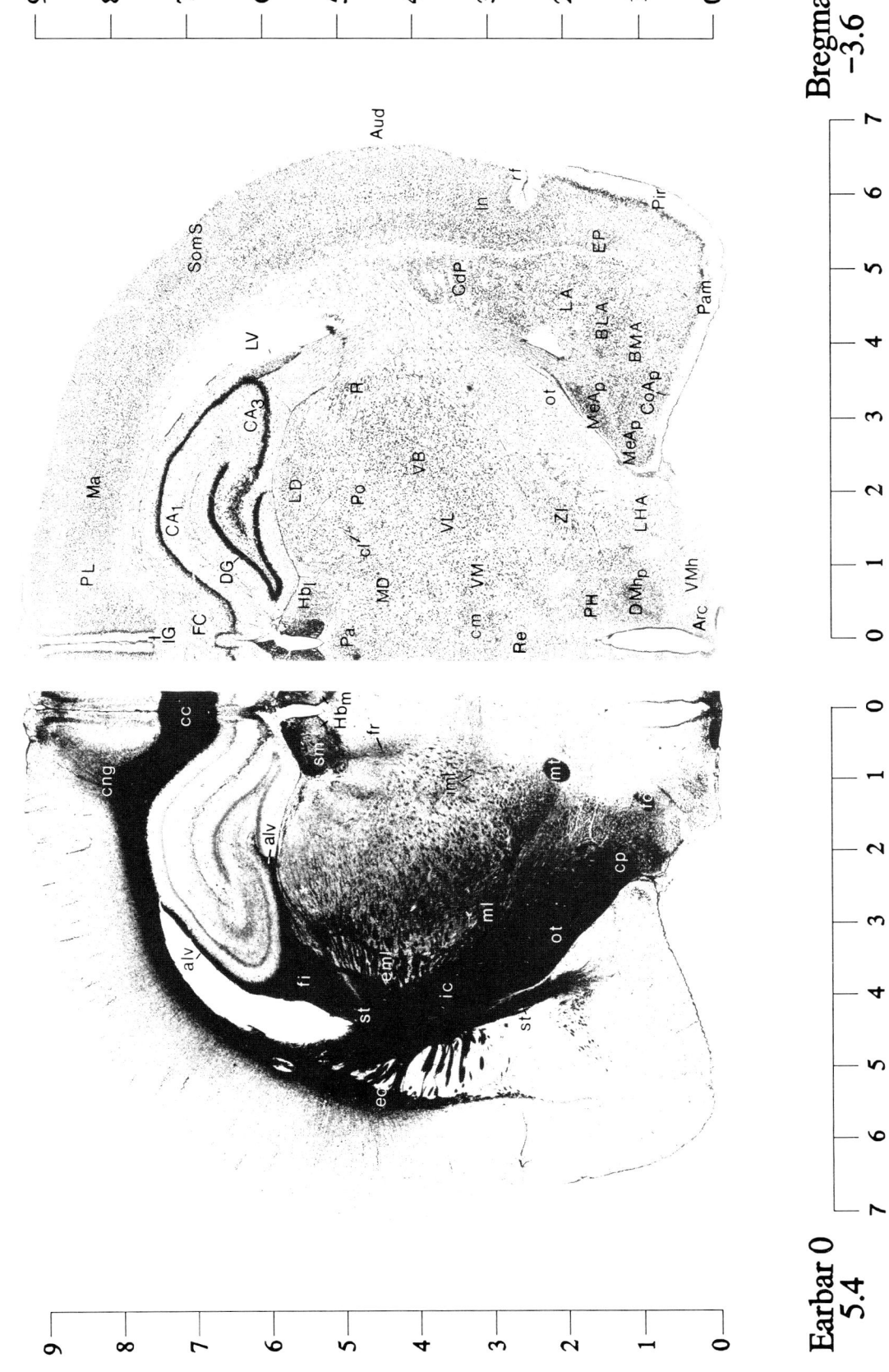

P&W -3.6

Bregma
-3.6

Earbar 0
5.4

Plate 21

AHZ Amygdalo-Hippocampal Area
alv Alveus
Arc Arcuate Nucleus (Hypothalamus)
BLA Basolateral Nucleus of the Amygdala
BMA Basomedial Nucleus of the Amygdala
CA₁ Hippocampal Area CA₁ (Ammon's Horn)
CA₃ Hippocampal Area CA₃ (Ammon's Horn)
cc Corpus Callosum
CdP Caudoputamen
cl Central Lateral Nucleus (Thalamus)
cm Central Medial Nucleus (Thalamus)
cng Cingulum Bundle
CoA_p Cortical Nucleus of the Amygdala (posterior)
cp Cerebral Peduncle
DG Dentate Gyrus
dhc Dorsal Hippocampal Commissure
DMh Dorsomedial Nucleus (Hypothalamus)
ec External Capsule
eml External Medullary Lamina
EP Endopiriform Nucleus
FC Fasciola Cinerea
fi Fimbria
fo Fornix
H₁, H₂ H Fields of Forel
Hb_l Habenula Nucleus (lateral) (Thalamus)
Hb_m Habenula Nucleus (medial) (Thalamus)

IG Induseum Griseum
iml Internal Medullary Lamina
In Insular Area (Cortex)
LA Lateral Nucleus of the Amygdala
LD Lateral Dorsal Nucleus (Thalamus)
LGd Dorsal Lateral Geniculate Nucleus (Thalamus)
LGv Ventral Lateral Geniculate Nucleus (Thalamus)
LHA Lateral Hypothalamic Area
MD Mediodorsal Nucleus (Thalamus)
mfb Medial Forebrain Bundle
ml Medial Lemniscus
mt Mammillothalamic Tract
ot Optic Tract
Pa Paraventricular Nucleus (Thalamus)
Pc Paracentral Nucleus (Thalamus)
PH Posterior Hypothalamic Nucleus
Pir Piriform Area (Cortex)
PL Posterior Limbic Area (Cortex)
Po Posterior Group (Thalamus)
Re Nucleus Reuniens (Thalamus)
rf Rhinal Fissure
sm Stria Medullaris
SomS Somatosensory Cortex
ST Subthalamic Nucleus
st Stria Terminalis
VB Ventrobasal Nuclear Complex (Thalamus)
VL Ventrolateral Nucleus (Thalamus)
VM Ventromedial Nucleus (Thalamus)
ZI Zona Incerta

Po: Posterior Group (Thalamus). Subsuming a group of architecturally different nuclei into a single taxon can provide new levels of confusion of immense proportion, and the posterior *group* serves as an excellent example. Various authors have designated a posterior *nucleus*, and some refer to a posterior *complex*, using it as synonymous with the posterior *group* of Rose and Woolsey ('58). The latter authors introduced the modern usage in tentative fashion as the sector of caudal thalamus that lacked an "essential" projection to cortex (i.e., specific cortical sector dependency) but rather projected to broader expanses of cortex and was thus defined as "sustaining" on the basis of its pattern of retrograde atrophy. This includes the region interposed between the more easily recognized main large nuclei and consists of small- and large-celled aggregates. The large-celled (medial) component of the medial geniculate body, the suprageniculate nucleus (of the embryonic dorsal thalamus, not to be confused with the suprageniculate of the pretectal group), the nucleus limitans, and the sector below the pulvinar called *nucleus posterior* by certain earlier workers were included in Rose and Woolsey's definition. The usage was expanded further in Paxinos and Watson's first edition, but they withdrew apologetically in the second edition (Paxinos and Watson, '86), recognizing the ambiguity that resulted from abuse of nomenclatural principles. Critical but somewhat confusing accounts of the evolving usage of Po can be found in Jones and Burton ('74), Berman and Jones ('82), and Jones ('85), but these refer largely to nonrodent species. Jones ('85) advocates recognition of two principal components: a suprageniculate-limitans aggregate (the latter term referring to the linear aggregate bordering the pretectal group), which is darkly stained and thus distinctive; and a posterior nucleus comprising the lightly stained, small-celled transitional region behind the ventrobasal complex, below the pulvinar and medial to the medial geniculate body. He advocates calling this heterogeneous collection the posterior *complex*, apparently rejecting Rose and Woolsey's distinction between *group* and *complex*. A more comprehensive reappraisal for the rat can be found in Feldman and Kruger ('80), Faull and Mehler ('86), and Fabri and Burton ('91).

P&W -3.8

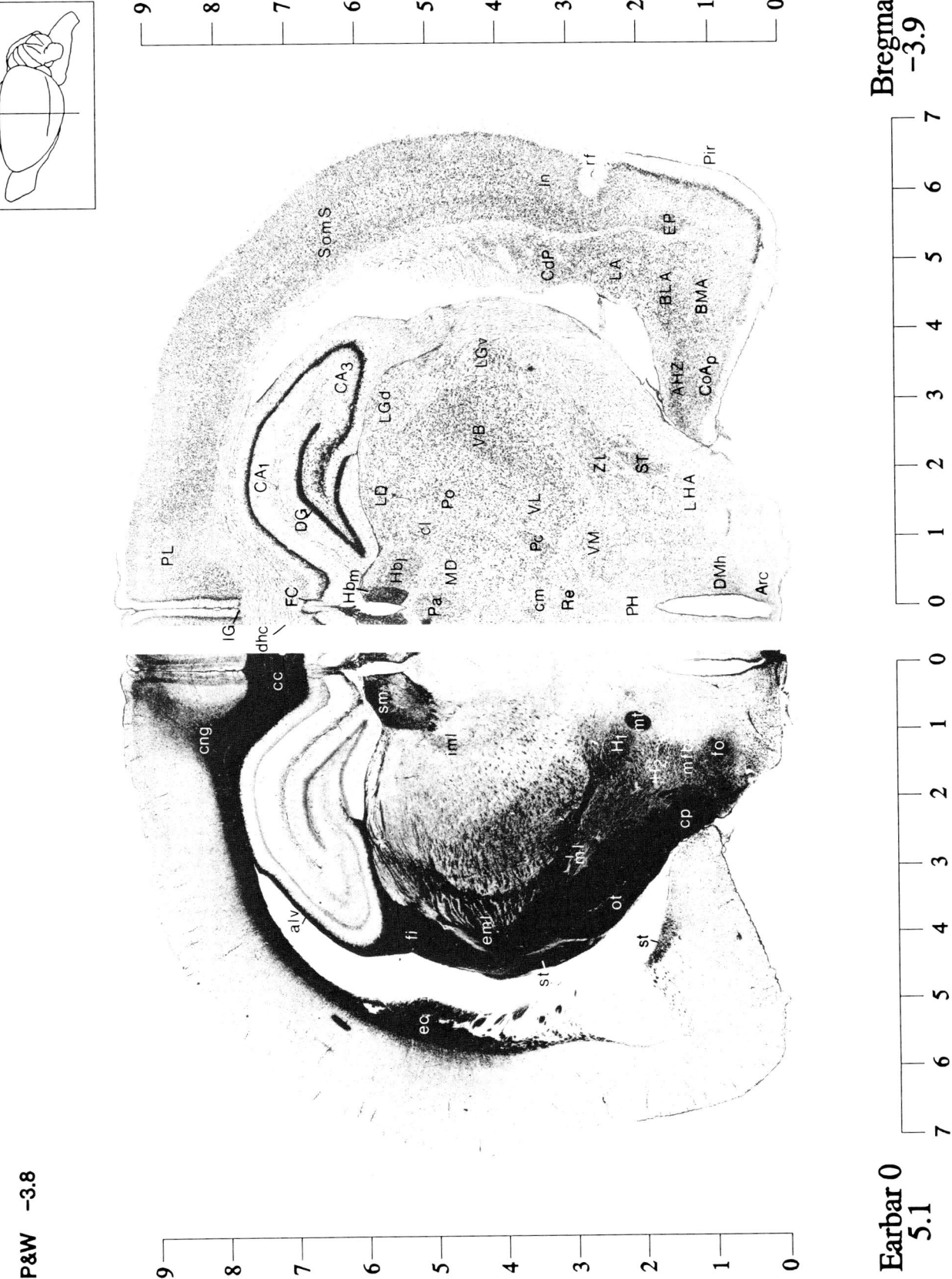

9
8
7
6
5
4
3
2
1
0

SomS

In

rf

Pir

CdP

LA

EP

BLA

BMA

LGv

AHZ

CoAp

CA3

LGd

VB

CA1

DG

LD

Po

cl

ST

Zl

LHA

FC

Hbm

Hbl

Pc

VL

PL

IG

dhc

Pa

MD

VM

DMh

cm

Re

PH

Arc

Bregma
-3.9

7
6
5
4
3
2
1
0

cc

cng

sm

iml

mt

Hi

fr

alv

ml

io

fi

eml

cp

ot

st

st

ef

0
1
2
3
4
5
6
7

Earbar 0
5.1

Plate 22

AHZ Amygdalo-Hippocampal Area
alv Alveus
Aud Auditory Area (Cortex)
BLA Basolateral Nucleus of the Amygdala
BMA Basomedial Nucleus of the Amygdala
bsc Brachium of Superior Colliculus
CA₁ Hippocampal Area CA₁ (Ammon's Horn)
CA₃ Hippocampal Area CA₃ (Ammon's Horn)
cc Corpus Callosum
CdP Caudoputamen
cl Central Lateral Nucleus (Thalamus)
CM Centromedial Nucleus (Thalamus)
cm Central Medial Nucleus (Thalamus)
cng Cingulum Bundle
CoA_p Cortical Nucleus of the Amygdala (posterior)
cp Cerebral Peduncle
DG Dentate Gyrus
dhc Dorsal Hippocampal Commissure
ec External Capsule
EP Endopiriform Nucleus
FC Fasciola Cinerea
fi Fimbria
fo Fornix
H₁,H₂ H Fields of Forel
Hb_l Habenula Nucleus (lateral) (Thalamus)
Hb_m Habenula Nucleus (medial) (Thalamus)
hf Hippocampal Fissure
IG Induseum Griseum
LA Lateral Nucleus of the Amygdala

LGd Dorsal Lateral Geniculate Nucleus (Thalamus)
LGv Ventral Lateral Geniculate Nucleus (Thalamus)
LHA Lateral Hypothalamic Area
LP Lateral Posterior Nucleus (Thalamus)
LV Lateral Ventricle
MD Mediodorsal Nucleus (Thalamus)
mfb Medial Forebrain Bundle
mr Mammillary Recess
mt Mammillothalamic Tract
ot Optic Tract
Pa Paraventricular Nucleus (Thalamus)
Pe_p Periventricular Nucleus (posterior) (Hypothalamus)
Pf Parafascicular Nucleus (Thalamus)
PH Posterior Hypothalamic Nucleus
Pir Piriform Area (Cortex)
PL Posterior Limbic Area (Cortex)
PM_v Premammillary Nucleus (ventral) (Hypothalamus)
Po Posterior Group (Thalamus)
R Reticular Nucleus (Thalamus)
Re Nucleus Reuniens (Thalamus)
rf Rhinal Fissure
sm Stria Medullaris
SomS Somatosensory Cortex
ST Subthalamic Nucleus
st Stria Terminalis
V3 Third Ventricle
VB Ventrobasal Nuclear Complex (Thalamus)
VM Ventromedial Nucleus (Thalamus)
VM_p Ventromedial Nucleus (posterior) (Thalamus)
ZI Zona Incerta

CM: Centromedial Nucleus (Thalamus) (also called Nucleus Centrum Medianum and Nucleus Centre Médian). The CM is arguably homologous with the prominent encapsulated nucleus of this name bearing the eponym of Luys in the primate thalamus. The homology is difficult on strictly architectural grounds in many animals, including cat, rabbit, and rat, and it has been suggested that it can be subsumed together with the parafascicular nucleus (e.g., Swanson, '92) following Rose and Woolsey's ('43) "postmedial group" concept. An ovoid nucleus, which is the rostral extension of the parafascicular nucleus (Pf), is outlined by Paxinos and Watson ('86), who indicate that it belongs to the paracentral nucleus and insist "it should not be mistaken for the oral pole of the parafascicular nucleus." Swanson ('92) takes the deliberate decision of including it as the anterior part of the parafascicular nucleus. It might also be argued that extensions of "parafascicular" beyond the nucleus that surrounds the fasciculus retroflexus (and from which the name is derived) should be avoided, an argument that also applies to the lateral extension called subparafascicular, whose position belies the name.

A universally satisfactory solution for this nomenclatural dilemma has not been reached, but this nucleus is sufficiently recognizable in rat to warrant separate status. Its position and form are consistent with the centre médian of Luys and yet it is distinct from the functionally linked parafascicular nucleus. Outlining distinct nuclei and yet giving them the same "paracentral" name (Paxinos and Watson, '86) is surprising from authors who have indulged in more extensive subdivisions than most descriptions. Thus, we tentatively retain the tenuous CM usage here, cautioning the user that the homology is uncertain and also that this should not be confused with the midline central medial (cm) nucleus.

cp: Cerebral Peduncle. The meaning of this term is somewhat confused by the earlier connotation that it consists of the midbrain segment of the corticospinal tract (the crus cerebri), along with the tegmentum, ventral to the tectum. In the rat the term now refers just to the crus cerebri lateral to the hypothalamus and midbrain, constituting the main compact descending tract from the forebrain.

P&W −4.16

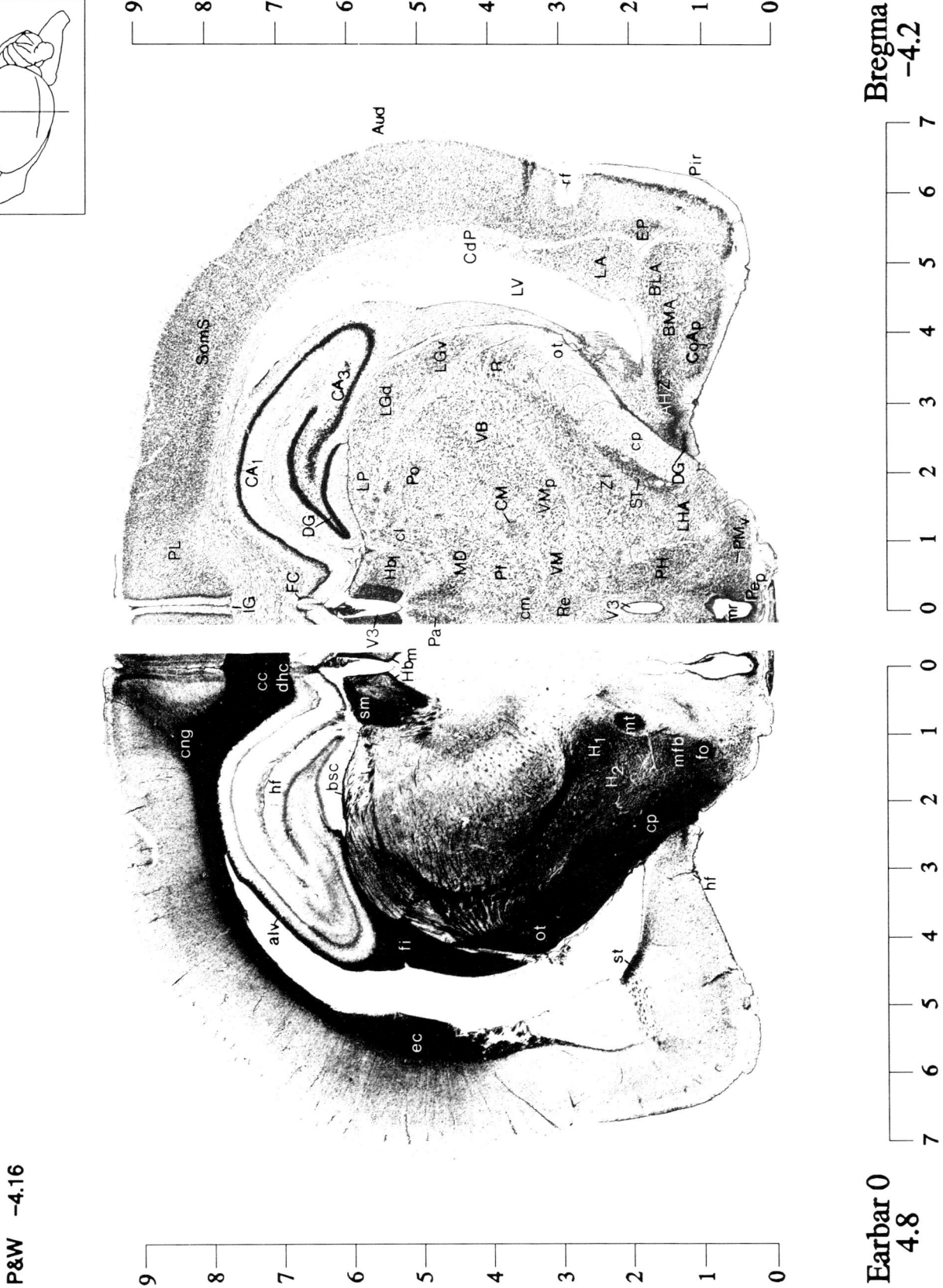

Bregma
−4.2

Earbar 0
4.8

Plate 23

AHZ Amygdalo-Hippocampal Area
alv Alveus
BLA Basolateral Nucleus of the Amygdala
BMA Basomedial Nucleus of the Amygdala
bsc Brachium of Superior Colliculus
CA₁ Hippocampal Area CA₁ (Ammon's Horn)
CA₃ Hippocampal Area CA₃ (Ammon's Horn)
cc Corpus Callosum
CM Centromedial Nucleus (Thalamus)
cm Central Medial Nucleus (Thalamus)
cng Cingulum Bundle
CoAₚ Cortical Nucleus of the Amygdala (posterior)
cp Cerebral Peduncle
DG Dentate Gyrus
dhc Dorsal Hippocampal Commissure
dMR Deep Mesencephalic Reticular Nucleus (Thalamus)
ec External Capsule
eml External Medullary Lamina
EP Endopiriform Nucleus
FC Fasciola Cinerea
fi Fimbria
fr Fasciculus Retroflexus
Hb₁ Habenula Nucleus (lateral) (Thalamus)
Hbₘ Habenula Nucleus (medial) (Thalamus)
hf Hippocampal Fissure
IG Induseum Griseum
Ig Intermediate Geniculate Nucleus (leaflet) (Thalamus)
iml Internal Medullary Lamina
is Infundibular Stalk
LA Lateral Nucleus of the Amygdala

LGd Dorsal Lateral Geniculate Nucleus (Thalamus)
LGv Ventral Lateral Geniculate Nucleus (Thalamus)
LHA Lateral Hypothalamic Area
LP Lateral Posterior Nucleus (Thalamus)
mfb Medial Forebrain Bundle
ml Medial Lemniscus
MM Medial Mammillary Nucleus (Hypothalamus)
mr Mammillary Recess
mt Mammillothalamic Tract
ot Optic Tract
Paₚ Paraventricular Nucleus (posterior) (Thalamus)
Peₚ Periventricular Nucleus (posterior) (Hypothalamus)
Pf Parafascicular Nucleus (Thalamus)
PH Posterior Hypothalamic Nucleus
Pir Piriform Area (Cortex)
PMd Premammillary Nucleus (dorsal) (Hypothalamus)
PMᵥ Premammillary Nucleus (ventral) (Hypothalamus)
Po Posterior Group (Thalamus)
R Reticular Nucleus (Thalamus)
rf Rhinal Fissure
sm Stria Medullaris
sPf Subparafascicular Nucleus (Thalamus)
ST Subthalamic Nucleus
st Stria Terminalis
TM Tuberomammillary Nucleus (Hypothalamus)
VB Ventrobasal Nuclear Complex (Thalamus)
VM Ventromedial Nucleus (Thalamus)
VMₚ Ventromedial Nucleus (posterior) (Thalamus)
ZI Zona Incerta

is: Infundibular Stalk (Pituitary Stalk). The portion of the infundibular stalk containing a central recess is preserved in some sections, although most of the remainder of the hypophysis is torn away.

Pf: Parafascicular Nucleus (Thalamus) (also called Nucleus Parafascicularis). The tightly packed small neurons surrounding the fasciculus retroflexus form a distinctive, easily recognized nucleus in cell and fiber architecture in most mammals such that there is general agreement in various descriptions and atlases. The anterior expansion of this nuclear group, which we have tentatively designated the *centromedial nucleus* or *n. centrum medianum* (CM), probably deserves distinct status because it is encapsulated, but the ventromedial border as it merges with the central gray is difficult to define, and Paxinos and Watson ('86) designate this region the *rostral continuation of the interstitial nucleus of Cajal*, associated with the parvicellular component of the oculomotor complex. The boundary between diencephalon and mesencephalon is indeed ambiguous, but the lateral extension presents a special problem. See discussion at sPf: Subparafascicular Nucleus (Plate 24).

PMd,ᵥ: Premammillary Nucleus (Hypothalamus). Although a single name is employed on topographic grounds, subdivision on architectural grounds into dorsal and ventral components is supported by connectional evidence, indicating that PMd is a component of the mammillary body (Canteras and Swanson, '92) and PMᵥ is related to the ventromedial nucleus (Canteras et al., '92b).

ST: Subthalamic Nucleus. In human, and many other primate brains, the ST bearing the eponym of Luys (*corpus Luysii*) is a distinct, large, encapsulated nucleus. Its homologue in the rat based on form and position below the zona incerta, is somewhat tenuous. Our designation is slightly smaller than that of Paxinos and Watson ('86) and that of Swanson ('92). In the apparent absence of experimental data on its presumptive striatal projection in the rat (Heimer et al., '85), revision may be warranted.

P&W −4.52

Bregma
−4.5

Earbar 0
4.5

CA1, DG, CA3, FC, IG, LP, Hbl, Pt, Pap, cm, CM, VM, sPf, PH, VMp, Po, VB, Zl, dMR, ST, LHA, MM, mr, is, Pep, PMv, PMd, TM, LGv, LGd, Ig, R, ot, cp, DG, hf, CoAp, BMA, BLA, LA, EP, Pir, fi, f

cng, cc, dhc, sm, bsc, eml, ot, alv, st, ec, fr, iml, ml, cp, Hbm, mt, mfb

Plate 24

sPf: Subparafascicular Nucleus (Thalamus). An elongate nucleus extending from the lateral edge of the parafascicular nucleus to the caudolateral margin of the diencephalon overlying the cerebral peduncle. The sPf might be identified as a singular, related entity because of the similarity of its spindle-shaped neurons, but this is not evident in Nissl-stained material. Bodian (39) employed the term *subparafascicular n.* to this lateral expansion in the opossum, using cytological criteria of similarity to Pf but specifying that it was obviously not "below the fasciculus" from which the name derives. Certain sectors of this continuous chain of cells have been given separate positional names. For example, the zone above the cerebral peduncle is called the *peripeduncular or suprapeduncular nucleus* in some descriptions. The intermediate region extending from the central gray to the peduncle has been called the *parvicellular subparafascicular nucleus* by Paxinos and Watson ('86), and because this band displays distinctive immunoreactivity using antibodies to calcitonin-gene-related peptide (CGRP), Kruger et al. ('88b) have proposed using this criterion for naming this nucleus, a somewhat unsatisfactory means of resolving a confusing nomenclature. The nuclear group extending along the caudal diencephalic boundary probably should not be encumbered with numerous names attached to heterogeneous nuclei. One logical resolution would be to follow Winer's private suggestion of calling this the *posterior intralaminar nucleus* as the lateral extension of the *posterior medial group* of Rose and Woolsey, provided this does not lead to further confusion by incorrect implication of association with the intralaminar wing embedded in the internal medullary lamina. Unfortunately, this "subparafascicular" lateral wing does not coincide with the *posterior intralaminar nucleus* indicated by Le Doux et al. ('85). A new terminology for this region would be welcome, but we demur from attempting resolution by imposing another ineffective nomenclature (see Winer et al., '88, for an excellent account in opossum). Mehler ('80) has summarized this literature and offered some interesting speculation suggesting that this caudal thalamic boundary constitutes the "chief diencephalic alar plate relay to limbic structures" by linking his subparafascicular nucleus with the paraventricular thalamic nuclei (see Kruger et al., '88b, for discussion). The designation *subparafascicular n.* is now generally accepted and is clearly depicted in diagrams by Swanson ('92); thus we employ it here despite the illogical descriptive term. The medial region, where sPf adjoins the parafascicular nucleus and central gray, is clearly complex, and subnuclei are plainly evident in sagittal and horizontal sections. Paxinos and Watson ('86) designate an *ethmoid* and a *scaphoid nucleus* in this region, terms we have failed to adopt largely because they are not generally accepted. The need for a new nomenclature for this region is profoundly evident.

AHZ Amygdalo-Hippocampal Area
alv Alveus
Aud Auditory Area (Cortex)
BLA Basolateral Nucleus of the Amygdala
bsc Brachium of Superior Colliculus
CA$_1$ Hippocampal Area CA$_1$ (Ammon's Horn)
CA$_3$ Hippocampal Area CA$_3$ (Ammon's Horn)
cc Corpus Callosum
cl Central Lateral Nucleus (Thalamus)
cng Cingulum Bundle
CoA$_p$ Cortical Nucleus of the Amygdala (posterior)
cp Cerebral Peduncle
DG Dentate Gyrus
dhc Dorsal Hippocampal Commissure
dMR Deep Mesencephalic Reticular Nucleus (Thalamus)
ec External Capsule
EP Endopiriform Nucleus
FC Fasciola Cinerea
fr Fasciculus Retroflexus
Hb$_l$ Habenula Nucleus (lateral) (Thalamus)
Hb$_m$ Habenula Nucleus (medial) (Thalamus)
hc Habenular Commissure
IG Induseum Griseum
LGd Dorsal Lateral Geniculate Nucleus (Thalamus)
LGv Ventral Lateral Geniculate Nucleus (Thalamus)
LHA Lateral Hypothalamic Area
LM Lateral Mammillary Nucleus
LP Lateral Posterior Nucleus (Thalamus)
LV Lateral Ventricle
ml Medial Lemniscus
MM Medial Mammillary Nucleus (Hypothalamus)
mr Mammillary Recess
mt Mammillothalamic Tract
ot Optic Tract
Pa$_p$ Paraventricular Nucleus (posterior) (Thalamus)
Pf Parafascicular Nucleus (Thalamus)
PH Posterior Hypothalamic Nucleus
Pir Piriform Area (Cortex)
PL Posterior Limbic Area (Cortex)
Po Posterior Group (Thalamus)
Pto Olivary Pretectal Nucleus (Thalamus)
rf Rhinal Fissure
sPf Subparafascicular Nucleus (Thalamus)
st Stria Terminalis
SUM$_l$ Supramammillary Nucleus (lateral)
VB Ventrobasal Nuclear Complex (Thalamus)
VM$_p$ Ventromedial Nucleus (posterior) (Thalamus)
ZI Zona Incerta

P&W -4.8

Bregma
-4.8

Earbar 0
4.2

Aud

LV

CA3

CA1

DG

PL

IG

FC

Hb

Pfo

Pap

LP

CA3

LGd

LGv

PO VB

VMp

PJ

sPf

cp

ZI

dMR

LHA

SuM

EML

MM

PH

PH

r f

Pir

EP

BLA

CeAp

AHZ

DG

cng

cc

dhc

hc

Hbmi

fr

ml

mt

cp

bsc

ot

ec

LV

st

alv

Plate 25

H₁,H₂: H Fields of Forel. The term derives from the split of myelinated fascicles in a form crudely approximating the letter H in the human brain. The H fields are generally divided into H₁ for the thalamic fascicle entering the rostral portion of the thalamic ventral group dorsal to the zona incerta. Fibers emerging from the globus pallidus of the lenticular nucleus run below the zona incerta in the dorsal part of the ansa lenticularis to form the lenticular fascicle called field H₂. The counterparts and limits of these tracts are clearly less pronounced in the rat brain, although their homology and position are evident as indicated on the plates.

Ptm: Medial Pretectal Area (Thalamus). This region is difficult to recognize as a nucleus because it fuses with the rostral superior colliculus. It was thus called *"area" praetectalis medialis* by J. E. Rose and Woolsey ('43) and *n. praebigeminalis medialis* by M. Rose ('35) in rabbit, but it is not recognized by Scalia ('72), who subsumes a portion within his olivary pretectal nucleus. The designation here is consistent with the useful illustrations of Siminoff et al. ('67) and especially those in Swanson ('92).

STa: Subthalamic Nucleus (accessory part). This group of neurons scattered in the anteromedial sector of the cerebral peduncle is known as the *accessory nucleus of Luys* and is nicely described and illustrated in the three cardinal planes by Watanabe and Kawana ('82). Although the subthalamic nucleus proper in humans has long been known by its eponymic *corpus Luysii*, we employ a new term hesitantly but deliberately to avoid the eponym.

AHZ Amygdalo-Hippocampal Area
alv Alveus
Aq Cerebral Aqueduct
BLA_p Basolateral Nucleus of the Amygdala (posterior)
bsc Brachium of Superior Colliculus
CA₁ Hippocampal Area CA₁ (Ammon's Horn)
CA₃ Hippocampal Area CA₃ (Ammon's Horn)
cc Corpus Callosum
CG Central Gray
cng Cingulum Bundle
CoA_p Cortical Nucleus of the Amygdala (posterior)
cp Cerebral Peduncle
DG Dentate Gyrus
dhc Dorsal Hippocampal Commissure
dMR Deep Mesencephalic Reticular Nucleus (Thalamus)
ec External Capsule
EP Endopiriform Nucleus
FC Fasciola Cinerea
fr Fasciculus Retroflexus
H₁,H₂ H Fields of Forel
hc Habenula Commissure
IG Induseum Griseum
LGd Dorsal Lateral Geniculate Nucleus (Thalamus)
LGv Ventral Lateral Geniculate Nucleus (Thalamus)
LM Lateral Mammillary Nucleus
LP Lateral Posterior Nucleus (Thalamus)
LV Lateral Ventricle
ml Medial Lemniscus
MM Medial Mammillary Nucleus (Thalamus)
mp Mammillary Peduncle
mt Mammillothalamic Tract
ot Optic Tract
pc Posterior Commissure
pcn Nucleus of Posterior Commissure
Pf Parafascicular Nucleus (Thalamus)
PH Posterior Hypothalamic Nucleus
Pir Piriform Area (Cortex)
Po Posterior Group (Thalamus)
Pta Anterior Pretectal Nucleus (Thalamus)
Ptm Medial Pretectal Area (Thalamus)
Pto Olivary Pretectal Nucleus (Thalamus)
rf Rhinal Fissure
sco Subcommissural Organ
sPf Subparafascicular Nucleus (Thalamus)
st Stria Terminalis
STa Subthalamic Nucleus (accessory part)
Sub Subiculum
SUM_l Supramammillary Nucleus (lateral)
SUM_m Supramammillary Nucleus (medial)
sumx Supramammillary Decussation
TM Tuberomammillary Nucleus (Hypothalamus)
V3 Third Ventricle
VB Ventrobasal Nuclear Complex (Thalamus)
ZI Zona Incerta

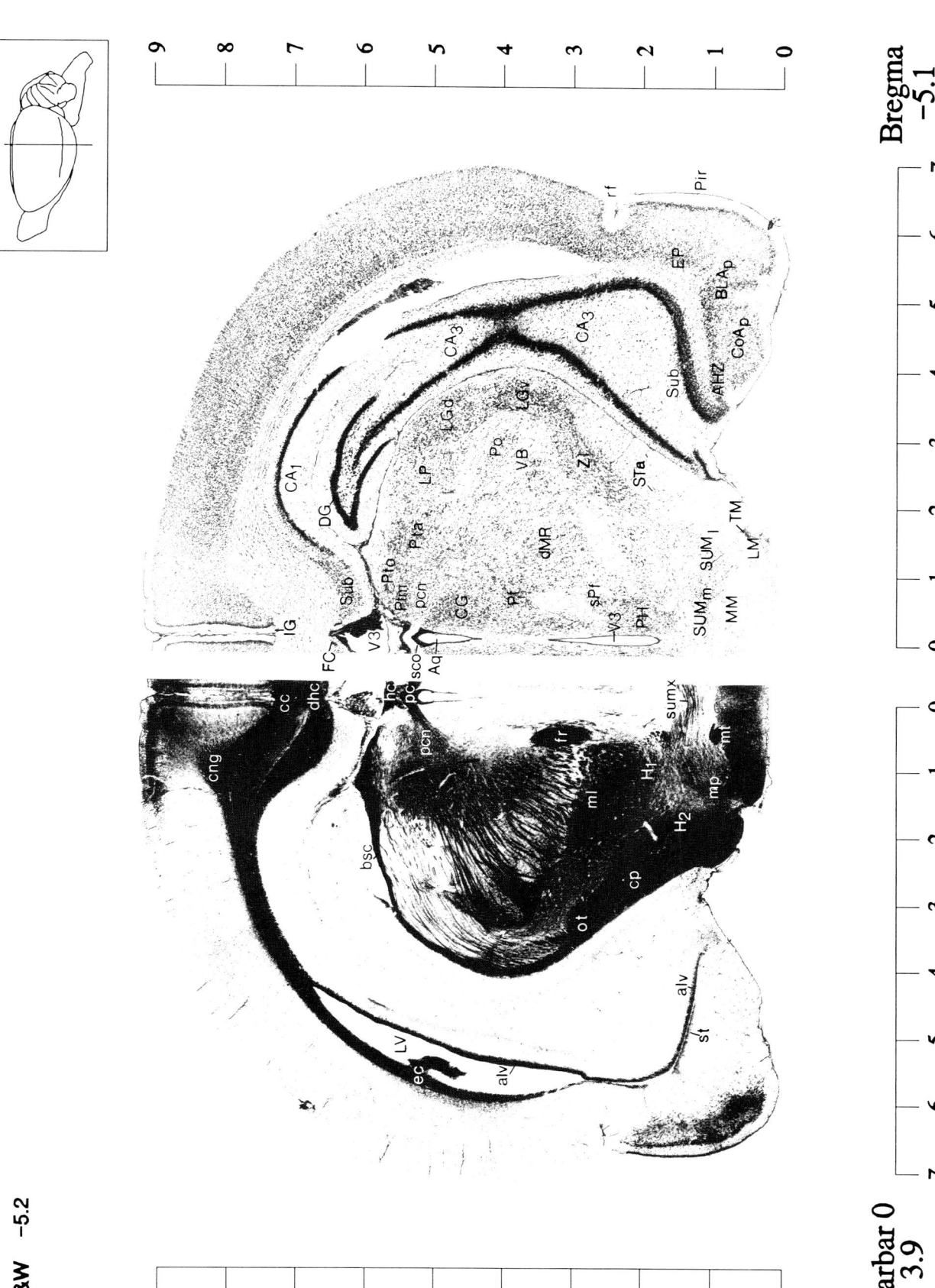

P&W −5.2

Bregma
−5.1

Earbar 0
3.9

Plate 26

AE: Amygdalo-Entorhinal Transition Area. Other than its having an input from the lateral olfactory tract, little is known about this rather expansive part of the temporal lobe, which has been referred to variously as the "area interposed between area entorhinalis caudally and the pyriform cortex and posterior pole of the amygdala rostrally" (Haug, '76), the ventromedial part of the lateral entorhinal area (Krettek and Price, '77b), the amygdalopiriform transition area (De Olmos et al., '85; Paxinos and Watson, '86), and the postpiriform transition area (Swanson, '92).

bsc: Brachium of the Superior Colliculus. A mixed tract, it contains numerous myelinated optic tract axons that continue past the lateral geniculate body lying on the dorsolateral surface of the caudal thalamus, en route to the superior colliculus, principally into the stratum opticum.

Ig: Intermediate Geniculate Nucleus (Thalamus) (also called Intergeniculate Leaflet). This small nucleus, interposed between the ventral nucleus of the lateral geniculate (LGv) and the principal nucleus of the medial geniculate body (MGp), can be seen as an encapsulated nucleus in fiber-stained sections (Plate 26) and is extensively illustrated in transverse and horizontal fiber and Nissl material together with autoradiographic tracing by Feldman and Kruger ('80) in the rat thalamus.

Ot: Nucleus of the Optic Tract (Thalamus). This nucleus lies at the junction of thalamus and midbrain. Indeed it is sometimes considered part of the mesencephalon (Swanson, '92), and it has been called the *n. lentiformis mesencephali magnocellularis* (Kuhlenbeck and Miller, '42; Shintani, '59) and was designated the γ (gamma) *subnucleus of the n. praebigeminalis lateralis* of the rabbit by M. Rose ('35). It was considered the *pretectal nucleus* by Nauta and van Straaten ('47). Scalia ('72), Gregory ('85), and Giolli et al. ('89) provide useful accounts in the rat.

AE Amygdalo-Entorhinal Transition Area
Aq Cerebral Aqueduct
BLA$_p$ Basolateral Nucleus of the Amygdala (posterior)
bsc Brachium of Superior Colliculus
CA$_1$ Hippocampal Area CA$_1$ (Ammon's Horn)
CA$_3$ Hippocampal Area CA$_3$ (Ammon's Horn)
CG Central Gray
cng Cingulum Bundle
CoA$_p$ Cortical Nucleus of the Amygdala (posterior)
cp Cerebral Peduncle
DG Dentate Gyrus
dMR Deep Mesencephalic Reticular Nucleus (Thalamus)
ENT Entorhinal Area
fr Fasciculus Retroflexus
H$_1$,H$_2$ H Fields of Forel
Ig Intermediate Geniculate Nucleus (leaflet) (Thalamus)
LGd Dorsal Lateral Geniculate Nucleus (Thalamus)
LGv Ventral Lateral Geniculate Nucleus (Thalamus)

LM Lateral Mammillary Nucleus
LV Lateral Ventricle
MGm Magnocellular Medial Geniculate Nucleus (Thalamus)
ml Medial Lemniscus
MM Medial Mammillary Nucleus (Hypothalamus)
mp Mammillary Peduncle
Ot Nucleus of Optic Tract
pc Posterior Commissure
PH Posterior Hypothalamic Nucleus
Pi Pineal Gland (Epiphysis)
PL Posterior Limbic Area (Cortex)
Po Posterior Group (Thalamus)
pp Peripeduncular Nucleus
Pta Anterior Pretectal Nucleus (Thalamus)
Pto Olivary Pretectal Nucleus (Thalamus)
rf Rhinal Fissure
sco Subcommissural Organ
SNc Substantia Nigra (compact)
SNr Substantia Nigra (reticular)
Sub Subiculum
SUM Supramammillary Nucleus
sumx Supramammillary Decussation
TM Tuberomammillary Nucleus (Hypothalamus)
VTA Ventral Tegmental Area (of Tsai)
ZI Zona Incerta

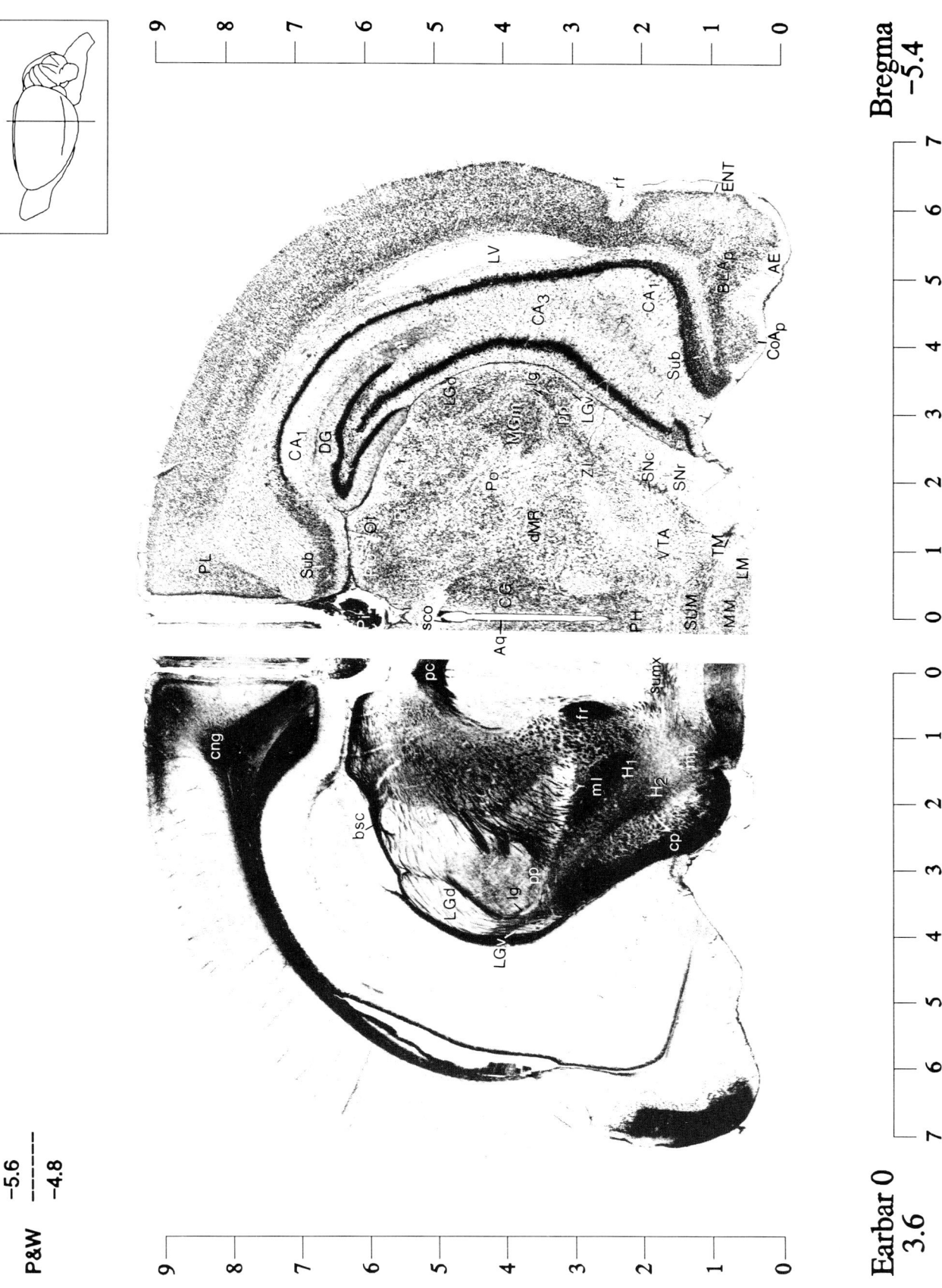

Bregma
−5.4

Earbar 0
3.6

Plate 27

AE Amygdalo-Entorhinal Transition Area
bsc Brachium of Superior Colliculus
CA₁ Hippocampal Area CA$_1$ (Ammon's Horn)
CA₃ Hippocampal Area CA$_3$ (Ammon's Horn)
CG Central Gray
cp Cerebral Peduncle
DG Dentate Gyrus
dMR Deep Mesencephalic Reticular Nucleus
fr Fasciculus Retroflexus
LGd Dorsal Lateral Geniculate Nucleus (Thalamus)
IT Terminal Nuclei of the Accessory Optic Root (basal root)
MGm Magnocellular Medial Geniculate Nucleus (Thalamus)
MGp Principal/Parvicellular Medial Geniculate Nucleus (Thalamus)
ml Medial Lemniscus
MM Medial Mammillary Nucleus (Hypothalamus)
mp Mammillary Peduncle
mT Medial Terminal Nucleus of Accessory Optic Tract
Ot Nucleus of Optic Tract

pc Posterior Commissure
PH Posterior Hypothalamic Nucleus
PL Posterior Limbic Area (Cortex)
Po Posterior Group (Thalamus)
pp Peripeduncular Nucleus
Pta Anterior Pretectal Nucleus (Thalamus)
Ptm Medial Pretectal Area (Thalamus)
Pto Olivary Pretectal Nucleus (Thalamus)
RL Nucleus Raphé Linearis (rostral)
SC Superior Colliculus
sco Subcommissural Organ
Sg Suprageniculate Nucleus (Thalamus)
sgd Deep Gray Layer of Superior Colliculus
sgi Intermediate Gray Layer of Superior Colliculus
sgs Superficial Gray Layer of Superior Colliculus
SN Substantia Nigra
SNc Substantia Nigra (compact)
SNr Substantia Nigra (reticular)
so Optic Layer of Superior Colliculus
Sub Subiculum
SUM Supramammillary Nucleus
sumx Supramammillary Decussation
swi Intermediate White Layer of Superior Colliculus
V3 Third Ventricle
VTA Ventral Tegmental Area (of Tsai)
zo Zonal Lamina of Superior Colliculus

mp: Mammillary Peduncle. This fiber tract is most evident near the base of the hypothalamus just lateral to the lateral mammillary nucleus. It can then be traced caudally into the midbrain where it disappears between the fasciculus retroflexus and the medial tip of the cerebral peduncle (see Shibata, '87).

mT: Medial Terminal Nucleus of the Accessory Optic Tract (also called Nucleus Opticus Tegmenti). At low magnification it is difficult to recognize the organization of this small nucleus lateral to the mammillary body. It is generally separated into dorsal and ventral subnuclei because it is partially separated by optic axons and also because each subnucleus displays distinctive architecture. Excellent experimental studies should be consulted for detailed and more useful illustrations (e.g., Hayhow et al., '60; Giolli et al., '84; Giolli et al., '85).

RL and RLc: Nucleus Raphé Linearis and Nucleus Raphé Linearis (central). These cell groups were named by Castaldi ('23) and do not contain serotonergic neurons. The nomenclature employed by various authors is especially confusing. The RL nucleus of our designation constitutes the rostral portion of the raphé system and is often called the *rostral linear nucleus raphé*; Swanson ('92) assigns the term *central linear nucleus raphé* to the ventral and caudal nucleus labeled RLc here. The most rostral member of the raphé system is also called the *caudal linear nucleus* (Törk, '85). The central part should not be confused with the nucleus raphé linearis caudalis, a term introduced by Castaldi ('23) for the superior central or median raphé nucleus.

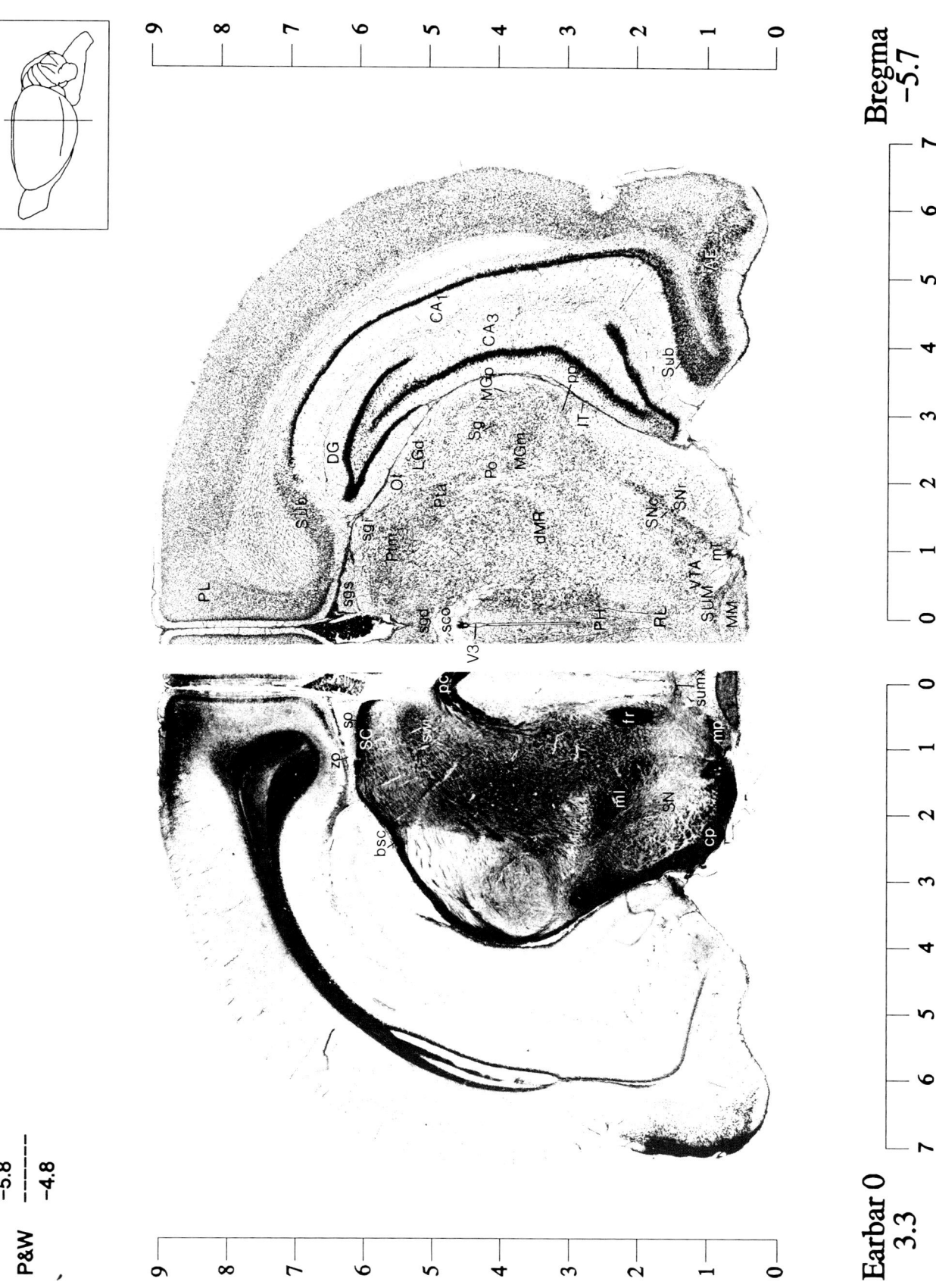

P&W
−5.8
−−−−−
−4.8

Bregma
−5.7

Earbar 0
3.3

9   8   7   6   5   4   3   2   1   0

7   6   5   4   3   2   1   0

0   1   2   3   4   5   6   7

9   8   7   6   5   4   3   2   1   0

PL

Sub

sgd
sgs

sco

scd

V3

Pm1

sgr

Ot

DG

Pta

LGd

Po

Sg

MGm

dMR

MGp

CA1

CA3

pp

IT

Sub

ME

PH

SNc

SNr

RL

SUM

VTA

MM

mt

ml

ZO

SO

SC

bsc

sSN

PC

fr

SN

mt

cp

MM

sumx

mp

Plate 28

AE   Amygdalo-Entorhinal Transition Area
aot   Accessory Optic Tract (basal optic root)
Aq   Cerebral Aqueduct
Ast   Area Striata (Cortex)
Aud   Auditory Area (Cortex)
bsc   Brachium of Superior Colliculus
CA_1 Hippocampal Area CA_1 (Ammon's Horn)
CA_3 Hippocampal Area CA_3 (Ammon's Horn)
CG   Central Gray
cp   Cerebral Peduncle
D   Nucleus of Darkschewitsch
DG   Dentate Gyrus
dMR   Deep Mesencephalic Reticular Nucleus
ENT_l Entorhinal Area (lateral)
fr   Fasciculus Retroflexus
hf   Hippocampal Fissure
IF   Interfascicular Nucleus
Inc   Interstitial Nucleus of Cajal
LGd   Dorsal Lateral Geniculate Nucleus (Thalamus)
LV   Lateral Ventricle
MGp   Principal/Parvicellular Medial Geniculate Nucleus (Thalamus)
ml   Medial Lemniscus
mp   Mammillary Peduncle
mT   Medial Terminal Nucleus of Accessory Optic Tract
pc   Posterior Commissure
PL   Posterior Limbic Area (Cortex)
pp   Peripeduncular Nucleus
Pta   Anterior Pretectal Nucleus (Thalamus)
Ptp   Posterior Pretectal Nucleus (Thalamus)
rf   Rhinal Fissure
RL   Nucleus Raphé Linearis (rostral)
SC   Superior Colliculus
sgd   Deep Gray Layer of Superior Colliculus
sgi   Intermediate Gray Layer of Superior Colliculus
sgs   Superficial Gray Layer of Superior Colliculus
SN   Substantia Nigra
SNc   Substantia Nigra (compact)
SNr   Substantia Nigra (reticular)
so   Optic Layer of Superior Colliculus
Sub   Subiculum
swi   Intermediate White Layer of Superior Colliculus
VTA   Ventral Tegmental Area (of Tsai)
zo   Zonal Lamina of Superior Colliculus
3n   Oculomotor Nerve

D: Nucleus of Darkschewitsch. This small-celled cluster embedded in the central gray has been classed among the "accessory oculomotor nuclei," although there is little anatomical evidence of its participation in oculomotor regulation. The separation of this nucleus from the medially contiguous cluster of parasympathetic preganglionic neurons constituting the nucleus of Edinger-Westphal and from the underlying and slightly caudal interstitial nucleus of Cajal is readily identifiable in myelin-stained sections of the human brain, but in rat the separation can be tenuous although identifiable. When relying solely on cytoarchitecture, it is tempting to recognize a parvicellular component of the oculomotor group (e.g., Paxinos and Watson, '86) because of the indistinct caudal boundary of the nucleus of Darkschewitsch, but the high neuronal packing density in its rostral extent renders acceptable its identification as a distinct nucleus in the rat. Although it may contribute axons to the posterior commissure, it should not be confused with the indistinct larger-celled nucleus of the posterior commissure.

IF: Interfascicular (Raphé) Nucleus. This tiny midline nucleus lies along the base of the midbrain, just caudal to the mammillary body and rostral to the interpeduncular nucleus, between the fasciculus retroflexus on either side. It was identified by Berman ('68) in the cat, and does not contain serotonergic neurons. Whether it should be regarded as part of the raphé nuclear group or as a rostral extension of the interpeduncular nucleus remains undetermined.

VTA: Ventral Tegmental Area (of Tsai). The VTA (of Tsai) is characterized by a dense accumulation of dopaminergic neurons (A_{10} of Dahlström and Fuxe, '64) projecting mainly to limbic regions of the telencephalon, in contrast to the immediately adjacent dopaminergic neurons in the compact part of the substantia nigra that project mainly to the caudoputamen (Phillipson, '79; Swanson, '82; Björklund and Lindvall, '84). There is no boundary between the VTA and the contiguous compact part of the substantia nigra and the cells are too scattered to constitute a nucleus.

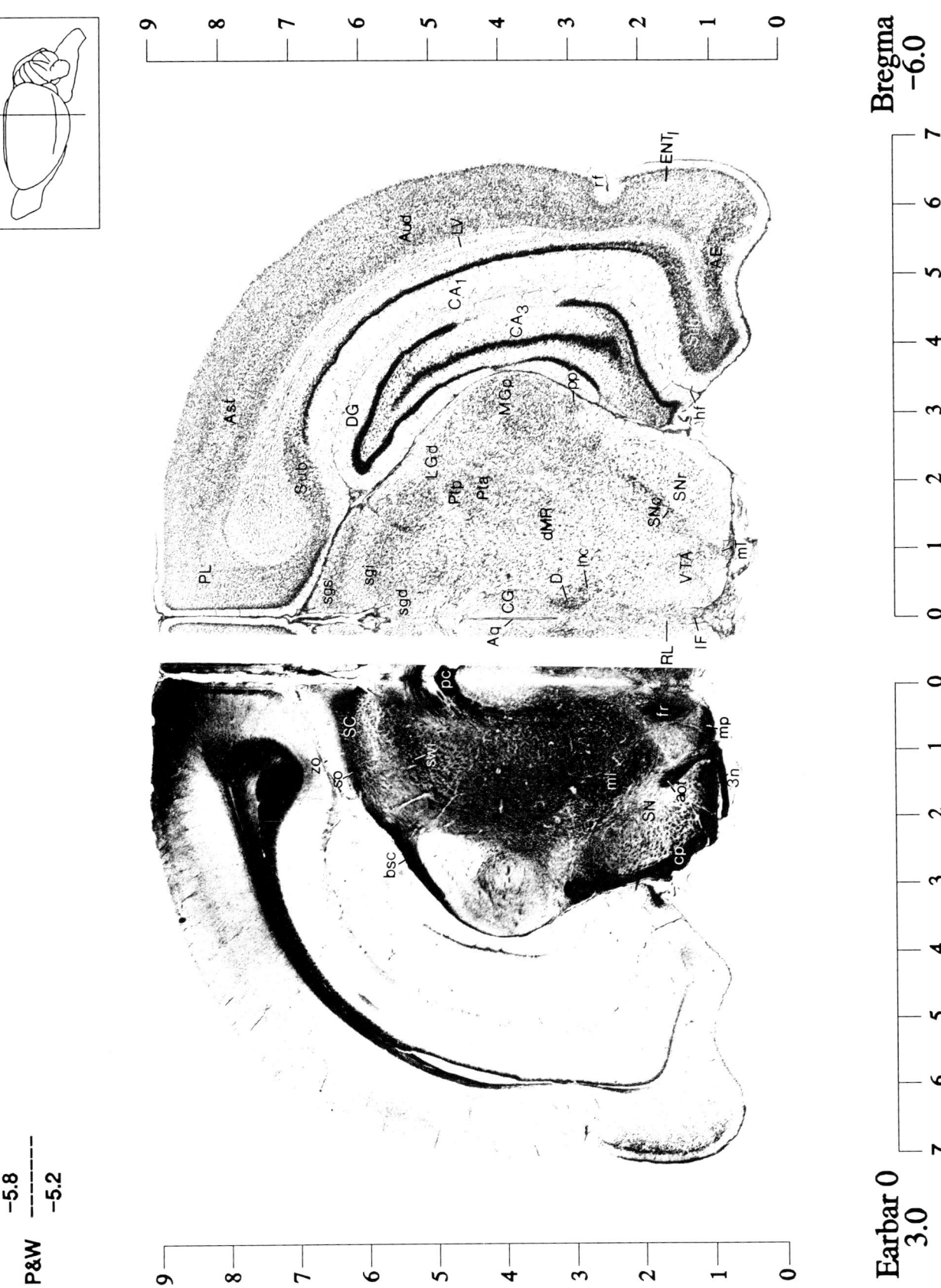

P&W

−5.8
−−−−
−5.2

Bregma
−6.0

Earbar 0
3.0

Plate 29

Aq   Cerebral Aqueduct
Ast   Area Striata (Cortex)
Aud   Auditory Area (Cortex)
CA₁   Hippocampal Area CA$_1$ (Ammon's Horn)
CA₃   Hippocampal Area CA$_3$ (Ammon's Horn)
CG   Central Gray
cp   Cerebral Peduncle
D   Nucleus of Darkschewitsch
DG   Dentate Gyrus
dMR   Deep Mesencephalic Reticular Nucleus
ENT$_l$ Entorhinal Area (lateral)
ENT$_m$ Entorhinal Area (medial)
fr   Fasciculus Retroflexus
hf   Hippocampal Fissure
IF   Interfascicular Nucleus
Inc   Interstitial Nucleus of Cajal
MGp   Principal/Parvicellular Medial Geniculate Nucleus (Thalamus)
ml   Medial Lemniscus

mp   Mammillary Peduncle
pc   Posterior Commissure
PL   Posterior Limbic Area (Cortex)
pp   Peripeduncular Nucleus
Ptp   Posterior Pretectal Nucleus (Thalamus)
RL   Nucleus Raphé Linearis (rostral)
Rup   Red Nucleus (parvicellular)
SC   Superior Colliculus
sgd   Deep Gray Layer of Superior Colliculus
sgi   Intermediate Gray Layer of Superior Colliculus
Sgp   Suprageniculate Nucleus of Pretectal Group
sgs   Superficial Gray Layer of Superior Colliculus
SN   Substantia Nigra
SNc   Substantia Nigra (compact)
SNr   Substantia Nigra (reticular)
so   Optic Layer of Superior Colliculus
Sub   Subiculum
swi   Intermediate White Layer of Superior Colliculus
VTA   Ventral Tegmental Area (of Tsai)
zo   Zonal Lamina of Superior Colliculus
3n   Oculomotor Nerve

CG: Central Gray. It has long been recognized that a poorly differentiated region of gray matter immediately surrounds the ventricular system of the spinal cord as well as of the midbrain and rostral pons. In the spinal cord this region is known as the *substantia gelatinosa centralis, central gray,* or *area X,* and may be involved preferentially in visceral functions. The brain-stem central gray is commonly divided for descriptive purposes into mesencephalic or periaqueductal and pontine regions. A major branch of the periventricular fiber system courses through the central gray. Cajal ('09-'11) referred to it as the *periependymal longitudinal fascicle,* and indicated that one of its components is the dorsal longitudinal fascicle (of Schütz). Some workers have regarded the periaqueductal gray as a dorsal component of the reticular formation, whereas others indicate that at least part of it forms the deepest layer of the superior colliculus. A number of smaller, relatively well defined nuclei lie within the brain-stem central gray, including (from rostral to caudal) the nucleus of Darkschewitsch, dorsal tegmental nucleus of Gudden, laterodorsal tegmental nucleus, locus ceruleus, and supragenual nucleus.

Inc: Interstitial Nucleus of Cajal. A rostral midbrain loose aggregate of mostly large neurons just caudal to the fasciculus retroflexus and ventrolateral to the central (periaqueductal) gray, from which it is separated by the fibers of the medial longitudinal fascicle (mlf). It is easily distinguished from the adjacent dense, small-celled nucleus of Darkschewitsch and other parvicellular nuclei associated with the oculomotor nuclear complex. Rutherford and Gwyn ('82) provide useful photos of coronal sections. In order to avoid confusion with other interstitial nuclei the eponym of Cajal is routinely employed.

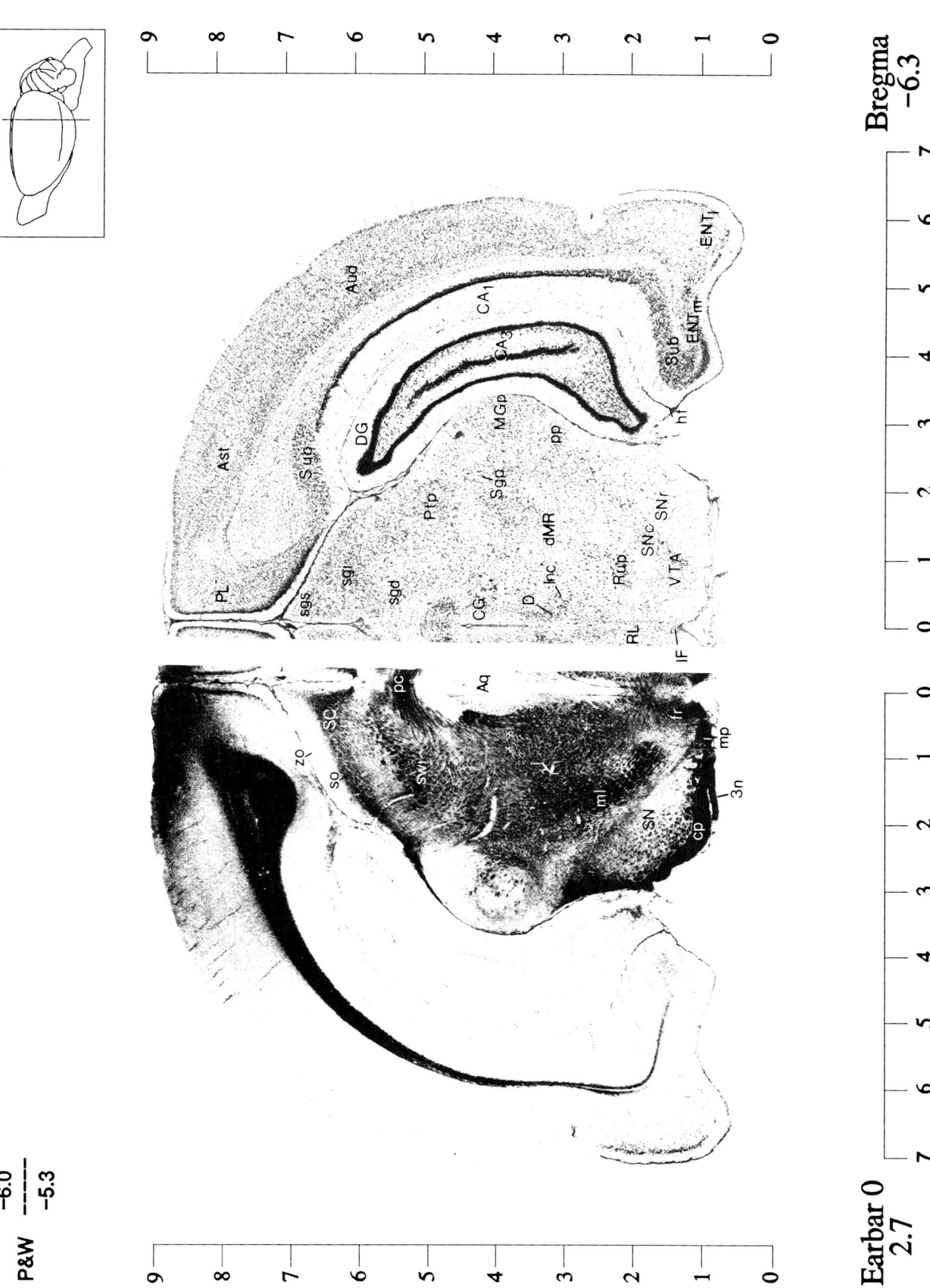

P&W

−6.0
——
−5.3

9
8
7
6
5
4
3
2
1
0

Aud
CA1
Ast
Sub
DG
CA3
MGp
pp
PL
sgi
Prp
Sgp
sgd
dMR
CG
D
Inc
R4p
SNc
SNr
sgs
RL
VTA
IF
Sub
ENTm
hf
ENTl

zo
SC
pc
so
SWi
Aq
ml
V
SN
fr
cp
mp
3n

9
8
7
6
5
4
3
2
1
0

Bregma
−6.3

7
6
5
4
3
2
1
0
0
1
2
3
4
5
6
7

Earbar 0
2.7

Plate 30

Sgp: Suprageniculate Nucleus of Pretectal Group. A small nucleus caudolateral to the posterior pretectal nucleus, recognized as the *nucleus suprageniculatis praetectalis* by Kuhlenbeck and Miller ('42) and J. E. Rose ('42a) and illustrated in horizontal sections in the rat by Siminoff et al. ('67), who found that it is not activated by visual stimuli. It appears to correspond to M. Rose's ('35) *nucleus dorsalis corporis geniculati interni* in rabbit. This should not be confused with the suprageniculate nucleus of the posterior group derived from the embryological dorsal thalamus. J. E. Rose ('42a) included it in the thalamic pretectal group by tracing its origin from the epithalamus, but it clearly lies in the midbrain and because it is not easily recognized in transverse Nissl-stained sections there are numerous discrepancies. In reviewing the nomenclature, Scalia ('72) opines that J. E. Rose's ('42a) nucleus Sgp is the part of his nucleus of the optic tract that migrates caudolaterally, and this is probably correct. The original literature is recommended only to the courageous scholar.

Ast Area Striata (Cortex)
Aud Auditory Area (Cortex)
CA₁ Hippocampal Area CA₁ (Ammon's Horn)
CG Central Gray
cp Cerebral Peduncle
DG Dentate Gyrus
dMR Deep Mesencephalic Reticular Nucleus
ENT Entorhinal Area
IP Interpeduncular Nucleus
MGp Principal/Parvicellular Medial Geniculate Nucleus (Thalamus)
ml Medial Lemniscus
mp Mammillary Peduncle
pc Posterior Commissure
pp Peripeduncular Nucleus
Ptp Posterior Pretectal Nucleus (Thalamus)
rf Rhinal Fissure
RL Nucleus Raphé Linearis (rostral)
Rsp Retrosplenial Area (Cortex)
Rum Red Nucleus (magnocellular)
Rup Red Nucleus (parvicellular)
SC Superior Colliculus
sgd Deep Gray Layer of Superior Colliculus
sgi Intermediate Gray Layer of Superior Colliculus
Sgp Suprageniculate Nucleus of Pretectal Group
sgs Superficial Gray Layer of Superior Colliculus
SN Substantia Nigra
SNc Substantia Nigra (compact)
SNl Substantia Nigra (lateral)
SNr Substantia Nigra (reticular)
so Optic Layer of Superior Colliculus
Sub Subiculum
swi Intermediate White Layer of Superior Colliculus
VTA Ventral Tegmental Area (of Tsai)
vtd Ventral Tegmental Decussation
zo Zonal Lamina of Superior Colliculus

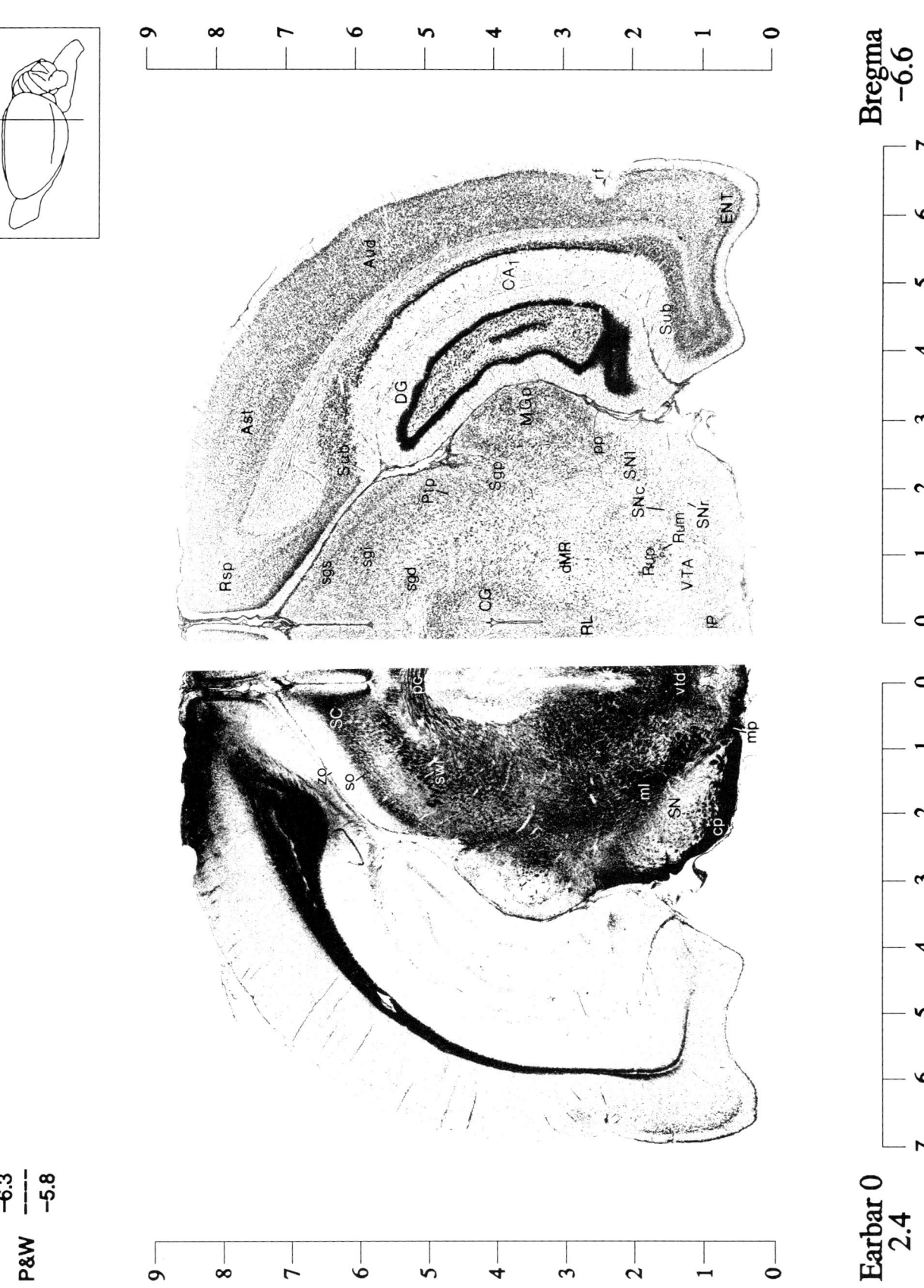

P&W
−6.3
—
−5.8

Bregma
−6.6

Earbar 0
2.4

Plate 31

Aq: Cerebral Aqueduct. This channel, bearing the eponym of Sylvius, provides the cerebrospinal fluid communication between the third and fourth ventricles. It consists of a cuboidal-to-columnar layer of ependymal cells surrounding a central canal formed by the development of the neural tube.

IP: Interpeduncular Nucleus. This bilaterally symmetrical, unpaired nucleus forms the midline base of the midbrain along with the rostrally adjacent interfascicular nucleus. Its major input arrives from the habenula via the fasciculus retroflexus, and it can be divided into six subnuclei (Groenewegen et al., '86).

alv Alveus
Aq Cerebral Aqueduct
Ast Area Striata (Cortex)
Aud Auditory Area (Cortex)
CG Central Gray
cp Cerebral Peduncle
DG Dentate Gyrus
dhc Dorsal Hippocampal Commissure
dMR Deep Mesencephalic Reticular Nucleus
ENT Entorhinal Area
EW Edinger-Westphal Nucleus
IF Interfascicular Nucleus
IP Interpeduncular Nucleus
MGp Principal/Parvicellular Medial Geniculate Nucleus (Thalamus)
ml Medial Lemniscus
mp Mammillary Peduncle
pp Peripeduncular Nucleus
RL Nucleus Raphé Linearis (rostral)
RLc Nucleus Raphé Linearis (central)
Rsp Retrosplenial Area (Cortex)
Rum Red Nucleus (magnocellular)
Rup Red Nucleus (parvicellular)
SC Superior Colliculus
sgd Deep Gray Layer of Superior Colliculus
sgi Intermediate Gray Layer of Superior Colliculus
sgs Superficial Gray Layer of Superior Colliculus
SN Substantia Nigra
SNc Substantia Nigra (compact)
SNl Substantia Nigra (lateral)
SNr Substantia Nigra (reticular)
so Optic Layer of Superior Colliculus
Sub Subiculum
swi Intermediate White Layer of Superior Colliculus
VTA Ventral Tegmental Area (of Tsai)
vtd Ventral Tegmental Decussation
zo Zonal Lamina of Superior Colliculus
5s Trigeminal Nerve Root (sensory or major division)

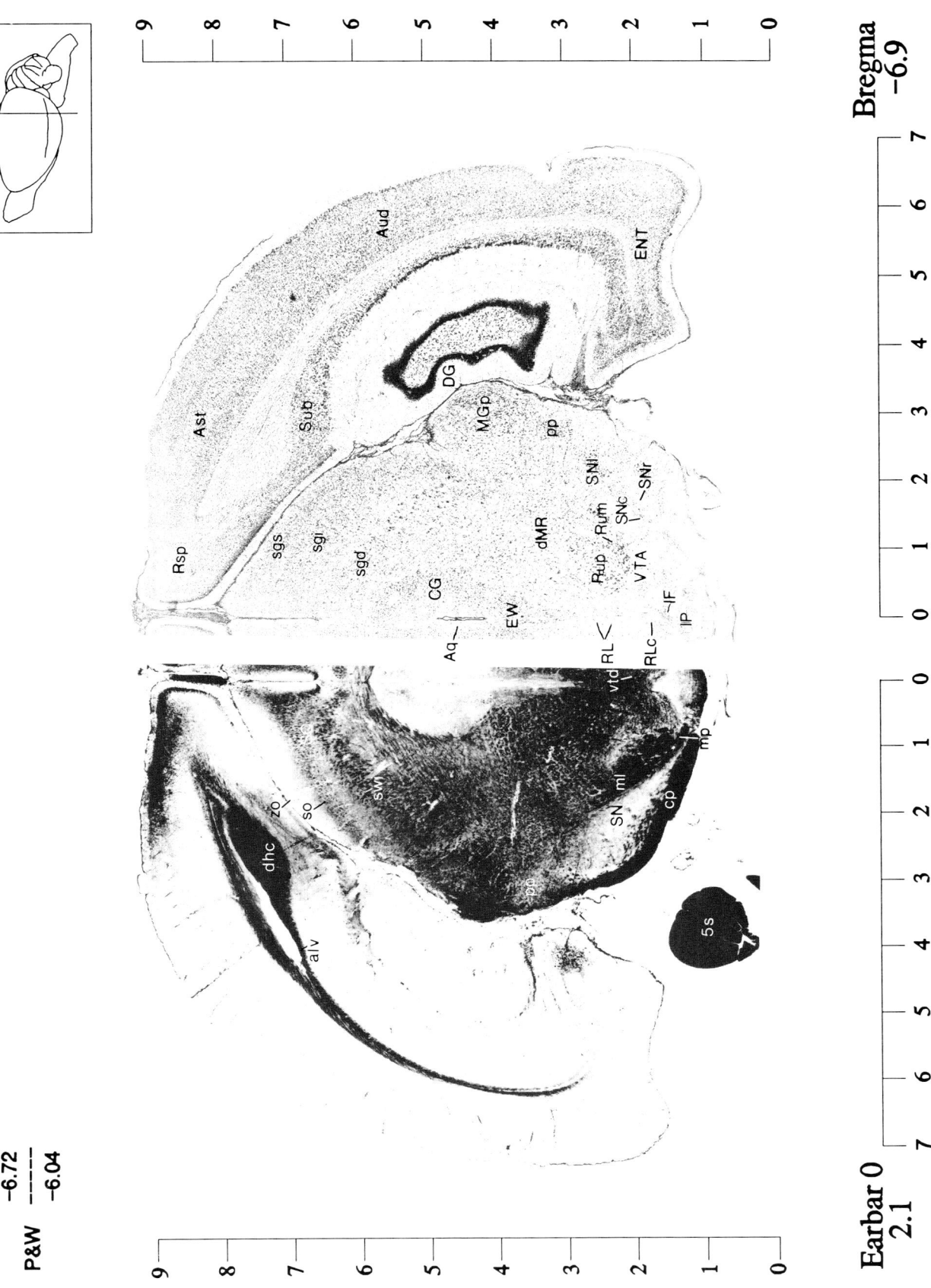

P&W —6.72
      ——— —6.04

Bregma
−6.9

Earbar 0
2.1

Plate 32

ab Angular Bundle
Aq Cerebral Aqueduct
Ast Area Striata (Cortex)
BI Nucleus of the Brachium Inferior Colliculus
bic Brachium of Inferior Colliculus
BIs Nucleus of the Brachium Inferior Colliculus (subbrachial sector)
CG Central Gray
cp Cerebral Peduncle
DG Dentate Gyrus
dMR Deep Mesencephalic Reticular Nucleus
ENT₁ Entorhinal Area (lateral)
ENTₘ Entorhinal Area (medial)
EW Edinger-Westphal Nucleus
IF Interfascicular Nucleus
IP Interpeduncular Nucleus
ml Medial Lemniscus
mlf Medial Longitudinal Fascicle
Pi Pineal Gland (Epiphysis)
rf Rhinal Fissure
RLc Nucleus Raphé Linearis (central)

Rsp Retrosplenial Area (Cortex)
Rum Red Nucleus (magnocellular)
Rup Red Nucleus (parvicellular)
SC Superior Colliculus
sgd Deep Gray Layer of Superior Colliculus
sgi Intermediate Gray Layer of Superior Colliculus
sgs Superficial Gray Layer of Superior Colliculus
SN Substantia Nigra
SNc Substantia Nigra (compact)
SNl Substantia Nigra (lateral)
SNr Substantia Nigra (reticular)
so Optic Layer of Superior Colliculus
Sub Subiculum
swi Intermediate White Layer of Superior Colliculus
VTA Ventral Tegmental Area (of Tsai)
vtd Ventral Tegmental Decussation
zo Zonal Lamina of Superior Colliculus
3 Oculomotor Nucleus
5m Trigeminal Nerve Root (motor or minor division)
5s Trigeminal Nerve Root (sensory or major division)

BI(s): Nucleus of the Brachium Inferior Colliculus (s, subbrachial sector). A ventral component of the nucleus of the brachium of the inferior colliculus can be recognized and probably corresponds to the *subbrachial nucleus* of Paxinos and Watson ('86), but this region has been encumbered with a profusion of names and inadequate illustrations for resolving possible overlaps. Some authors follow Cajal ('11) in recognizing a *suprapeduncular nucleus* (e.g., Castaldi, '23, in guinea pig), probably distinct from the *nucleus sagulum* (Sag), which is a caudal extension of this region indicated as Sag by Swanson ('92), but this appears to be at least largely overlapping the *microcellular tegmental nucleus* of Paxinos and Watson ('86), a somewhat idiosyncratic usage. More serious discrepancies can be found in the indication of *nucleus sagulum* by these various authors, although this nucleus is easily identified lying just medial to the distinct, compact parabigeminal nucleus (Pbg).

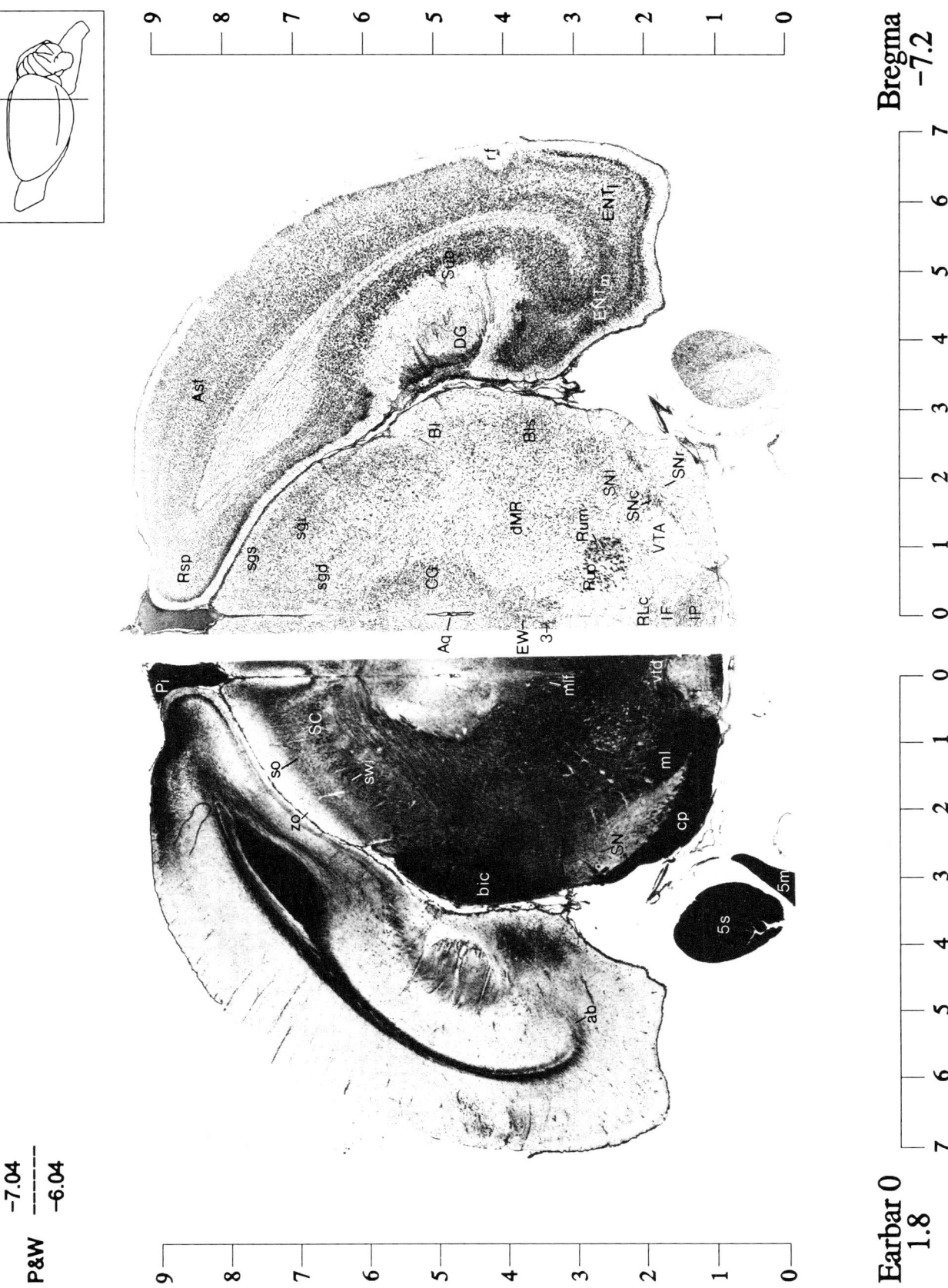

P&W

-7.04
-6.04

Bregma
-7.2

Earbar 0
1.8

Plate 33

ab Angular Bundle
Aq Cerebral Aqueduct
Ast Area Striata (Cortex)
Bl Nucleus of the Brachium Inferior Colliculus
bic Brachium of Inferior Colliculus
CG Central Gray
cp Cerebral Peduncle
dMR Deep Mesencephalic Reticular Nucleus
ENT$_l$ Entorhinal Area (lateral)
ENT$_m$ Entorhinal Area (medial)
IP Interpeduncular Nucleus
mcp Middle Cerebellar Peduncle
ml Medial Lemniscus
mlf Medial Longitudinal Fascicle
Pbg Parabigeminal Nucleus
Pi Pineal Gland (Epiphysis)
PpT Pedunculopontine Tegmental Nucleus
rf Rhinal Fissure
RLc Nucleus Raphé Linearis (central)
Rsp Retrosplenial Area (Cortex)
Rum Red Nucleus (magnocellular)
Sag Nucleus Sagulum
SC Superior Colliculus
sgd Deep Gray Layer of Superior Colliculus
sgi Intermediate Gray Layer of Superior Colliculus
sgs Superficial Gray Layer of Superior Colliculus
SN Substantia Nigra
so Optic Layer of Superior Colliculus
Sub Subiculum
swi Intermediate White Layer of Superior Colliculus
VTA Ventral Tegmental Area (of Tsai)
vtd Ventral Tegmental Decussation
zo Zonal Lamina of Superior Colliculus
3 Oculomotor Nucleus
5m Trigeminal Nerve Root (motor or minor division)
5s Trigeminal Nerve Root (sensory or major division)

Sag: Nucleus Sagulum. This can be recognized most easily in horizontal sections, where it lies caudal to the nucleus of the brachium of the inferior colliculus (Bl) and rostral to the nucleus of the lateral lemniscus (LL); it is capped laterally by a distinct dense parabigeminal nucleus (Pbg). In transverse sections the Sag of Swanson ('92) clearly includes the *microcellular tegmental nucleus* of Paxinos and Watson ('86), which we have been unable to identify as a distinct entity.

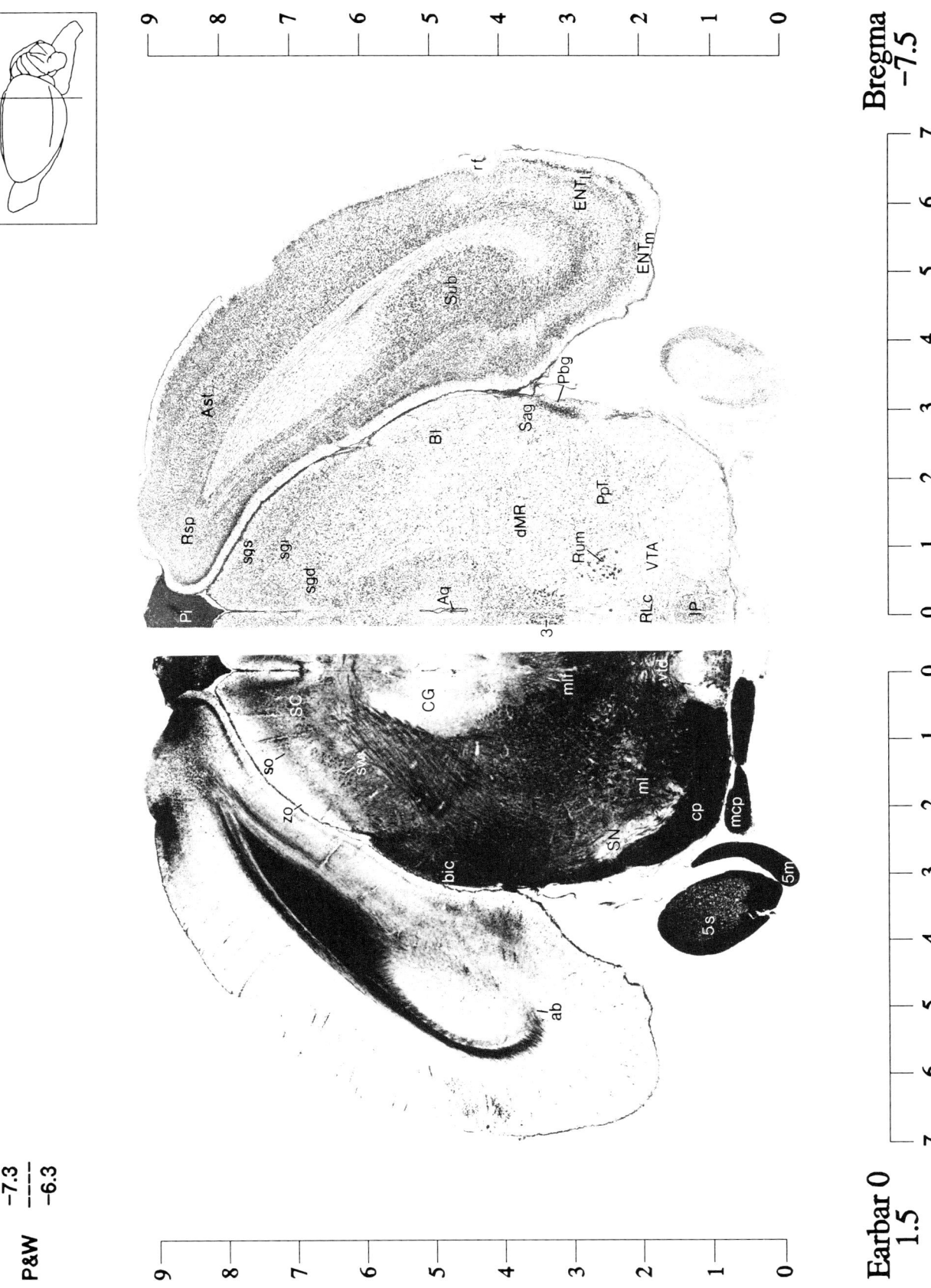

Bregma
−7.5

Earbar 0
1.5

Plate 34

Rsp: Retrosplenial Area (Cortex). The posterior limbic region was divided into an *area cingularis* (cingulate) and *area retrosplenialis* in the rabbit cortex by Rose and Woolsey (48), indicating dorsal and ventral fields, a nomenclature that can be applied to the rat cortex. The retrosplenial area surrounds the caudal part of the corpus callosum (the splenium) interposed between the cingulate field (area 23 of Brodmann) and the subicular fields of the hippocampal gyrus, displaying a gradient of thinning ventrally. It was given separate status by Rose and Woolsey, based partly on the dependency of the thalamic anterodorsal nucleus upon this cortical area, but is subsumed with the subicular fields by some anatomists and is partially coextensive with Brodmann's area 29 based on cytoarchitecture. *Retrosplenial* implies "behind the splenium" of the corpus callosum, but unfortunately various authors have extended this cortical field forward to occupy the zone overlying the caudal half of the body of the corpus callosum to meet the anterior limbic cortex (Krettek and Price, '77c). Thus Rsp and *posterior limbic* have been used for the same region.

Aq Cerebral Aqueduct
Ast Area Striata (Cortex)
BI Nucleus of the Brachium Inferior Colliculus
bic Brachium of Inferior Colliculus
CG Central Gray
dMR Deep Mesencephalic Reticular Nucleus
ENT Entorhinal Area
IP Interpeduncular Nucleus
LL_d Nucleus of Lateral Lemniscus (dorsal)
LL_v Nucleus of Lateral Lemniscus (ventral)
mcp Middle Cerebellar Peduncle
ml Medial Lemniscus
mlf Medial Longitudinal Fascicle
ms5 Mesencephalic Nucleus of the Trigeminal
Pbg Parabigeminal Nucleus
Pd Pontine Gray (Deep)
Pi Pineal Gland (Epiphysis)
PpT Pedunculopontine Tegmental Nucleus
RLc Nucleus Raphé Linearis (central)
Rsp Retrosplenial Area (Cortex)
SC Superior Colliculus
sgd Deep Gray Layer of Superior Colliculus
sgi Intermediate Gray Layer of Superior Colliculus
sgs Superficial Gray Layer of Superior Colliculus
so Optic Layer of Superior Colliculus
swi Intermediate White Layer of Superior Colliculus
VTA Ventral Tegmental Area (of Tsai)
vtd Ventral Tegmental Decussation
zo Zonal Lamina of Superior Colliculus
4 Trochlear Nucleus
5m Trigeminal Nerve Root (motor or minor division)
5s Trigeminal Nerve Root (sensory or major division)

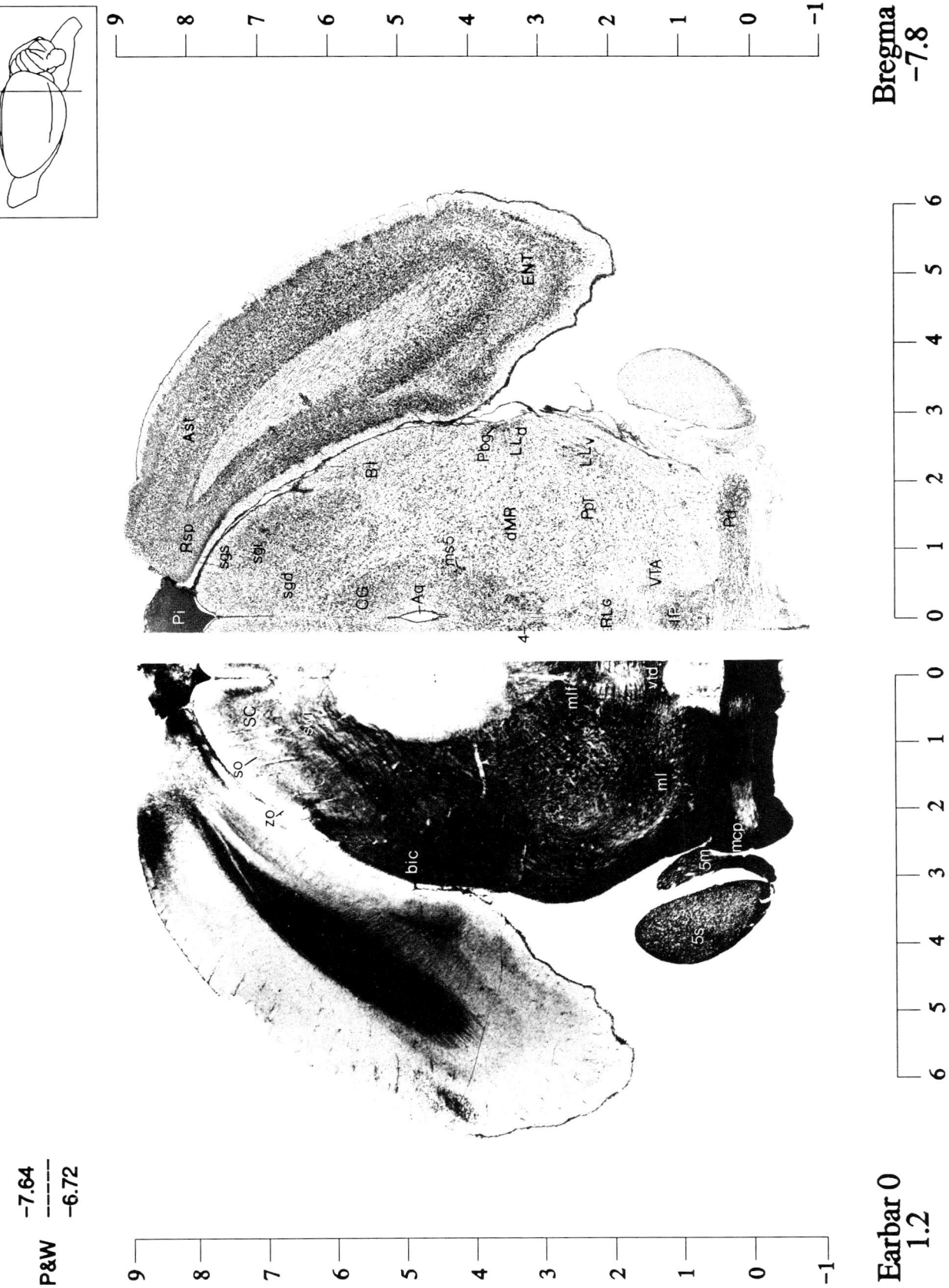

Plate 35

Rd: Dorsal Raphé Nucleus. The dorsal raphé nucleus presents a special problem because some authors recognize an unpaired median component surrounded by a more loosely packed lateral division, although these lateral wings probably warrant separate status and are readily susceptible to further subdivision. Some authors designate the lateral division of the dorsal raphé nucleus the *pars ventromedialis of the dorsal tegmental nucleus* (Morest, '61; Cowan et al., '64). Others recognize an *annular nucleus of the medial longitudinal fasciculus* (see Castaldi's ('23) account in guinea pig) for the condensed ventral component of the lateral expansion of the dorsal raphé nucleus overlying this conspicuous fiber tract. Paxinos and Watson ('86) outline the lateral wing as one of at least three subdivisions of the Rd, but do not employ their usual practice of providing names for distinct entities.

Rm: Median Raphé Nucleus (also called Nucleus Raphé Medianus, Medial Raphé Nucleus, Superior Central Nucleus, and Group B8). This is the most easily defined of the raphé nuclei. The zone can be detected in unstained sections by the presence of numerous small vessels at its lateral margins. Its separation from the rostrally contiguous interpeduncular nucleus and caudally from the dorsally contiguous dorsal raphé nucleus is nicely illustrated with serotonin immunohistochemistry by Törk ('85). The nucleus raphé pontis (RP) is a tiny, poorly defined cell group lying just caudal to the median raphé nucleus (e.g., see Brown '43).

Aq Cerebral Aqueduct
Ast Area Striata (Cortex)
bic Brachium of Inferior Colliculus
CG Central Gray
dMR Deep Mesencephalic Reticular Nucleus
ENT Entorhinal Area
IC Inferior Colliculus
IP Interpeduncular Nucleus
LL$_d$ Nucleus of Lateral Lemniscus (dorsal)
LL$_v$ Nucleus of Lateral Lemniscus (ventral)
ll Lateral Lemniscus
mcp Middle Cerebellar Peduncle
ml Medial Lemniscus
mlf Medial Longitudinal Fascicle
ms5 Mesencephalic Nucleus of the Trigeminal
Pbg Parabigeminal Nucleus
Pd Pontine Gray (Deep)
Pi Pineal Gland (Epiphysis)
PpT Pedunculopontine Tegmental Nucleus
Rd Dorsal Raphé Nucleus
RLc Nucleus Raphé Linearis (central)
Rm Median Raphé Nucleus
SC Superior Colliculus
scpx Decussation of the Superior Cerebellar Peduncle
sgd Deep Gray Layer of Superior Colliculus
sgi Intermediate Gray Layer of Superior Colliculus
sgs Superficial Gray Layer of Superior Colliculus
so Optic Layer of Superior Colliculus
swi Intermediate White Layer of Superior Colliculus
VTA Ventral Tegmental Area (of Tsai)
zo Zonal Lamina of Superior Colliculus
4 Trochlear Nucleus
5m Trigeminal Nerve Root (motor or minor division)
5s Trigeminal Nerve Root (sensory or major division)

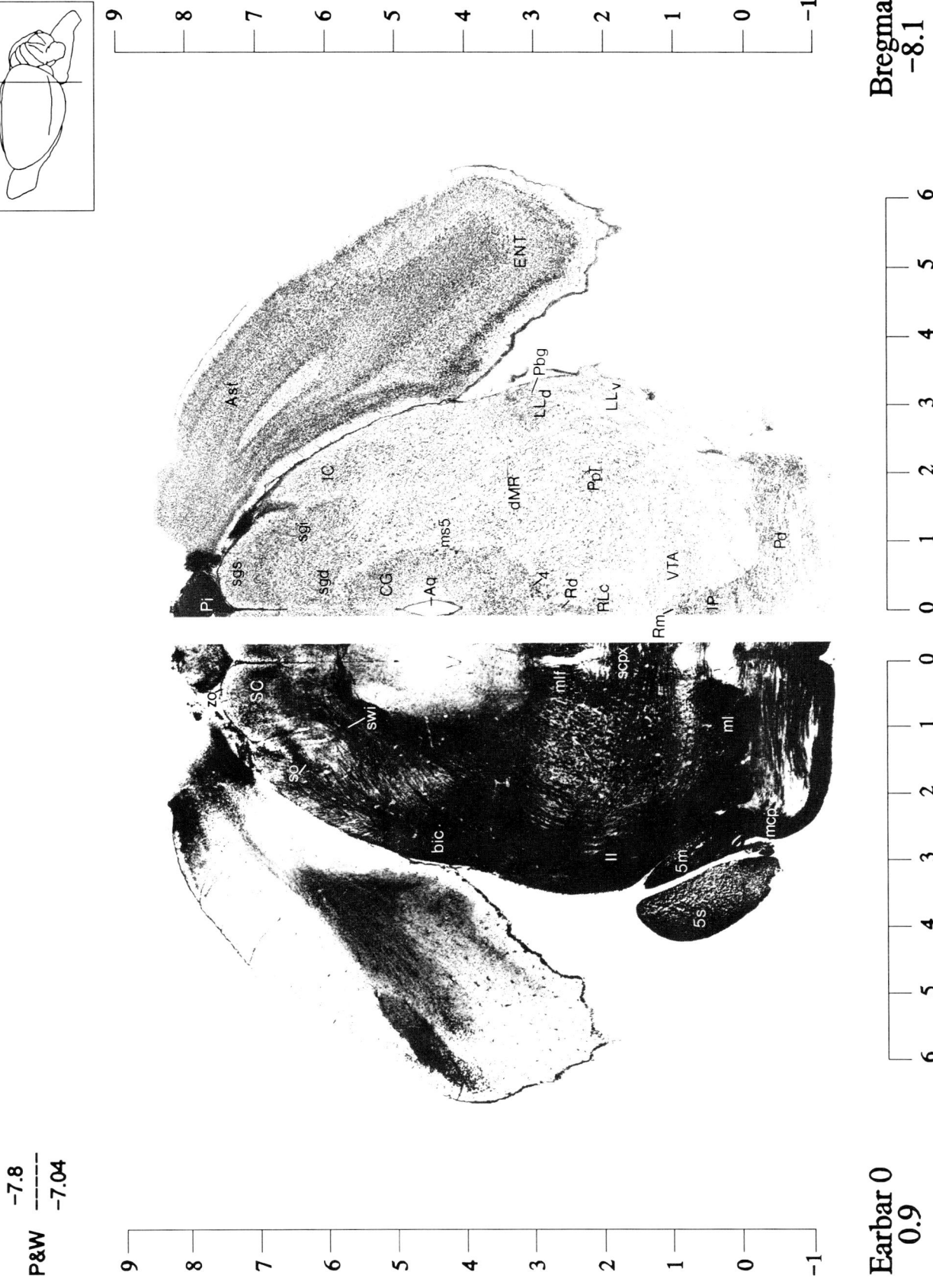

P&W −7.8
      ―――
      −7.04

Bregma
−8.1

Earbar 0
0.9

Plate 36

Aq Cerebral Aqueduct
Ast Area Striata (Cortex)
bic Brachium of Inferior Colliculus
CG Central Gray
Cun Cuneiform Nucleus
ENT$_l$ Entorhinal Area (lateral)
ENT$_m$ Entorhinal Area (medial)
IC Inferior Colliculus
IP Interpeduncular Nucleus
LL$_d$ Nucleus of Lateral Lemniscus (dorsal)
LL$_v$ Nucleus of Lateral Lemniscus (ventral)
ll Lateral Lemniscus
mcp Middle Cerebellar Peduncle
ml Medial Lemniscus
mlf Medial Longitudinal Fascicle
ms5 Mesencephalic Nucleus of the Trigeminal
Pbg Parabigeminal Nucleus
Pd Pontine Gray (Deep)
Pi Pineal Gland (Epiphysis)
PpT Pedunculopontine Tegmental Nucleus
Rd Dorsal Raphé Nucleus
Rm Median Raphé Nucleus
SC Superior Colliculus
scp Superior Cerebellar Peduncle
sgd Deep Gray Layer of Superior Colliculus
sgi Intermediate Gray Layer of Superior Colliculus
sgs Superficial Gray Layer of Superior Colliculus
so Optic Layer of Superior Colliculus
swi Intermediate White Layer of Superior Colliculus
VTA Ventral Tegmental Area (of Tsai)
zo Zonal Lamina of Superior Colliculus
4 Trochlear Nucleus
5m Trigeminal Nerve Root (motor or minor division)
5s Trigeminal Nerve Root (sensory or major division)

Ast: Area Striata (Cortex). This commonly retained misnomer for the primary visual cortex in rodents derives its name from the horizontal stripe (stria) visible by naked eye in the visual cortex of higher primates. It was thereby recognized as a morphologically distinct cortical area by Gennari in 1776. The homologous area in rat cortex lacks the distinctive horizontal stripe and thus its boundary is less accurately determined; but there is little controversy concerning the general location, which was identified by several independent experimental tracing methods, including constituting the principal projection field of the dorsal lateral geniculate nucleus.

Pd: Pontine Gray (Deep or basal). The neurons in this transverse bridge across the brain stem are separated into groups by longitudinal fascicles of the corticospinal tract and give rise to the middle cerebellar peduncle.

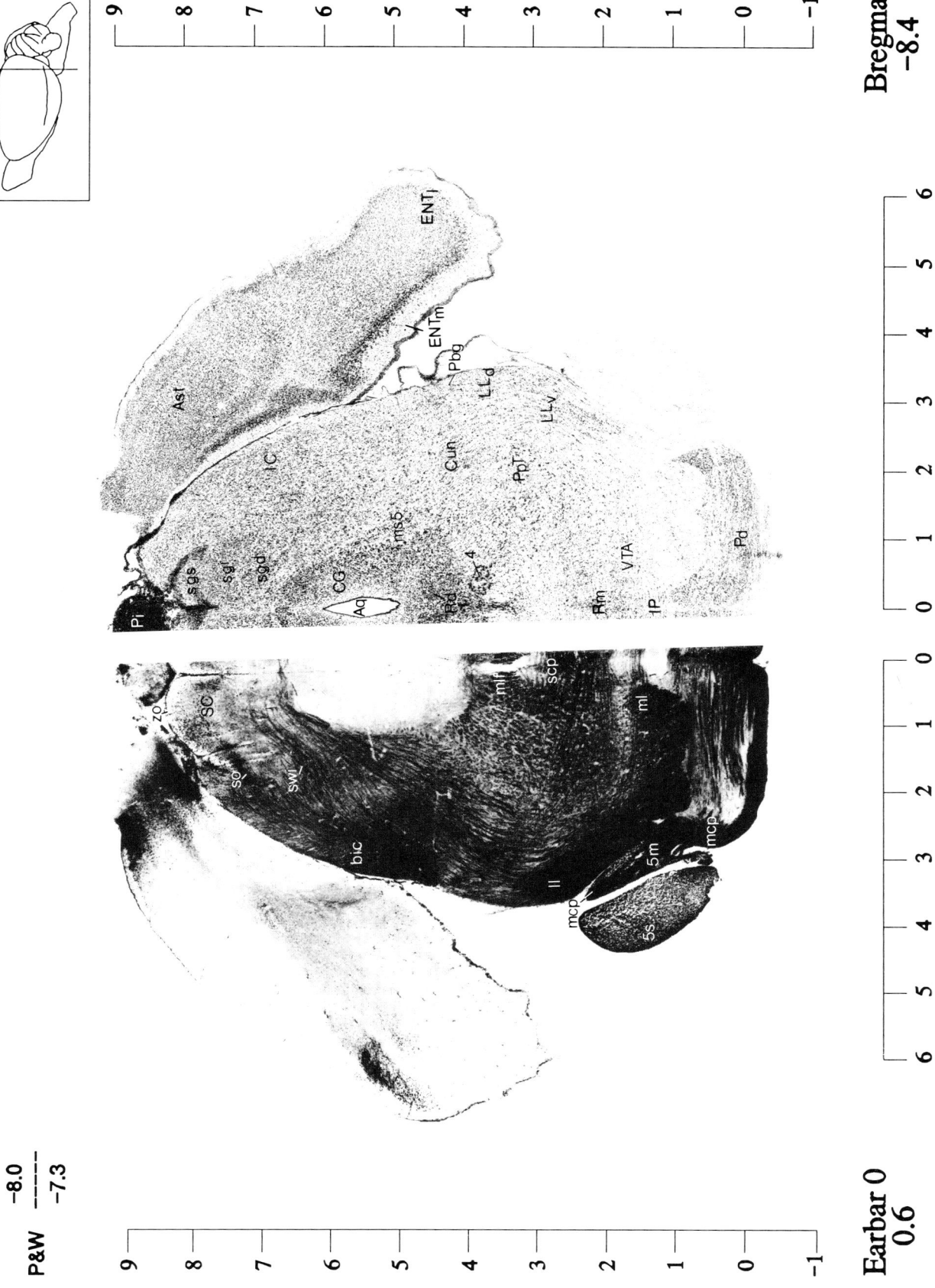

P&W
−8.0
—
−7.3

Bregma
−8.4

Earbar 0
0.6

Plate 37

Aq Cerebral Aqueduct
Ast Area Striata (Cortex)
bic Brachium of Inferior Colliculus
CG Central Gray
Cun Cuneiform Nucleus
ENT_m Entorhinal Area (medial)
IC Inferior Colliculus
ICc Central Nucleus of Inferior Colliculus
LL_d Nucleus of Lateral Lemniscus (dorsal)
LL_v Nucleus of Lateral Lemniscus (ventral)
ll Lateral Lemniscus
mcp Middle Cerebellar Peduncle
ml Medial Lemniscus
mlf Medial Longitudinal Fascicle
ms5 Mesencephalic Nucleus of the Trigeminal
Pd Pontine Gray (Deep)
PpT Pedunculopontine Tegmental Nucleus
PR_o Pontine Reticular Nucleus (oral)
py Pyramidal Tract
Rd Dorsal Raphé Nucleus
Rm Median Raphé Nucleus
x Needle Track
5s Trigeminal Nerve Root (sensory or major division)

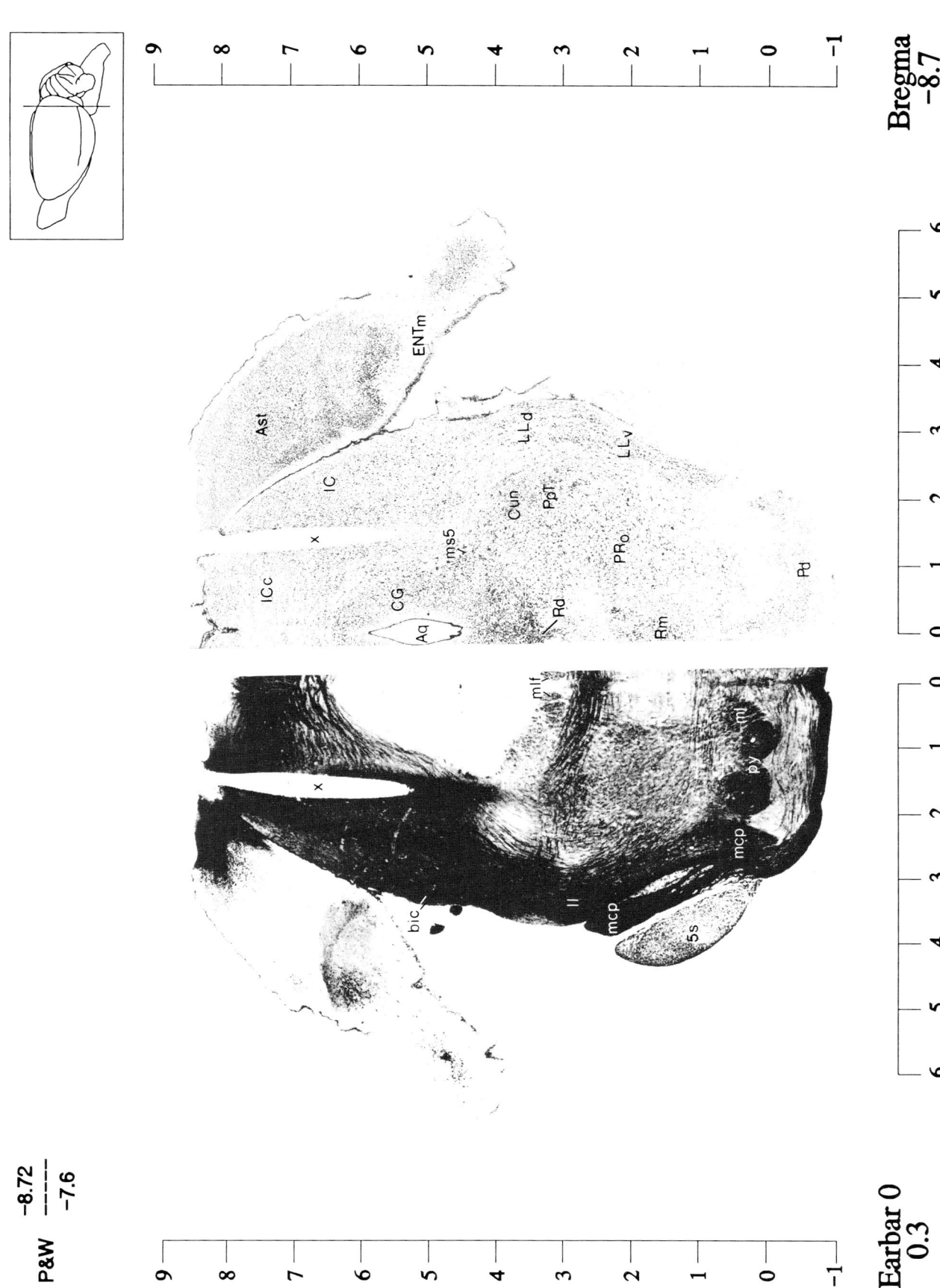

P&W −8.72 ——— −7.6

Bregma −8.7

Earbar 0 0.3

Plate 38

Ast Area Striata (Cortex)
bic Brachium of Inferior Colliculus
Cbl Cerebellum
CG Central Gray
Cun Cuneiform Nucleus
IC Inferior Colliculus
ICc Central Nucleus of Inferior Colliculus
LDT Laterodorsal Tegmental Nucleus
LL$_d$ Nucleus of Lateral Lemniscus (dorsal)
LL$_v$ Nucleus of Lateral Lemniscus (ventral)
ll Lateral Lemniscus
mcp Middle Cerebellar Peduncle
ml Medial Lemniscus
mlf Medial Longitudinal Fascicle
ms5 Mesencephalic Nucleus of the Trigeminal
Pd Pontine Gray (Deep)
PpT Pedunculopontine Tegmental Nucleus
PR$_o$ Pontine Reticular Nucleus (oral)
py Pyramidal Tract
Rd Dorsal Raphé Nucleus
Rm Median Raphé Nucleus
Rpm Paramedian Raphé Nucleus
scp Superior Cerebellar Peduncle
TR Tegmental Reticular Nucleus
x Needle Track
4n Trochlear Nerve
5s Trigeminal Nerve Root (sensory or major division)

Cun: Cuneiform Nucleus. This term has been applied in different ways to the rat brain stem, and we follow the usage described by Castaldi ('26), Olszewski and Baxter ('54), and Swanson et al. ('84). Thus, it is the relatively cell-dense region just ventral to the inferior colliculus and dorsal to the parabrachial nucleus and pedunculopontine tegmental nucleus from which it is readily distinguished.

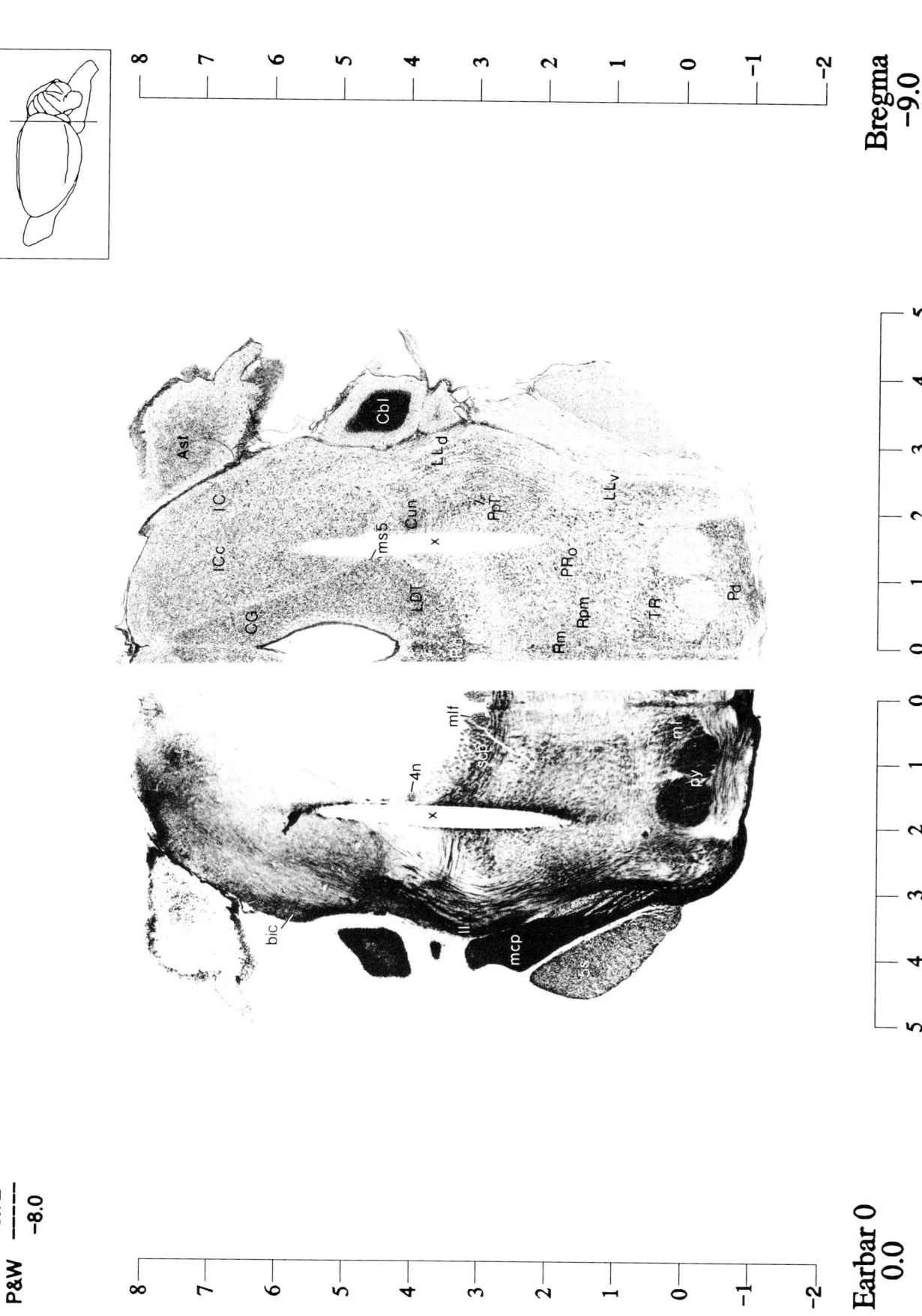

P&W
−8.72
−8.0

Earbar 0
0.0

8
7
6
5
4
3
2
1
0
−1
−2

5
4
3
2
1
0

0
1
2
3
4
5

8
7
6
5
4
3
2
1
0
−1
−2

Ast
IC
ICc
CG
LDT
Cbl
LLd
Cun
ms5
PpT
7
LLv
PRo
Rm
Rpm
TR
Pd
Po

4n
mlf
Su3
ll
bic
mcp
x
ml
py

Plate 39

Cun Cuneiform Nucleus
DT Dorsal Tegmental Nucleus
ICc Central Nucleus of Inferior Colliculus
KF Kölliker-Fuse Nucleus
LDT Laterodorsal Tegmental Nucleus
LL$_v$ Nucleus of Lateral Lemniscus (ventral)
ll Lateral Lemniscus
mcp Medial Cerebellar Peduncle
ml Medial Lemniscus
mlf Middle Longitudinal Fascicle
ms5 Mesencephalic Nucleus of the Trigeminal
Olep External Periolivary Nucleus
PB$_l$ Parabrachial Nucleus (lateral)
PpT Pedunculopontine Tegmental Nucleus
PR$_o$ Pontine Reticular Nucleus (oral)
py Pyramidal Track
Rd Dorsal Raphé Nucleus
Rm Median Raphé Nucleus
Rpm Paramedian Raphé Nucleus
scp Superior Cerebellar Peduncle
TR Tegmental Reticular Nucleus
vsct Ventral Spinocerebellar Tract
VT Ventral Tegmental Nucleus
x Needle Track
5s Trigeminal Nerve Root (sensory or major division)

Olep: External Periolivary Nucleus. Ventral to the superior olivary complex lies a scattered group of cells (called the *external periolivary nucleus* by Harrison and Warr, '62) corresponding to the lateral, and perhaps part of the ventral, subnucleus of the trapezoid body in rabbit according to Meessen and Olszewski ('49). Paxinos and Watson ('86) recognize a medioventral and a lateral periolivary nucleus here but do not draw a line of separation. We have elected to designate this region the *external periolivary nucleus* (Olep) in order to avoid conflict with multiple previous usages and also because we cannot subdivide these nuclei in a consistent manner. Paxinos and Watson ('86) designate a separate rostral component capping the head of the superior olivary complex that they call the *rostral periolivary region*. Although their distinction is recognized and can be seen as a separate entity in the sagittal plane (Plate 111), we have included this region in the external periolivary nucleus (Olep). The portion of this rostral external periolivary nucleus external to the lateral lemniscus (Plate 39) is called the *paralemniscal nucleus* by Paxinos and Watson ('86).

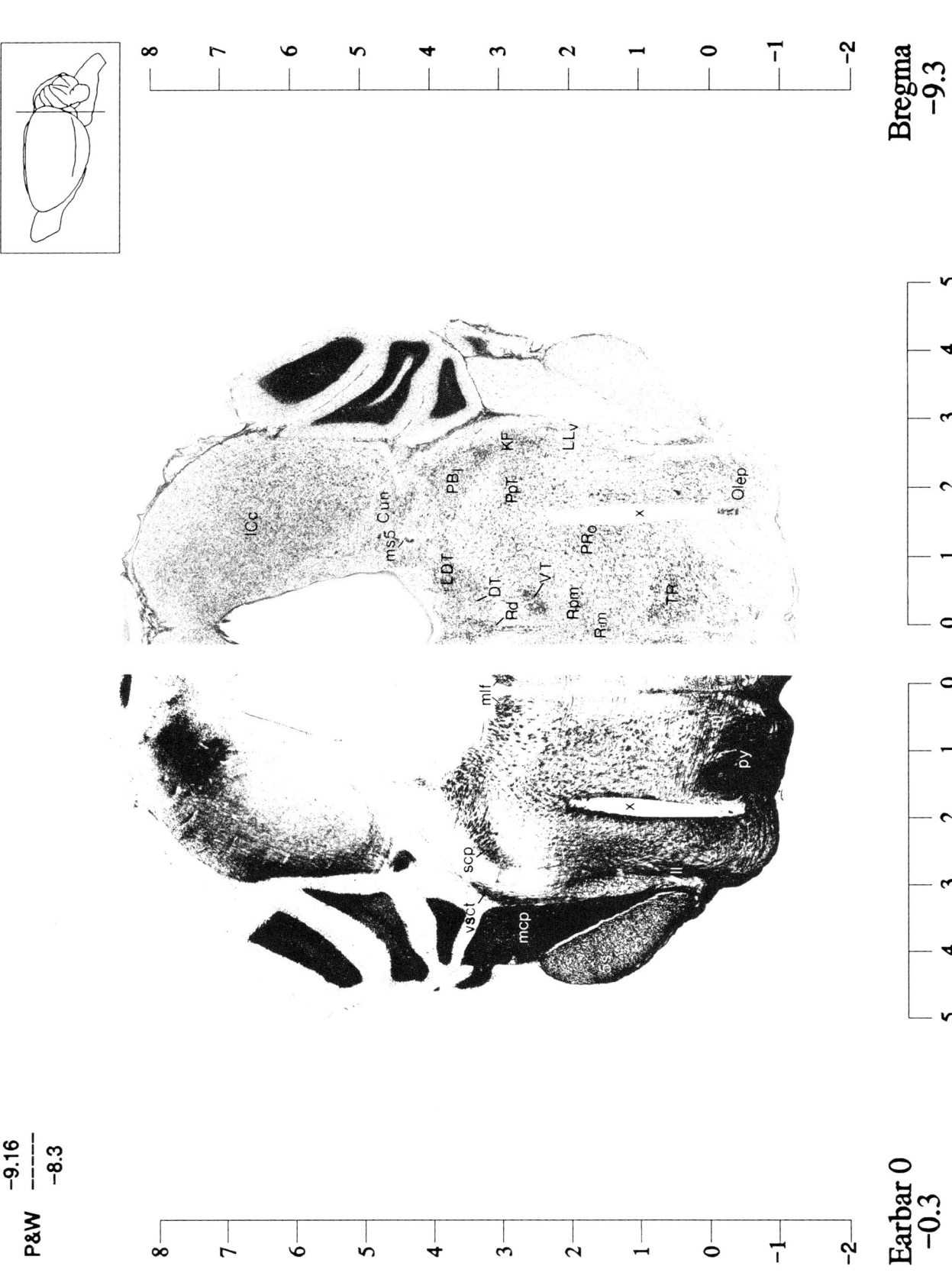

P&W
—9.16
—8.3

Bregma
−9.3

Earbar 0
−0.3

Plate 40

DT$_d$ Dorsal Tegmental Nucleus (dorsal)
DT$_v$ Dorsal Tegmental Nucleus (ventral)
IC Inferior Colliculus
KF Kölliker-Fuse Nucleus
LDT Laterodorsal Tegmental Nucleus
ll Lateral Lemniscus
mcp Middle Cerebellar Peduncle
ml Medial Lemniscus
mlf Medial Longitudinal Fascicle
ms5 Mesencephalic Nucleus of the Trigeminal
ms5t Mesencephalic Tract (root) of the Trigeminal
Olep External Periolivary Nucleus
Olsp Superior Paraolivary Nucleus
PB$_l$ Parabrachial Nucleus (lateral)
PB$_m$ Parabrachial Nucleus (medial)
Pr5 Principal Sensory Trigeminal Nucleus
PR$_o$ Pontine Reticular Nucleus (oral)
py Pyramidal Tract
Rd Dorsal Raphé Nucleus
Rdl Dorsal Raphé Nucleus (lateral extension)
RP Nucleus Raphé Pontis
scp Superior Cerebellar Peduncle
SO$_l$ Superior Olivary Complex (lateral)
TR Tegmental Reticular Nucleus
vsct Ventral Spinocerebellar Tract
VT Ventral Tegmental Nucleus
5s Trigeminal Nerve Root (sensory or major division)
7n Facial Nerve

LDT: Laterodorsal Tegmental Nucleus (also called Dorsolateral Tegmental Nucleus). This darkly stained aggregate embedded in the ventrolateral central gray of the tegmentum of the pons is most easily recognized with Nissl stains and has been variously subdivided. Tohyama et al. ('78) advocate recognizing a magnocellular central core with a parvicellular surround and identify a separate lateral subdivision but rely heavily on hypothalamic connections of this region (also see Cornwall et al., '90; Swanson, '92). A distinct sublaterodorsal nucleus lying ventral to LDT, dorsal to the pedunculopontine nucleus, and medial to the medial (ventral) parabrachial nucleus is clearly illustrated by Swanson ('92) and apparently includes the ventral LDT nucleus and a portion of the nucleus subcoeruleus of Paxinos and Watson ('86). The nucleus indicated as the LDT in this atlas can be selectively labeled by nitric oxide synthase immunoreactivity excluding the adjacent structures (Onstott et al., '93).

scp: Superior Cerebellar Peduncle (also called the Brachium Conjunctivum). This large bundle can be seen as a distinct crescent-shaped tract ventrolateral to the fourth ventricle, containing the principal efferent path from the cerebellum, derived from the deep cerebellar nuclei. Most of the axons can be traced to their decussation in the base of the caudal midbrain (bcx).

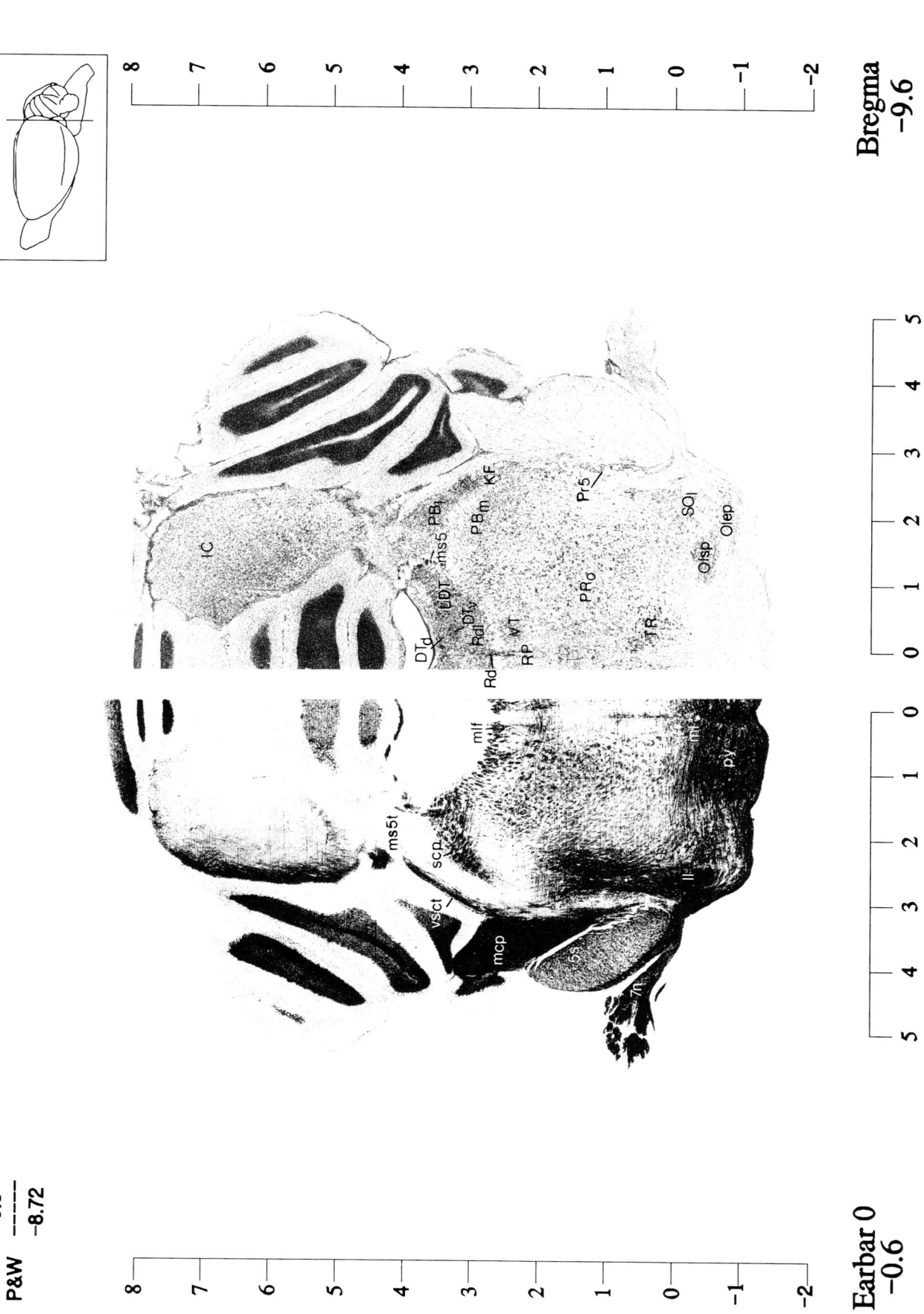

P&W
−9.3
−−−
−8.72

8

7

6

5

4

3

2

1

0

−1

−2

Earbar 0
−0.6

Bregma
−9.6

8

7

6

5

4

3

2

1

0

−1

−2

5

4

3

2

1

0

5

4

3

2

1

0

−1

−2

Plate 41

COd Cochlear Nucleus (dorsal)
COv Cochlear Nucleus (ventral anterior)
DTd Dorsal Tegmental Nucleus (dorsal)
DTv Dorsal Tegmental Nucleus (ventral)
KF Kölliker-Fuse Nucleus
LDT Laterodorsal Tegmental Nucleus
m5 Motor Trigeminal Nucleus
mcp Middle Cerebellar Peduncle
ml Medial Lemniscus
mlf Medial Longitudinal Fascicle
ms5 Mesencephalic Nucleus of the Trigeminal
Olep External Periolivary Nucleus
Olsp Superior Paraolivary Nucleus
PBl Parabrachial Nucleus (lateral)
PBm Parabrachial Nucleus (medial)
Pr5 Principal Sensory Trigeminal Nucleus
PRc Pontine Reticular Nucleus (caudal)
PRo Pontine Reticular Nucleus (oral)
PRv Pontine Reticular Nucleus (ventral)
py Pyramidal Tract
Rd Dorsal Raphé Nucleus
Rdll Dorsal Raphé Nucleus (lateral extension)
RP Nucleus Raphé Pontis
S5 Supratrigeminal Nucleus
scp Superior Cerebellar Peduncle
SOl Superior Olivary Complex (lateral)
SOm Superior Olivary Complex (medial)
TB Nucleus of the Trapezoid Body
tb Trapezoid Body
TR Tegmental Reticular Nucleus
5r Trigeminal Root
7n Facial Nerve
8n Vestibular-Cochlear (Acoustic) Nerve

DT: Dorsal Tegmental Nucleus. The dorsal tegmental nucleus (of von Gudden) can be subdivided into dorsal and ventral subnuclei based on their projections to the mammillary body region. Historical precedent might dictate acknowledging Gudden's contribution of retrograde atrophy of the cell bodies of these nuclei in young rabbits after hypothalamic lesions, but Gudden's nomenclature of a and b subdivisions has not endured. Modern descriptions generally recognize dorsal and ventral sectors, the latter containing larger and more scattered cells. The description employed by Hayakawa and Zyo ('83) in several mammals is followed in this atlas, but the dorsal and ventral subnuclei are not readily discernible throughout and thus are labeled only in a few selected sections (Plates 40 and 41). Separation on the basis of projection (the ventral and dorsal components principally to the medial and lateral mammillary nuclei respectively) has been elaborated by degeneration and retrograde labeling methods (Cowan et al., '64, and Petrovicky, '85). A more detailed subdivision is offered by Morest ('61). Paxinos and Butcher ('85) also recognize a small wedge-shaped *spheroid nucleus* dorsal to the dorsal tegmental nucleus. A densely packed nucleus at the lower margin is called the *lateral extension of the dorsal raphé nucleus* (see Rd: Dorsal Raphé Nucleus at Plate 35) and an *anterior extension of the central gray* by some authors, including Paxinos and Watson ('86), who further subdivide the DT into central and pericentral regions. Also called *Gudden a* in rabbit, the nucleus proprius of the central gray substance can be subdivided (see Morest '61 and Cowan et al., '64) into anterior and posterior divisions, readily apparent in horizontal Plate 88 and sagittal Plates 100 and 102. The ventromedial division probably constitutes the lateral division of the dorsal raphé nucleus of some descriptions.

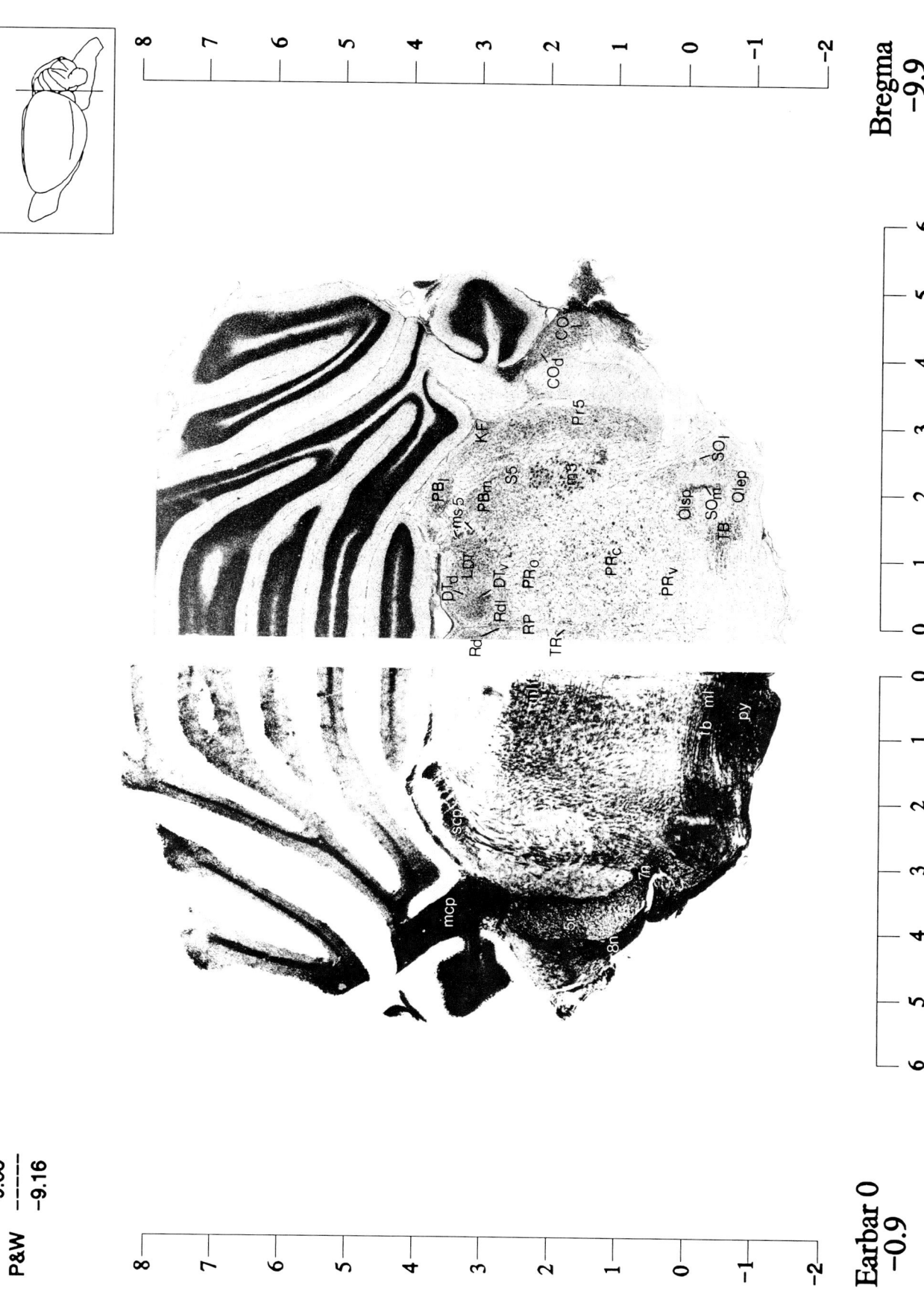

P&W −9.68
      −9.16

Bregma
−9.9

Earbar 0
−0.9

Plate 42

KF: Kölliker-Fuse Nucleus. A loose cluster of large multipolar neurons contiguous with the ventrolateral corner of the lateral (dorsal) parabrachial nucleus capping the superior cerebellar peduncle (see Fulwiler and Saper, '84).

TB: Nucleus of the Trapezoid Body. This compact nucleus, overlying the pyramidal tract and embedded in fibers of the trapezoid body, is sometimes divided into medial and ventral subdivisions following Meessen and Olszewski ('49) in rabbit. But general usage for the rat refers to a single, easily recognized nucleus. We have employed the term *trapezoid body* (tb) for the gross structure and the fibers associated with it.

TR: Tegmental Reticular Nucleus. This nucleus is susceptible to various interpretations and subdivisions and lies lateral to the midline fibers of the middle cerebellar peduncle and above the medial lemniscus and corticospinal fibers. It contains large, deeply stained neurons, but lateral and ventral components contain smaller scattered cells that are sometimes included with the contiguous deep pontine gray (Pd) or given separate status as a parvicellular tegmental reticular nucleus (Torigoe et al., '86). Our designation should be regarded as conservative and tentative, anticipating further anatomical analysis and consensus based on experimental studies.

CO$_d$	Cochlear Nucleus (dorsal)
CO$_v$	Cochlear Nucleus (ventral anterior)
DT	Dorsal Tegmental Nucleus
KF	Kölliker-Fuse Nucleus
LC	Locus Ceruleus
LDT	Laterodorsal Tegmental Nucleus
m5	Motor Trigeminal Nucleus
ml	Medial Lemniscus
mlf	Medial Longitudinal Fascicle
Olep	External Periolivary Nucleus
Olsp	Superior Paraolivary Nucleus
PB$_l$	Parabrachial Nucleus (lateral)
PB$_m$	Parabrachial Nucleus (medial)
Pr5	Principal Sensory Trigeminal Nucleus
PR$_c$	Pontine Reticular Nucleus (caudal)
PR$_o$	Pontine Reticular Nucleus (oral)
PR$_v$	Pontine Reticular Nucleus (ventral)
py	Pyramidal Tract
Rd	Dorsal Raphé Nucleus
Rdl	Dorsal Raphé Nucleus (lateral extension)
RM	Nucleus Raphé Magnus
RP	Nucleus Raphé Pontis
S5	Supratrigeminal Nucleus
scp	Superior Cerebellar Peduncle
SO$_l$	Superior Olivary Complex (lateral)
SO$_m$	Superior Olivary Complex (medial)
TB	Nucleus of the Trapezoid Body
tb	Trapezoid Body
TR	Tegmental Reticular Nucleus
5t	Trigeminal tract
7n	Facial Nerve
8n	Vestibular-Cochlear (Acoustic) Nerve

P&W

−10.04
−−−−−
−9.3

Bregma
−10.2

Earbar 0
−1.2

Plate 43

CO_d Cochlear Nucleus (dorsal)
CO_v Cochlear Nucleus (ventral anterior)
DT Dorsal Tegmental Nucleus
KF Kölliker-Fuse Nucleus
LC Locus Ceruleus
LDT Laterodorsal Tegmental Nucleus
m5 Motor Trigeminal Nucleus
ml Medial Lemniscus
mlf Medial Longitudinal Fascicle
ms5 Mesencephalic Nucleus of Trigeminal
Olep External Periolivary Nucleus
Olsp Superior Paraolivary Nucleus
PB_m Parabrachial Nucleus (medial)
Pr5 Principal Sensory Trigeminal Nucleus
PR_c Pontine Reticular Nucleus (caudal)
PR_v Pontine Reticular Nucleus (ventral)
py Pyramidal Tract
Rdl Dorsal Raphé Nucleus (lateral extension)
RM Nucleus Raphé Magnus
RP Nucleus Raphé Pontis
S5 Supratrigeminal Nucleus
scp Superior Cerebellar Peduncle
SO_l Superior Olivary Complex (lateral)
SO_m Superior Olivary Complex (medial)
TB Nucleus of the Trapezoid Body
tb Trapezoid Body
TR Tegmental Reticular Nucleus
5t Trigeminal Tract
7n Facial Nerve
8n Vestibular-Cochlear (Acoustic) Nerve

Barrington's Nucleus. We have not designated this nucleus, implicated on functional grounds in cardiorespiratory regulation, because of uncertainty in recognizing it on strictly cytoarchitectonic grounds. However, its location below the laterodorsal tegmental nucleus at the ventral margin of the central gray has been mapped by electrophysiological methods, and it can be shown to belong to the parasympathetic control system by retroviral labeling (Haxhiu et al., '93).

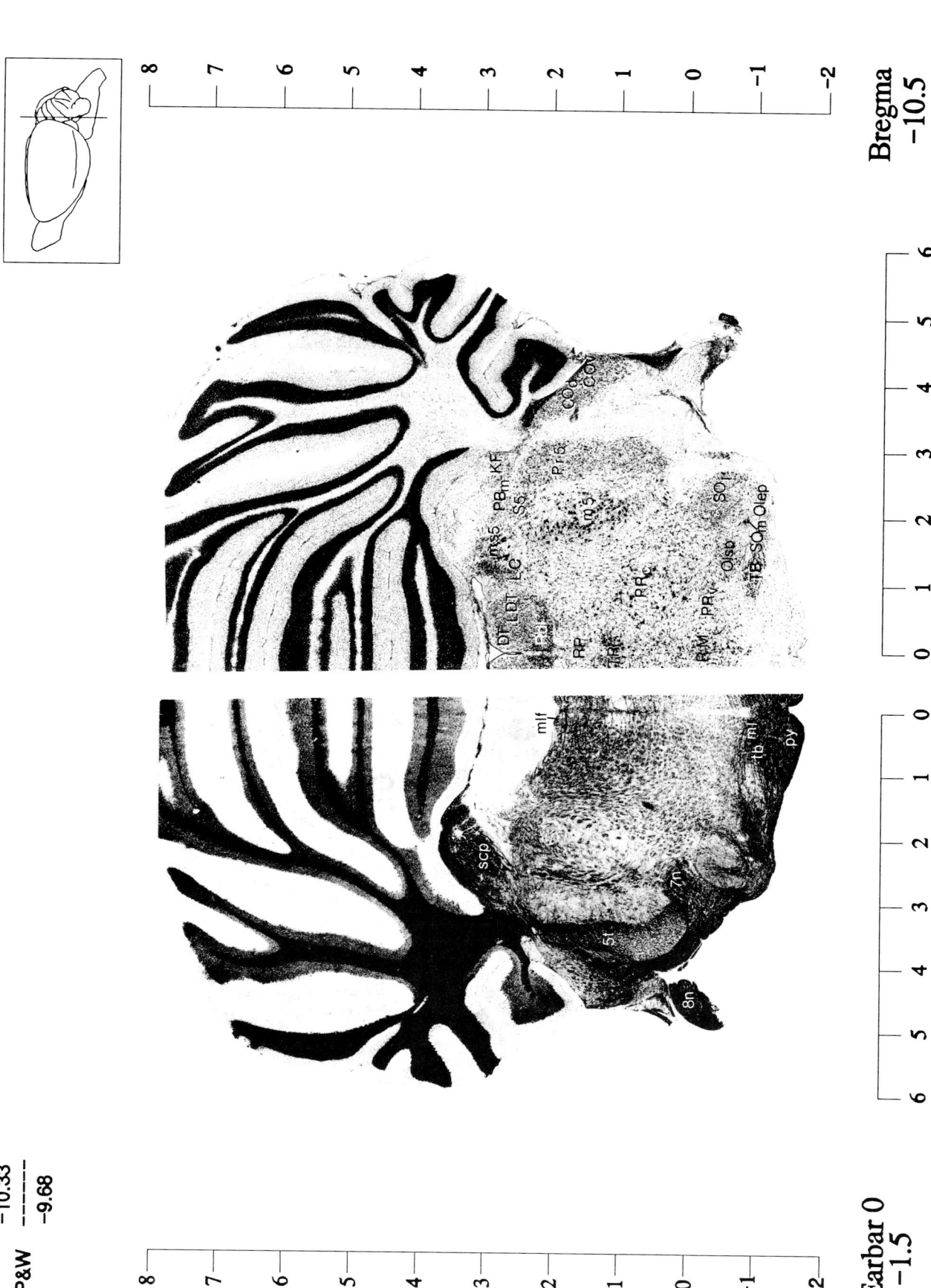

P&W
−10.33
−9.68

Bregma −10.5

Earbar 0 −1.5

8
7
6
5
4
3
2
1
0
−1
−2

6
5
4
3
2
1
0

0
1
2
3
4
5
6

Plate 44

CO$_d$ Cochlear Nucleus (dorsal)
CO$_v$ Cochlear Nucleus (ventral anterior)
DT Dorsal Tegmental Nucleus
LC Locus Ceruleus
LDT Laterodorsal Tegmental Nucleus
m5 Motor Trigeminal Nucleus
ml Medial Lemniscus
mlf Medial Longitudinal Fascicle
Olep External Periolivary Nucleus
Olsp Superior Paraolivary Nucleus
Pr5 Principal Sensory Trigeminal Nucleus
PR$_c$ Pontine Reticular Nucleus (caudal)
PR$_v$ Pontine Reticular Nucleus (ventral)
py Pyramidal Tract
RM Nucleus Raphé Magnus
RPa Nucleus Raphé Pallidus
S5 Supratrigeminal Nucleus
scp Superior Cerebellar Peduncle
SO$_l$ Superior Olivary Complex (lateral)
TB Nucleus of the Trapezoid Body
tb Trapezoid Body
Vs Superior Vestibular Nucleus
5t Trigeminal Tract
7n Facial Nerve
8n Vestibular-Cochlear (Acoustic) Nerve

LC: Locus Ceruleus (or Coeruleus). This longitudinally oriented string of large neurons derives its name from the natural blue pigment evident with the naked eye in the human brain stem. In the rat it is an almost pure population of about 1,450 noradrenergic neurons in the dorsolateral pontine central gray, just medial to a part of the mesencephalic nucleus of the trigeminal nerve (Swanson, '76). It was originally described by Dahlström and Fuxe ('64) as consisting of separate A4 and A6 groups of catecholaminergic neurons; later work has shown, as mentioned above, that it is a continuous cell group that should be distinguished from the subceruleus area, which lies ventral and contains noradrenergic neurons scattered among other cell types. However, the subceruleus area has been variously defined by many workers, and little consensus has yet emerged. The original definition of Meessen and Olszewski's ('49) *nucleus subcoeruleus a* (for the rabbit) appears reasonable. Recently, it has become clear on the basis of corticotropin releasing hormone (CRH) expression, retrograde tracing experiments, and cytoarchitecture that a separate nucleus projecting (at least in part) to sacral parasympathetic preganglionic neurons lies immediately rostral and ventral to the locus ceruleus in the rat. It has been referred to as *Barrington's nucleus* because it probably corresponds to the long-recognized (on physiological grounds) "pontine micturition center" (see Imaki et al., '91). The lateral margin is bounded by the distinctive, large neurons of the mesencephalic nucleus of the trigeminal (mes 5). A group of large scattered neurons lies below the LC and mes 5 and has been designated the *n. subcoeruleus* by some authors, but other proposed subdivisions are discrepant, and the nicely illustrated presentations by Paxinos and Watson ('86) and Swanson ('92) differ substantially in naming and organization. Scattered neurons in this region displaying formaldehyde-induced fluorescence (Dahlström and Fuxe, '64) may belong to the LC on functional grounds. Extensive subdivisions based on architectural criteria are offered for the rabbit by Meessen and Olszewski ('49).

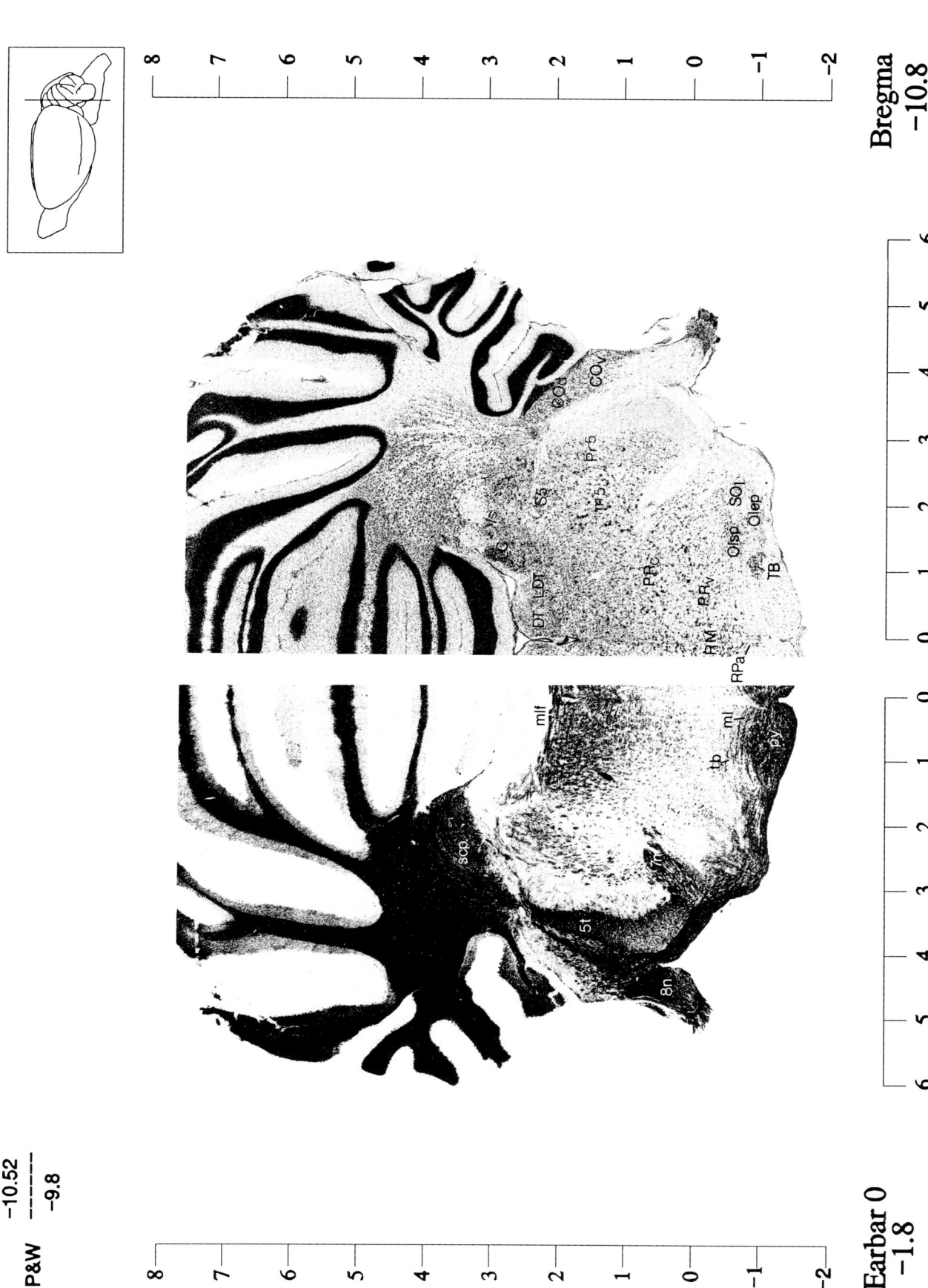

P&W
-10.52
-9.8

Bregma -10.8

Earbar 0 -1.8

Plate 45

A5 Noradrenergic Cell Group 5
CO_d Cochlear Nucleus (dorsal)
CO_v Cochlear Nucleus (ventral anterior)
De Dentate (Lateral) Cerebellar Nucleus
DT_p Dorsal Tegmental Nucleus (posterior)
GcR Gigantocellular Reticular Nucleus
icp Inferior Cerebellar Peduncle
Int Intermediate (Interpositus or Interposed) Cerebellar Nucleus
LC Locus Ceruleus
m5 Motor Trigeminal Nucleus
McR Magnocellular Reticular Nucleus (Medulla)
ml Medial Lemniscus
mlf Medial Longitudinal Fascicle
ocb Olivocochlear Bundle
Olep External Periolivary Nucleus
PcR Parvicellular Reticular Nucleus
Pr5 Principal Sensory Trigeminal Nucleus
py Pyramidal Tract
RM Nucleus Raphé Magnus
RPa Nucleus Raphé Pallidus
s5o Spinal Trigeminal Nucleus (oral)
Sge Supragenual Nucleus
SO_l Superior Olivary Complex (lateral)
tb Trapezoid Body
Vl Lateral Vestibular Nucleus
Vs Superior Vestibular Nucleus
5t Trigeminal Tract
6 Abducens Nucleus
7n Facial Nerve
8n Vestibular-Cochlear (Acoustic) Nerve

A1–A12. (These are designations without abbreviation.) In the first systematic application of the formaldehyde-induced fluorescence method for the cellular localization of biogenic amines to the central nervous system, Dahlström and Fuxe ('64) identified a number of cell groups that, for the most part, did not conform strictly to the better-known nuclei defined on the basis of cytoarchitecture. On the basis of emission characteristics and topography, they identified these cell groups with letters and numbers. Presumed catecholaminergic cell groups were assigned the letter A, whereas presumably indolaminergic cell groups, which tended to cluster near the brainstem raphé, were assigned the letter B, and numbering for each began caudally and ventrally. In some areas such as the ventrolateral medulla (A1 group), catecholaminergic neurons are scattered among a number of other cell types in a poorly defined part of the reticular formation, while in other areas (A4 and A6, together forming the locus ceruleus) they are virtually the only neuronal type present in a well-defined nucleus. The situation became more complicated when immunohistochemical methods clearly distinguished among three distinct types of catecholaminergic neurons: adrenergic, noradrenergic, and dopaminergic, leading to the distinction of adrenergic groups, apparently confined to the medulla, with the letter C (Hökfelt et al., '74). Swanson ('92) refers to recent literature on the distribution of these cell types. We have indicated an example (A5), the caudal portion of the pontine reticular nucleus, in order to guide the user in how this nomenclature has been applied. The medial nucleus of the solitary tract (TS_m) is indicated as A2, the retrorubral area of the deep mesencephalic reticular nucleus is A8, the compact portion of the substantia nigra is A9, the ventral tegmental area is A10, and the arcuate nucleus of the hypothalamus is A12 in this schema. The precedence of common usage prevails here as in most contemporary accounts.

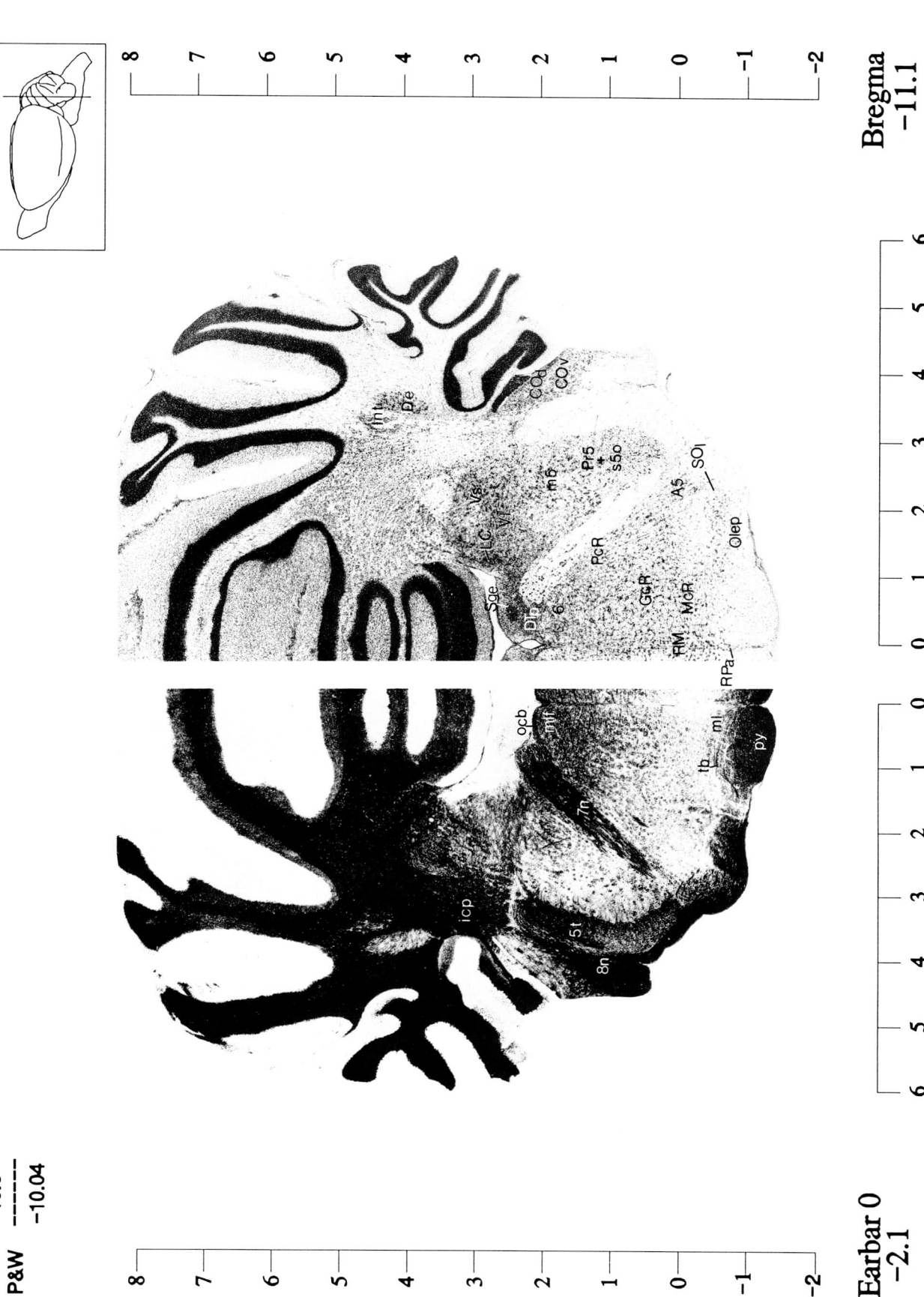

Plate 46

RM: Nucleus Raphé Magnus (also called groups B2 and B3). This long nucleus lying above and within the medial lemniscus in the rostral medulla extends laterally and is generally outlined as a triangular area in transverse sections. It is contiguous caudally with the nucleus raphé obscurus. With serotonin immunohistochemistry it appears as a distinctive large-celled, loose network (see Dahlström and Fuxe, '64, who introduced the nomenclature of B1–B9 in a caudal-to-rostral sequence).

RPa: Nucleus Raphé Pallidus (also called Group B1). A dense, small-celled, rod-shaped nucleus lying between the pyramids at the base of the medulla below the inferior olive. The cells stain poorly but are strikingly illustrated with serotonin immunohistochemistry by Törk ('85).

Sge: Supragenual Nucleus. This distinct small dark aggregate of neurons, lying dorsomedial to the genu (knee) of the facial nerve as it bends in the tegmental pons, is regrettably named for its location (Meessen and Olszewski, '49) and appears to be unrelated to the underlying facial nerve.

CO$_d$ Cochlear Nucleus (dorsal)
CO$_p$ Cochlear Nucleus (posterior)
De Dentate (Lateral) Cerebellar Nucleus
GcR Gigantocellular Reticular Nucleus
icp Inferior Cerebellar Peduncle
Int Intermediate (Interpositus or Interposed) Cerebellar Nucleus
m5 Motor Trigeminal Nucleus
McR Magnocellular Reticular Nucleus (Medulla)
ml Medial Lemniscus
mlf Medial Longitudinal Fascicle
ocb Olivocochlear Bundle
PcR Parvicellular Reticular Nucleus
Pr5 Principal Sensory Trigeminal Nucleus
py Pyramidal Tract
RM Nucleus Raphé Magnus
RPa Nucleus Raphé Pallidus
s5o Spinal Trigeminal Nucleus (oral)
Sge Supragenual Nucleus
V4 Fourth Ventricle
Vl Lateral Vestibular Nucleus
Vm Medial Vestibular Nucleus
Vs Superior Vestibular Nucleus
5t Trigeminal Tract
6 Abducens Nucleus
7 Facial Nucleus
7g Genu of Facial Nerve
8n Vestibular-Cochlear (Acoustic) Nerve

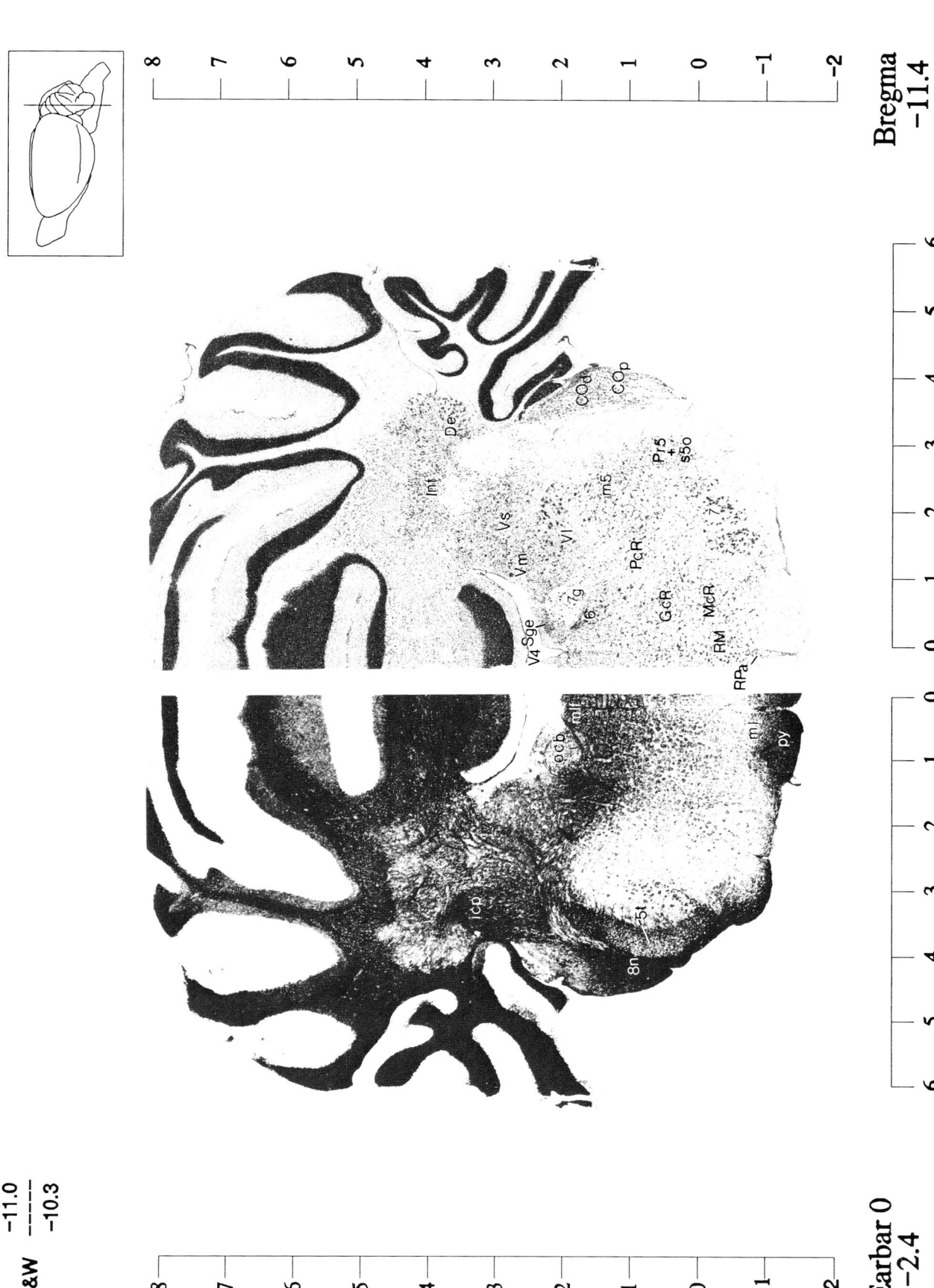

P&W

−11.0
———
−10.3

8
7
6
5
4
3
2
1
0
−1
−2

Earbar 0
−2.4

8
7
6
5
4
3
2
1
0
−1
−2

Bregma
−11.4

6
5
4
3
2
1
0

0
1
2
3
4
5
6

Int
De

Vs
Vm
V4
Sge
Vl
7g
6

m5
PcR
GcR
RM
McR
RPa
7

Pr5
+
s5o

COd
COp

icp
8n
5t

ocb
mlf
ml
py

Plate 47

CO_d   Cochlear Nucleus (dorsal)
CO_p   Cochlear Nucleus (posterior)
De   Dentate (Lateral) Cerebellar Nucleus
Fa   Fastigial (Medial) Cerebellar Nucleus
GcR   Gigantocellular Reticular Nucleus
icp   Inferior Cerebellar Peduncle
Int   Intermediate (Interpositus or Interposed) Cerebellar Nucleus
McR   Magnocellular Reticular Nucleus (Medulla)
ml   Medial Lemniscus
mlf   Medial Longitudinal Fascicle
PcR   Parvicellular Reticular Nucleus
PgR   Paragigantocellular Reticular Nucleus
Ph   Perihypoglossal Nucleus
py   Pyramidal Tract
RM   Nucleus Raphé Magnus
RPa   Nucleus Raphé Pallidus
s5o   Spinal Trigeminal Nucleus (oral)
Vl   Lateral Vestibular Nucleus
Vm   Medial Vestibular Nucleus
Vs   Superior Vestibular Nucleus
5t   Trigeminal Tract
7   Facial Nucleus
7g   Genu of Facial Nerve

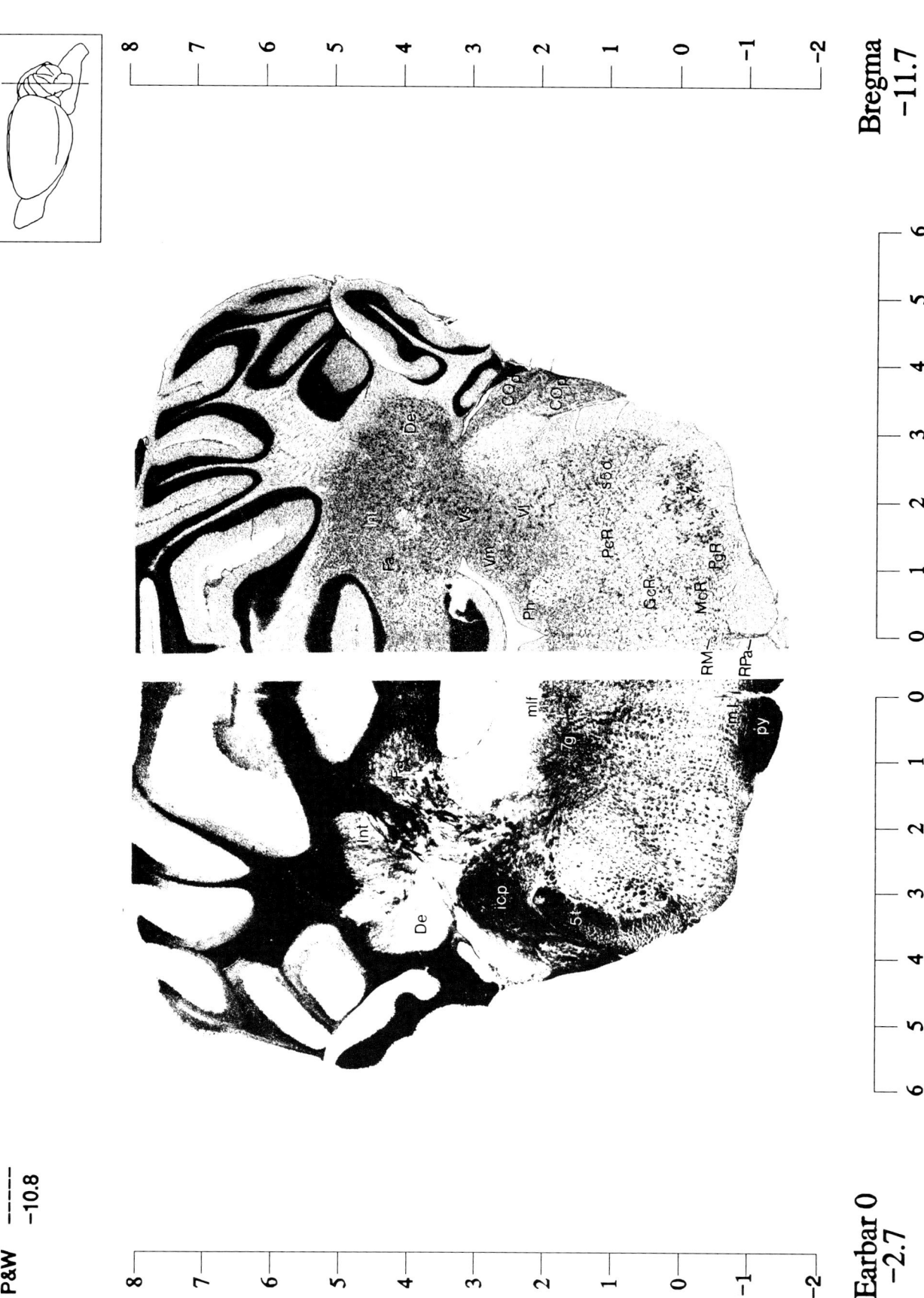

8
7
6
5
4
3
2
1
0
−1
−2

Bregma
−11.7

6
5
4
3
2
1
0

0
1
2
3
4
5
6

8
7
6
5
4
3
2
1
0
−1
−2

Earbar 0
−2.7

Plate 48

CO$_d$ Cochlear Nucleus (dorsal)
CO$_p$ Cochlear Nucleus (posterior)
De   Dentate (Lateral) Cerebellar Nucleus
Fa   Fastigial (Medial) Cerebellar Nucleus
GcR   Gigantocellular Reticular Nucleus
icp   Inferior Cerebellar Peduncle
Int   Intermediate (Interpositus or Interposed) Cerebellar Nucleus
McR   Magnocellular Reticular Nucleus (Medulla)
ml   Medial Lemniscus
mlf   Medial Longitudinal Fascicle
PcR   Parvicellular Reticular Nucleus
PgR   Paragigantocellular Reticular Nucleus
Ph   Perihypoglossal Nucleus
py   Pyramidal Tract
RM   Nucleus Raphé Magnus
RPa   Nucleus Raphé Pallidus
s5o   Spinal Trigeminal Nucleus (oral)
Vl   Lateral Vestibular Nucleus
Vm   Medial Vestibular Nucleus
Vsp   Spinal Vestibular Nucleus
Y   Nucleus Y
5t   Trigeminal Tract
7   Facial Nucleus

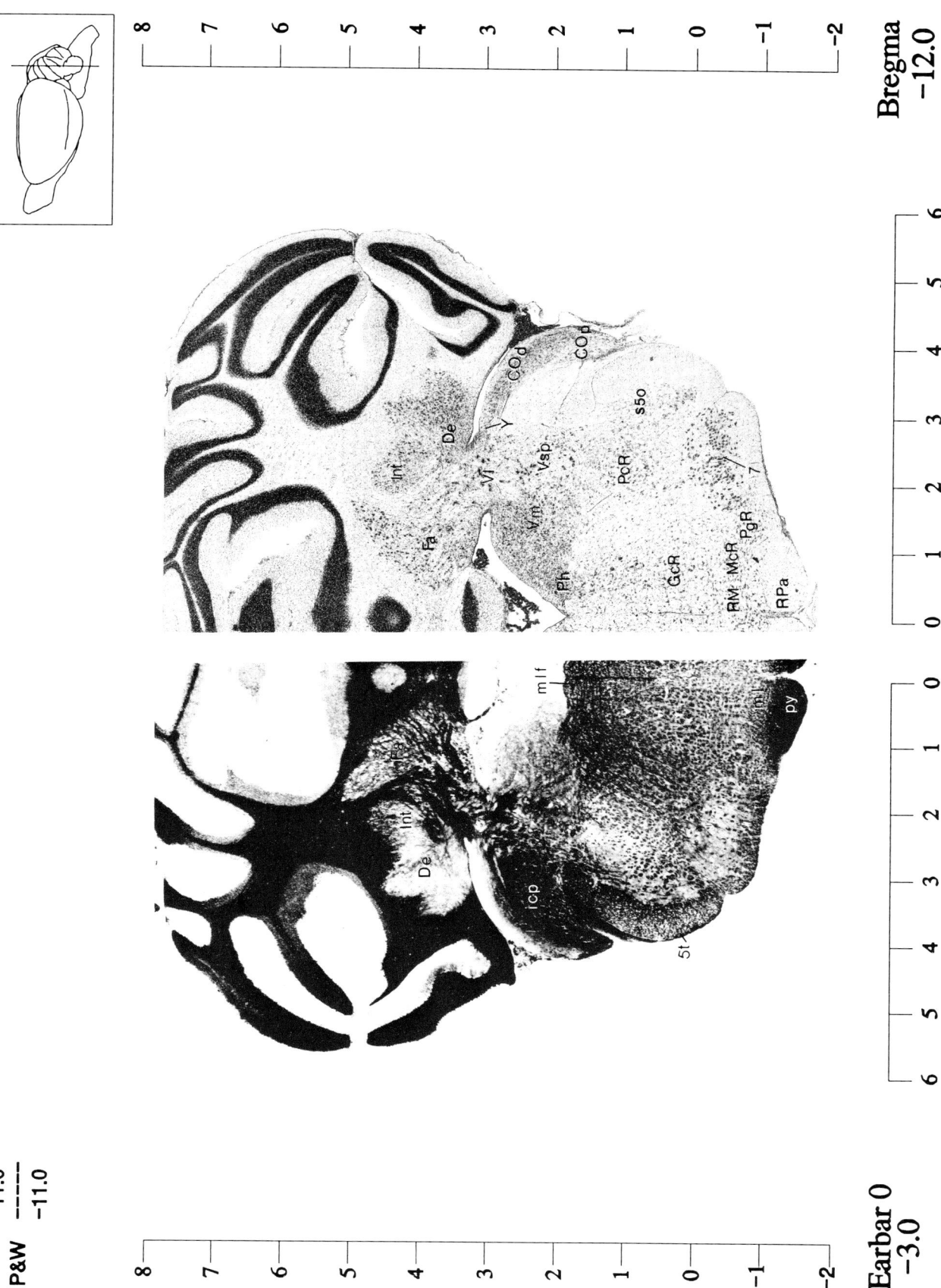

P&W

−11.6
−11.0

Bregma
−12.0

Earbar 0
−3.0

Plate 49

chp Choroid Plexus
CO_d Cochlear Nucleus (dorsal)
CO_p Cochlear Nucleus (posterior)
De Dentate (Lateral) Cerebellar Nucleus
dsct Dorsal Spinocerebellar Tract
Fa Fastigial (Medial) Cerebellar Nucleus
fl Foramen of Luschka
GcR Gigantocellular Reticular Nucleus
icp Inferior Cerebellar Peduncle
Int Intermediate (Interpositus or Interposed) Cerebellar Nucleus
McR Magnocellular Reticular Nucleus (Medulla)
ml Medial Lemniscus
mlf Medial Longitudinal Fascicle
PcR Parvicellular Reticular Nucleus
PgR Paragigantocellular Reticular Nucleus
Ph Perihypoglossal Nucleus
py Pyramidal Tract
RM Nucleus Raphé Magnus
RO Nucleus Raphé Obscurus
RPa Nucleus Raphé Pallidus
s5o Spinal Trigeminal Nucleus (oral)
Vl Lateral Vestibular Nucleus
Vm Medial Vestibular Nucleus
Vsp Spinal Vestibular Nucleus
X Nucleus X
Y Nucleus Y
5t Trigeminal Tract
7 Facial Nucleus

GcR(v): Gigantocellular Reticular Nucleus. The distinctive large neurons of this "core" medullary reticular nucleus have been encumbered with numerous subdivisions by various authors, but we have succumbed only to indicate the generally recognized ventral subsector. Some authors recognize an alpha sector and various surrounding zones described as paragigantocellular nuclei with topographic descriptors. Choosing among the alternatives in the vast literature on this subject seems impractical currently, and the interested reader is advised to employ key-word computer searches rather than rely on earlier reviews because new subdivisions are being devised based on "chemical neuroanatomy."

X: Nucleus X. An irregular aggregate of small neurons lying at the ventrolateral edge of the spinal vestibular nucleus and partially interspersed within the subjacent inferior cerebellar peduncle was identified in the cat by Brodal and Pompeiano ('57) as a nonvestibular nucleus receiving ascending spinal input. There is evidence that these neurons project upon the cerebellum in rat (see Mehler and Rubertone, '85), but we can only vaguely indicate its approximate location in an effort to achieve concordance with other atlases, although there is substantial discrepancy between the Paxinos and Watson ('86) and Swanson ('92) atlases.

Y: Nucleus Y. A small compact aggregate of neurons lying medially adjacent to the dorsal cochlear nucleus astride the inferior cerebellar peduncle has been illustrated in detail in the mouse brain stem together with experimental evidence of afferents from the saccule of the vestibular apparatus (Frederickson and Trune, '86). It is sometimes included in the infracerebellar nucleus and can be difficult to separate from the contiguous lateral and superior vestibular nuclei, especially in low-power magnification photomicrographs. This renders its depiction here somewhat dubious. Its identification in Nissl-stained sections is feasible at higher magnification and its location noted here is roughly consistent with the atlas designations of Paxinos and Watson ('86) and Swanson ('92), but we would not be able to outline them readily in this material.

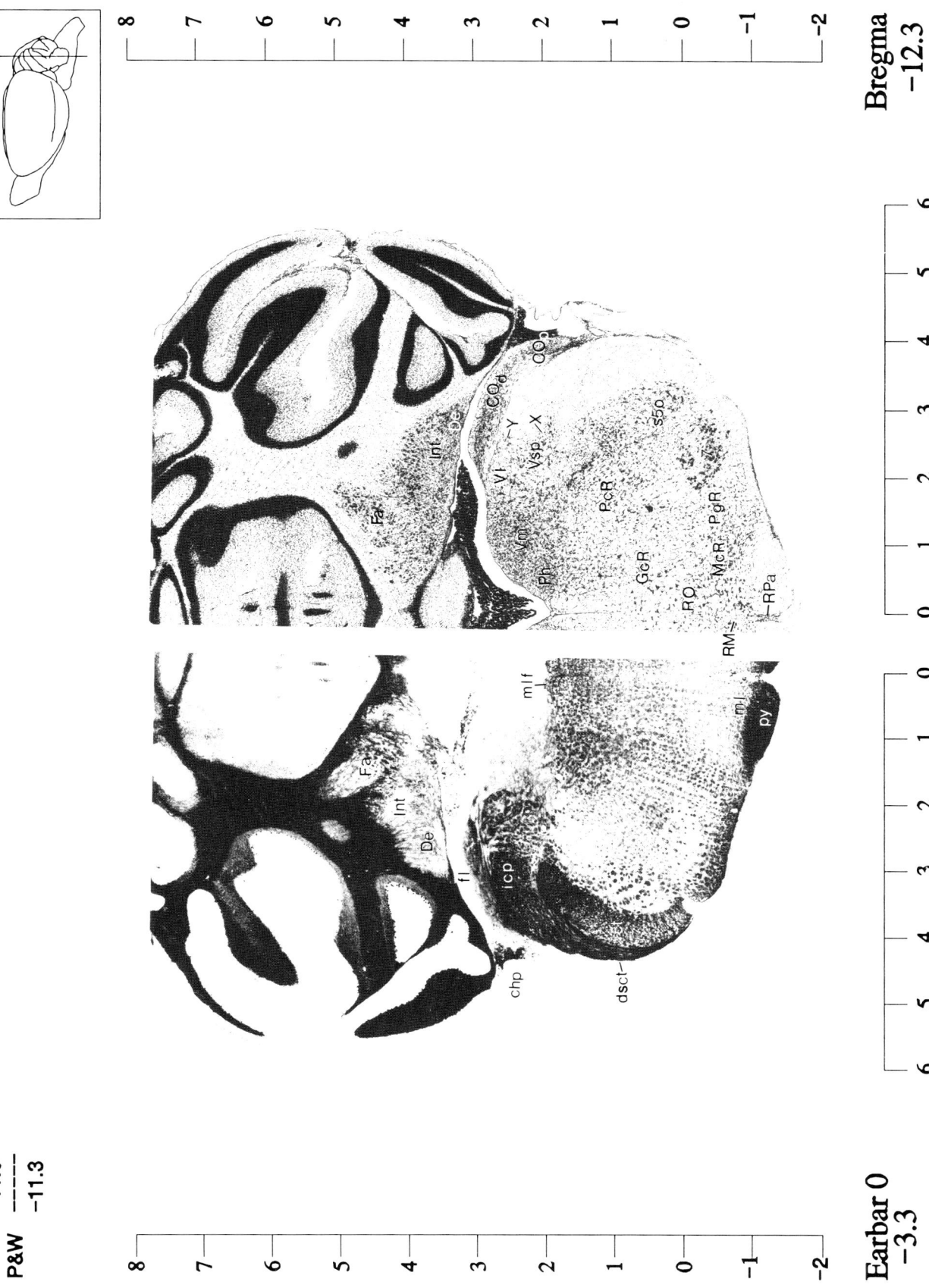

P&W
-11.6
-11.3

Bregma
-12.3

Earbar 0
-3.3

Plate 50

Salivatory Nuclei. An essentially vertical column of preganglionic parasympathetic neurons, the salivatory nuclei stretch in the rat between the facial nucleus (and rostral tip of the nucleus ambiguus) ventrally and the rostral end of the dorsal vagal complex dorsally. Its neurons innervate three major ganglia: pterygopalatine, submandibular, and otic. Axons to the first two ganglia course through the intermediate nerve (part of cranial nerve VII). Thus, the first two are innervated by the superior (or rostral) salivatory nucleus, while the third is innervated by the inferior (or caudal) salivatory nucleus. However, these nuclei are exceedingly difficult to identify in Nissl sections, and the problem has been evaded in most descriptions. Paxinos and Watson (86) indicate an approximate position without outlines for the superior salivatory nucleus. Only Swanson (92) attempts to define the form of both superior and inferior salivatory nuclei but he was obliged to resort to examination in the microscope at high magnification. The principles employed by Olszewski and Baxter ('54) and Meessen and Olszewski (49) in cytological recognition of these and other parasympathetic preganglionic neurons in the human and rabbit brain stem, are frankly intimidating in the small rat brain. Swanson (92) places the superior ventral to the inferior salivatory nucleus, and it seems clear from experimental studies (below) that his inferior salivatory nucleus contains neurons projecting to both the submandibular and otic ganglia (the salivatory nucleus proper). Experimental studies validate the approximate position, but the conflicts are significant and require reference to the original papers (Contreras et al., '80; Nicholson and Severin, '81; Eisenman and Azmitia, '82; Jansen et al., '92). Separate small and medium cell divisions of the inferior salivatory nucleus extending ventrolateral to the vagal nucleus are nicely illustrated by Matesz and Székely ('83). The labeling of the parasympathetic preganglionic "column" apparently has been achieved with a gene expression marker for axonal outgrowth (GAP-43 mRNA), but even with such clues, recognition in Nissl-stained sections is difficult (Kruger et al., '93). The preganglionic parasympathetic nasolacrimal nucleus is presumably smaller and may lie dorsolateral to the caudal end of the facial nucleus (Contreras et al., '80; Senba et al., '87). Nevertheless, it is likely that each of these nuclei is small, scattered, and cytologically similar to the nonbranchial portion of the nucleus ambiguus.

P&W −11.6

8
7
6
5
4
3
2
1
0
−1
−2

Int
Fa
chp
Ph
Vm
Vsp
COd
Y
X
GcR
PcR
McR
PgR
RPa
RM
RO
s5o
7
mlf
mlf
py
icp
dsct
chp
Fa
Int

Bregma
−12.6

6
5
4
3
2
1
0
1
2
3
4
5
6

Earbar 0
−3.6

8
7
6
5
4
3
2
1
0
−1
−2

Plate 51

AMB Nucleus Ambiguus
GcR Gigantocellular Reticular Nucleus
icp Inferior Cerebellar Peduncle
IO Inferior Olive
McR Magnocellular Reticular Nucleus (Medulla)
ml Medial Lemniscus
mlf Medial Longitudinal Fascicle
PcR Parvicellular Reticular Nucleus
PgR Paragigantocellular Reticular Nucleus
Ph Perihypoglossal Nucleus
py Pyramidal Tract
RO Nucleus Raphé Obscurus
RPa Nucleus Raphé Pallidus
s5i Spinal Trigeminal Nucleus (interpolar)
TS$_m$ Nucleus of the Solitary Tract (medial)
ts Solitary Tract
Vm Medial Vestibular Nucleus
Vsp Spinal Vestibular Nucleus
X Nucleus X
5t Trigeminal Tract

Vestibular Nuclear Group. The vestibular nuclear group is generally divided into four main nuclei recognized in a wide variety of mammals and consists of medial, superior, lateral and spinal vestibular nuclei (Vl, Vm, Vs, and Vsp). Mehler and Rubertone ('85) provide an illustrated review and also recognize the neurons interspersed with the vestibular portion of the eighth nerve root as the *interstitial nucleus of Cajal*. This may cause some confusion because there is an interstitial nucleus of Cajal associated with the oculomotor nuclear group as well as another in the spinal gray (there also are interstitial cells of Cajal in the peripheral nervous system, conspicuously evident in the gastrointestinal tract). The interstitial vestibular nucleus consisting of scattered nests of neurons within the vestibular nerve was indeed recognized as the *ganglion interstitel* by Cajal ('09), who attributed this to L. Sala. Cajal noted that this nucleus was poorly developed in mammals and believed it was probably the homologue of the *noyau tangentiel* (tangential nucleus) of fishes and birds. Some small associated groups described by Brodal and Pompeiano ('57) as nuclei x, y, and z in the cat and identified by some authors in the rat are considered separately. X: Nucleus X (Plate 103); Y: Nucleus Y (Plate 103); Z: Nucleus Z (Plate 58).

The superior vestibular nucleus (Vs) is most easily outlined in fiber preparations, wedged beneath the superior cerebellar peduncle and bounded laterally by the middle cerebellar peduncle. Its medial boundary is formed by a distinctive medial vestibular (Vm) nuclear complex readily susceptible to subdivision on cytoarchitectural grounds (see Mehler and Rubertone, '85). The ventrolateral portion of Vm contains scattered large neurons that merge with the lateral vestibular nucleus (Vl), the most easily identified nucleus because of its large multipolar neurons. The spinal vestibular nucleus (Vsp) (also called the *inferior* or *descending vestibular nucleus*) lies on the dorsal surface of the medulla for much of its extent after emerging from beneath the Vl; it is extensively pierced by rostrocaudal fiber bundles, providing an easily recognizable fiber architecture in contrast to the cytoarchitecture that merges indistinctly at various loci with the medial and lateral vestibular nuclei. The most useful description is probably that of Henkel and Martin ('77a) in the opossum; it contains excellent illustrations in the three principal planes and commentary on nuclear subdivision.

P&W −11.8

8
7
6
5
4
3
2
1
0
−1
−2

Bregma
−12.9

6
5
4
3
2
1
0

Vsp
X
Vm
S5
TSm
AMB
Ph
PgR
GcR
PgR
McR
RO
IO
RPa

mlf
ts
icp
py
g7

Earbar 0
−3.9

8
7
6
5
4
3
2
1
0
−1
−2

0
1
2
3
4
5
6

Plate 52

AMB Nucleus Ambiguus
CUe External Cuneate Nucleus
GcR Gigantocellular Reticular Nucleus
icp Inferior Cerebellar Peduncle
IO Inferior Olive
McR Magnocellular Reticular Nucleus
(Medulla)
ml Medial Lemniscus
mlf Medial Longitudinal Fascicle
PcR Parvicellular Reticular Nucleus
PgR Paragigantocellular Reticular Nucleus
Ph Perihypoglossal Nucleus
py Pyramidal Tract
RO Nucleus Raphé Obscurus
RPa Nucleus Raphé Pallidus
s5i Spinal Trigeminal Nucleus (interpolar)
TS_m Nucleus of the Solitary Tract (medial)
ts Solitary tract
Vm Medial Vestibular Nucleus
Vsp Spinal Vestibular Nucleus
X Nucleus X
5t Trigeminal Tract

s5i: Spinal Trigeminal Nucleus (interpolar) (also called Spinal Intermediate or Spinal V Nucleus Interpolaris). This distinct subnucleus is recognizable on several grounds but is most easily seen in horizontal sections, where it is almost almond-shaped. It contains slightly larger and more loosely packed neurons than its surrounding subnuclei. The caudal border is delineated by the gelatinosa of the caudal subnucleus, enabling subdivision by fiber architecture, the most striking criterion for seeing the nuclear outline. It also is prominent as a distinct functional entity by retrograde labeling from the thalamus and patterns of peptide immunoreactivity (Jacquin et al., '86; Kruger et al., '88a).

P&W −11.96

Bregma −13.2

Earbar 0 −4.2

8 7 6 5 4 3 2 1 0 −1 −2

6 5 4 3 2 1 0

0 1 2 3 4 5 6

CUe
X
Vsp
Vm
Ph
rSm
PcR
s5i
5t
AMB
GcR
RO MdR
PgR
IO
RPa
mlf
ml
py
ts
cp

Plate 53

AMB Nucleus Ambiguus
chp Choroid Plexus
CUe External Cuneate Nucleus
GcR Gigantocellular Reticular Nucleus
icp Inferior Cerebellar Peduncle
IO Inferior Olive
McR Magnocellular Reticular Nucleus
(Medulla)
ml Medial Lemniscus
mlf Medial Longitudinal Fascicle
pa5 Paratrigeminal Nucleus
PcR Parvicellular Reticular Nucleus
PgR Paragigantocellular Reticular Nucleus
Ph Perihypoglossal Nucleus
py Pyramidal Tract
RO Nucleus Raphé Obscurus
RPa Nucleus Raphé Pallidus
s5i Spinal Trigeminal Nucleus (interpolar)
TS$_m$ Nucleus of the Solitary Tract (medial)
ts Solitary Tract
Vm Medial Vestibular Nucleus
Vsp Spinal Vestibular Nucleus
X Nucleus X
5t Trigeminal Tract

CUe: External Cuneate Nucleus. A somatosensory relay nucleus lying dorsolateral to the "main" cuneate nucleus deserves separate status from the continuous cuneate-gracile complex because of (1) its distinctive cytoarchitecture containing larger, scattered neurons, (2) its receipt of input from somatotopically organized neck proprioceptors (Campbell et al., '74), and (3) its principal projection upon the cerebellum, in contrast to the somatotopically organized cutaneous projection to the cuneate nucleus, which in turn projects via the medial lemniscus to the ventrobasal (VB) thalamus.

pa5: Paratrigeminal Nucleus. A group of scattered neurons embedded in the dorsolateral portion of the spinal trigeminal tract overlying the spinal trigeminal nucleus extending from above the rostral portion of s5c forward to overlie a portion of s5i. It was called the *trigeminal extension of the lateral cervical nucleus* by Gobel et al. ('77), but recent usage of *paratrigeminal* has become standard.

Bregma
-13.5

Plate 54

AMB Nucleus Ambiguus
CU Cuneate Nucleus
CUe External Cuneate Nucleus
GcR Gigantocellular Reticular Nucleus
GR Gracile Nucleus
Ic Intercalated Nucleus (Medulla)
IO Inferior Olive
McR Magnocellular Reticular Nucleus (Medulla)
mlf Medial Longitudinal Fascicle
PcR Parvicellular Reticular Nucleus
PgR Paragigantocellular Reticular Nucleus
py Pyramidal Tract
RO Nucleus Raphé Obscurus
Rol Nucleus of Roller
RPa Nucleus Raphé Pallidus
s5i Spinal Trigeminal Nucleus (interpolar)
TS$_l$ Nucleus of the Solitary Tract (lateral)
TS$_m$ Nucleus of the Solitary Tract (medial)
ts Solitary Tract
5t Trigeminal Tract
12 Hypoglossal Nucleus

IO: Inferior Olive. This tortuous nucleus is considered a single functional entity whose neurons provide "climbing fibers" to the dendrites of cerebellar cortex Purkinje neurons as well as collaterals to the deep cerebellar nuclei. The IO consists of three main subnuclei: the principal, dorsal accessory, and medial accessory olive, the latter continuous with a dorsal cap of Kooy and a ventrolateral outgrowth. Some authors provide letter designations for the leaflets of the principal olive (e.g., Paxinos and Watson, '86). An illustrated account is provided by Gwyn et al. ('77).

PcR: Parvicellular Reticular Nucleus. As the name implies, this lateral part of the reticular formation is dominated by small neurons, and it lies medial to the sensory nuclei of the fifth, ninth, and tenth nerves throughout most of the length of the medulla and pons. Its boundaries with other parts of the reticular formation are indistinct.

PgR: Paragigantocellular Reticular Nucleus (also called Nucleus Paragigantocellularis Lateralis). A predominantly small-celled, poorly defined ventral sector of the rostral medulla easily separated from its medially adjacent gigantocellular reticular nucleus (GcR). It contains a heterogeneous variety of neurons, including large ones in its caudal sector, and is described in detail by Andrezik et al. ('81).

P&W   -12.72

Bregma
-13.8

Earbar 0
-4.8

Plate 55

AMB Nucleus Ambiguus
CU Cuneate Nucleus
CUe External Cuneate Nucleus
GcR Gigantocellular Reticular Nucleus
GR Gracile Nucleus
Ic Intercalated Nucleus (Medulla)
IO Inferior Olive
LR$_l$ Lateral Reticular Nucleus (lateral)
mlf Medial Longitudinal Fascicle
pa5 Paratrigeminal Nucleus
PcR Parvicellular Reticular Nucleus
PmR Paramedian Reticular Nucleus
py Pyramidal Tract
RO Nucleus Raphé Obscurus
Rol Nucleus of Roller
RPa Nucleus Raphé Pallidus
s5i Spinal Trigeminal Nucleus (interpolar)
s5o Spinal Trigeminal Nucleus (oral)
TS$_l$ Nucleus of the Solitary Tract (lateral)
TS$_m$ Nucleus of the Solitary Tract (medial)
ts Solitary Tract
5t Trigeminal Tract
12 Hypoglossal Nucleus

Ic: Intercalated Nucleus (Medulla) (also called Nucleus Intercalatus of Staderini). A small-celled mass lateral to the hypoglossal nucleus interposed between these motoneurons and the dorsal motor nucleus of the vagus. This is partially labeled by Paxinos and Watson ('86), but they add a discontinuous similar nucleus farther caudal, wedged between the hypoglossal and vagal nuclei, which they call the *intermedius nucleus of the medulla* bearing the eponym of Cajal, an identification attributed to Mehler. The intercalated nucleus is more easily recognized in larger mammalian brains, as in the cat (Berman, '68) and human (Jacobsohn, '09), and some authors deliberately avoid this separation (e.g., Swanson, '92). We are uncertain about the status of the intermedius nucleus and the origin of this usage, and caution the reader not to confuse it with the small-celled interstitial nucleus (of Cajal) near the oculomotor nuclear group.

P&W −12.8

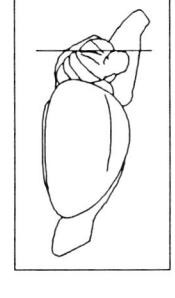

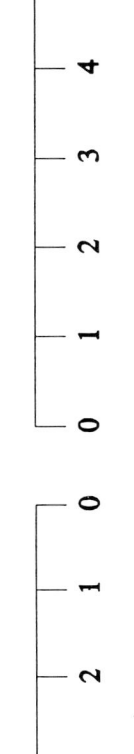

Bregma
−14.1

Earbar 0
−5.1

Plate 56

AMB   Nucleus Ambiguus
CU   Cuneate Nucleus
CUe   External Cuneate Nucleus
GcR   Gigantocellular Reticular Nucleus
GR   Gracile Nucleus
IO   Inferior Olive
LR$_l$ Lateral Reticular Nucleus (lateral)
LR$_m$ Lateral Reticular Nucleus (medial)
mlf   Medial Longitudinal Fascicle
pa5   Paratrigeminal Nucleus
PcR   Parvicellular Reticular Nucleus
PmR   Paramedian Reticular Nucleus
py   Pyramidal Tract
RO   Nucleus Raphé Obscurus
Rol   Nucleus of Roller
RPa   Nucleus Raphé Pallidus
s5i   Spinal Trigeminal Nucleus (interpolar)
TS$_l$ Nucleus of the Solitary Tract (lateral)
TS$_m$ Nucleus of the Solitary Tract (medial)
ts   Solitary Tract
Z   Nucleus Z
5t   Trigeminal Tract
10   Dorsal Motor Nucleus of Vagus
12   Hypoglossal Nucleus

RO: Nucleus Raphé Obscurus (also called Group B2). This nucleus consists of scattered cells in two paramedian lines caudal to the nucleus raphé magnus (RM) and above the inferior olive. Its distinctive large multipolar neurons are nicely visualized with serotonin immunohistochemistry (see Törk, '85), but its form is not prominent in Nissl-stained sections.

P&W −13.24

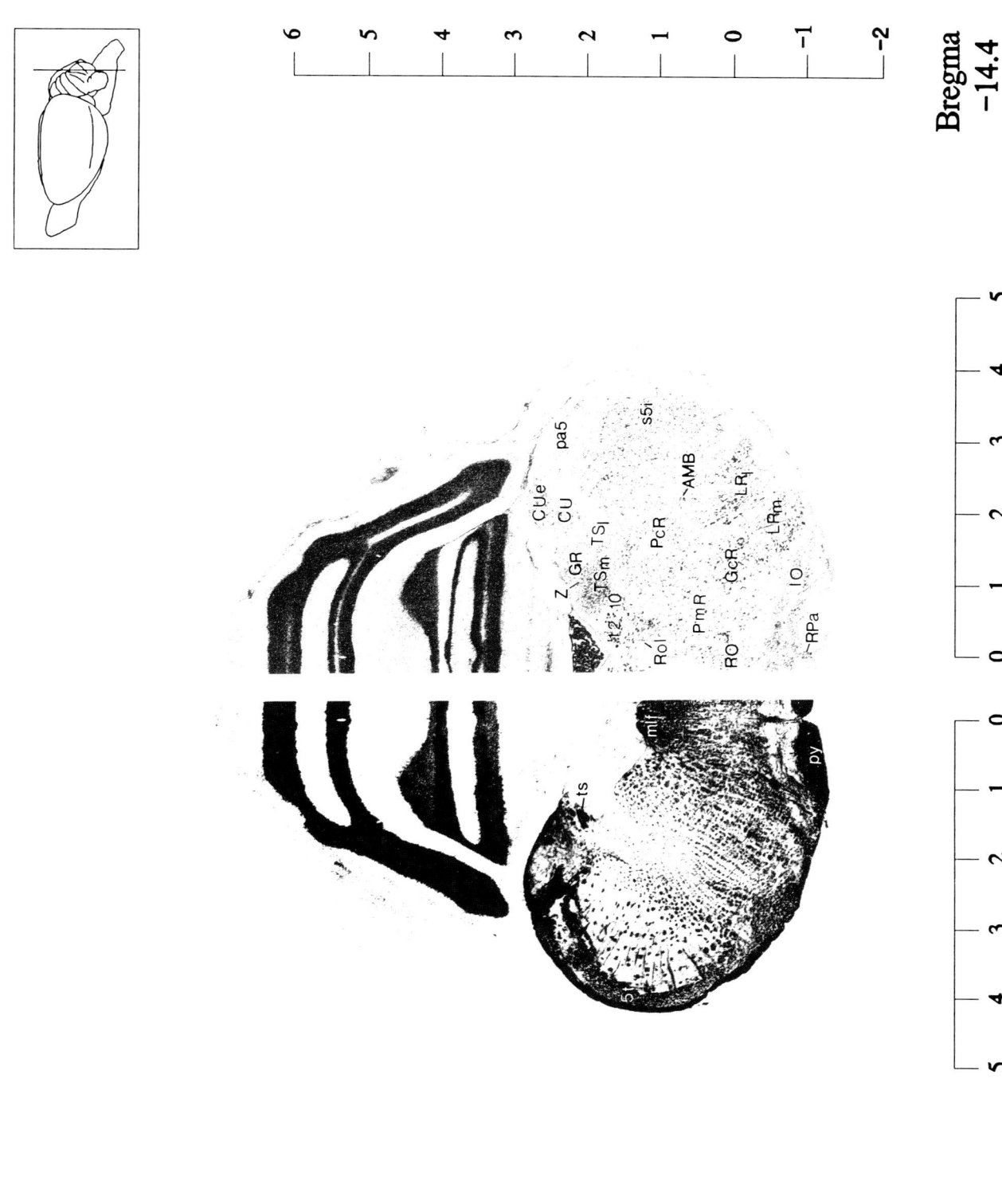

6
5
4
3
2
1
0
−1
−2

Bregma
−14.4

5
4
3
2
1
0

0
1
2
3
4
5

Earbar 0
−5.4

6
5
4
3
2
1
0
−1
−2

Plate 57

AMB Nucleus Ambiguus
CU Cuneate Nucleus
CUe External Cuneate Nucleus
GcR Gigantocellular Reticular Nucleus
GR Gracile Nucleus
IO Inferior Olive
LR$_l$ Lateral Reticular Nucleus (lateral)
LR$_m$ Lateral Reticular Nucleus (medial)
mlf Medial Longitudinal Fascicle
PcR Parvicellular Reticular Nucleus
PmR Paramedian Reticular Nucleus
py Pyramidal Tract
RO Nucleus Raphé Obscurus
RPa Nucleus Raphé Pallidus
s5i Spinal Trigeminal Nucleus (interpolar)
TS$_l$ Nucleus of the Solitary Tract (lateral)
TS$_m$ Nucleus of the Solitary Tract (medial)
ts Solitary Tract
VR Ventral Reticular Nucleus
Z Nucleus Z
5t Trigeminal Tract
10 Dorsal Motor Nucleus of Vagus
12 Hypoglossal Nucleus

P&W −14.33

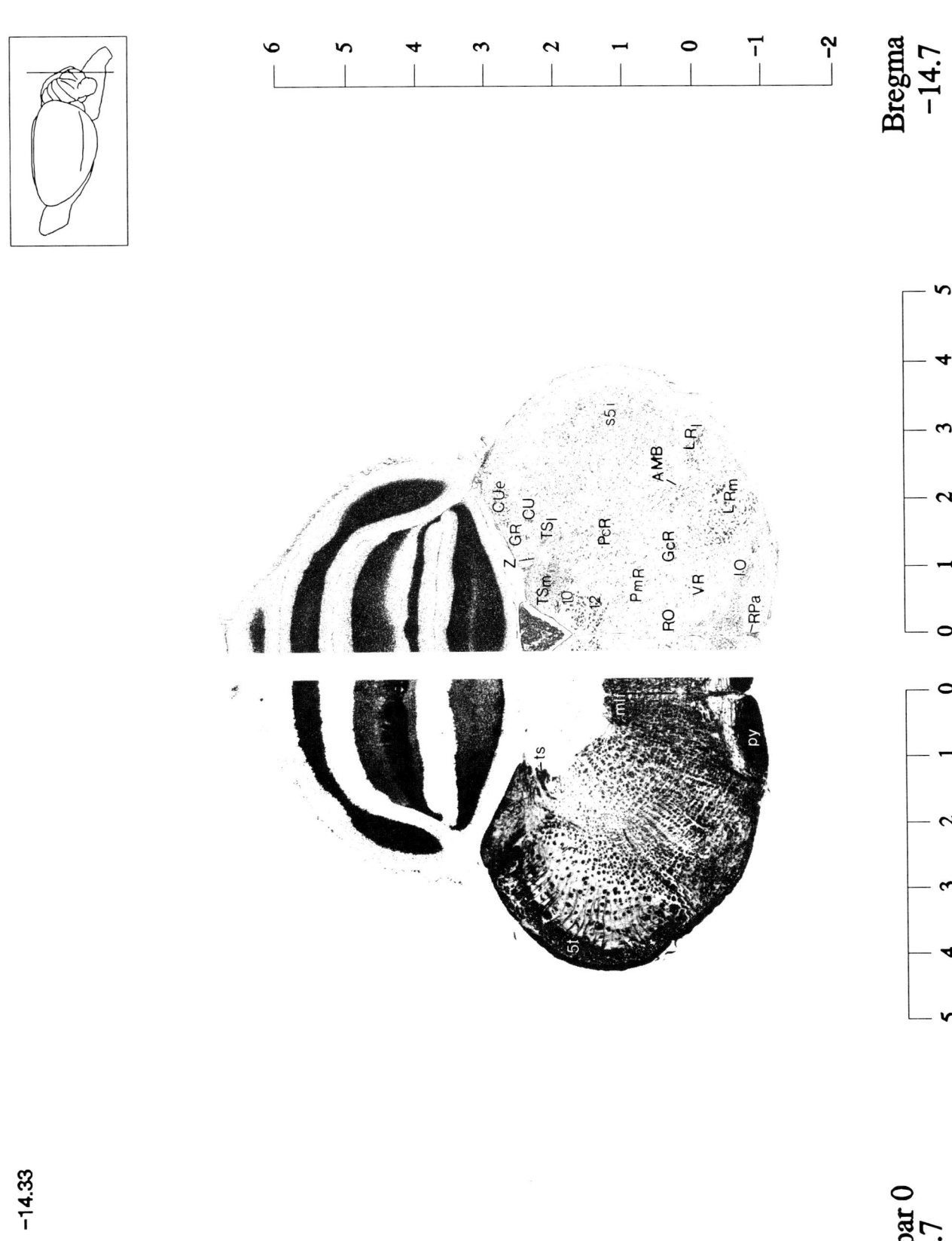

Bregma
−14.7

Earbar 0
−5.7

Plate 58

Z: Nucleus Z. A poorly defined small-celled aggregate contiguous with the rostrolateral pole of the gracile nucleus was identified in cat by Brodal and Pompeiano ('57). These neurons receive proprioceptive input from the dorsal nucleus (Clarke's column) spinal cord region known to contribute axons to the dorsal spinocerebellar tract. Nucleus Z has been identified in the rat by Low et al. ('86), who estimate that about 3% of all dorsal spinocerebellar tract projecting neurons contribute to terminals in this nucleus, which they have difficulty in defining in Nissl-stained sections. Unfortunately, nucleus Z and the gracile nucleus both project to the contralateral thalamus, so that retrograde neuronal labeling following injection of tracer into ventrobasal thalamus does not enable separation of these entities. In the absence of a clear discontinuity or cytoarchitectural difference, the results of such experimental labeling have been interpreted by some authors (e.g., Feldman and Kruger, '80) to indicate the presence of the gracile nucleus rostral to the obex into the region called nucleus Z. Attempts to outline this nucleus (e.g., Paxinos and Watson, '86, and Swanson, '92) can result in apparent discrepancies because it is difficult to identify, even with our access to sections in the three cardinal planes. Thus, our identification of the approximate location of this "nucleus" constitutes acquiescence to current usage and should be regarded as merely tentative; we do not imply that we can outline this structure or that experimental studies to date can satisfactorily resolve this problem.

AMB Nucleus Ambiguus
ap Area Postrema
CU Cuneate Nucleus
CUe External Cuneate Nucleus
GR Gracile Nucleus
IO Inferior Olive
LR Lateral Reticular Nucleus
mlf Medial Longitudinal Fascicle
PcR Parvicellular Reticular Nucleus
PmR Paramedian Reticular Nucleus
py Pyramidal Tract
RO Nucleus Raphé Obscurus
RPa Nucleus Raphé Pallidus
s5c Spinal Trigeminal Nucleus (caudal)
TS$_c$ Nucleus of the Solitary Tract (commissural)
TS$_l$ Nucleus of the Solitary Tract (lateral)
TS$_m$ Nucleus of the Solitary Tract (medial)
ts Solitary Tract
VR Ventral Reticular Nucleus
Z Nucleus Z
5t Trigeminal Tract
10 Dorsal Motor Nucleus of Vagus
12 Hypoglossal Nerve

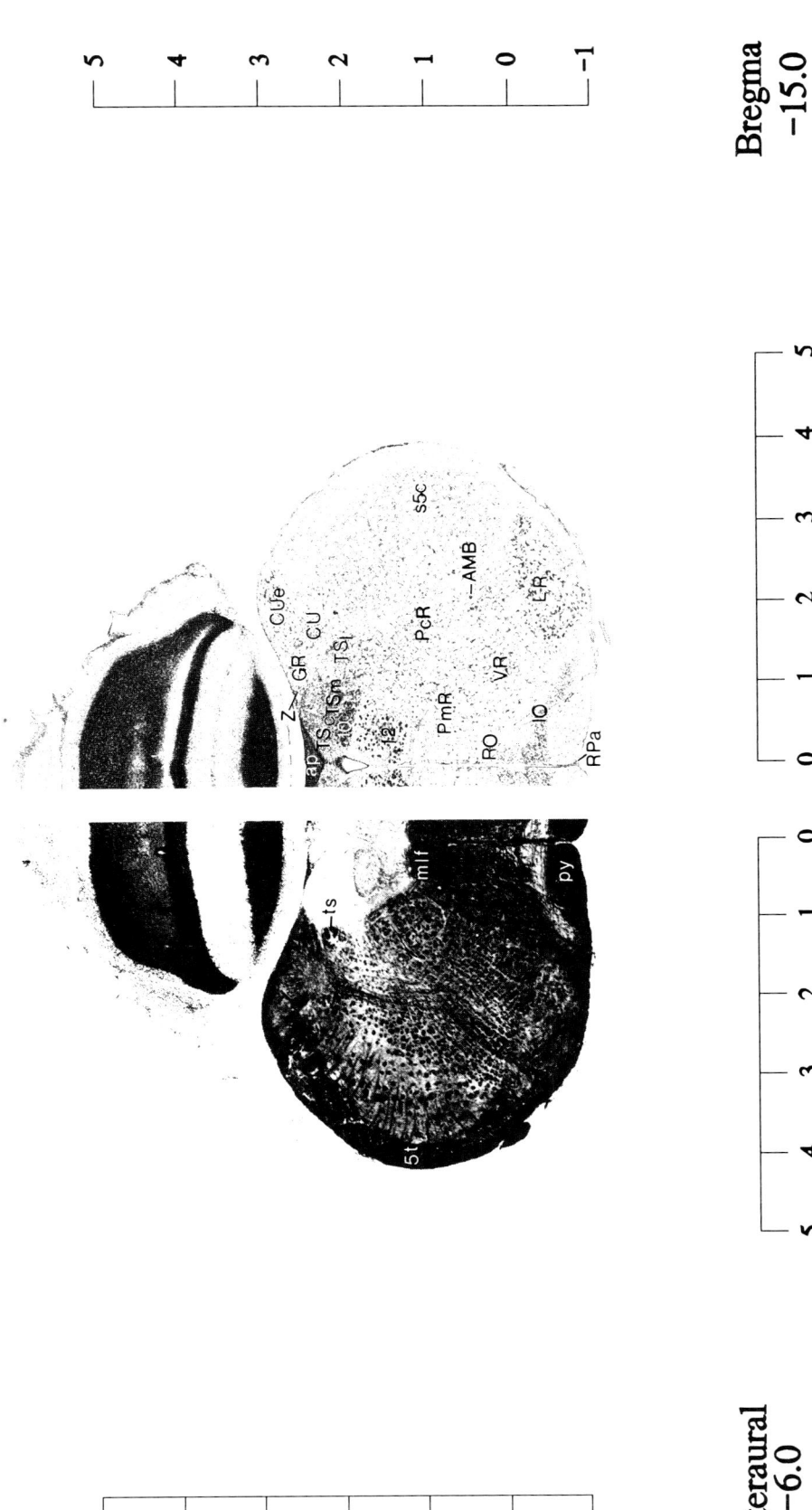

Plate 59

AMB Nucleus Ambiguus
ap Area Postrema
CU Cuneate Nucleus
CUe External Cuneate Nucleus
GR Gracile Nucleus
IO Inferior Olive
LR Lateral Reticular Nucleus
mlf Medial Longitudinal Fascicle
PcR Parvicellular Reticular Nucleus
PmR Paramedian Reticular Nucleus
py Pyramidal Tract
RO Nucleus Raphé Obscurus
RPa Nucleus Raphé Pallidus
s5c Spinal Trigeminal Nucleus (caudal)
TS$_c$ Nucleus of the Solitary Tract (commissural)
TS$_l$ Nucleus of the Solitary Tract (lateral)
TS$_m$ Nucleus of the Solitary Tract (medial)
ts Solitary Tract
VR Ventral Reticular Nucleus
5t Trigeminal Tract
10 Dorsal Motor Nucleus of Vagus
12 Hypoglossal Nerve
12r Hypoglossal Nerve Root

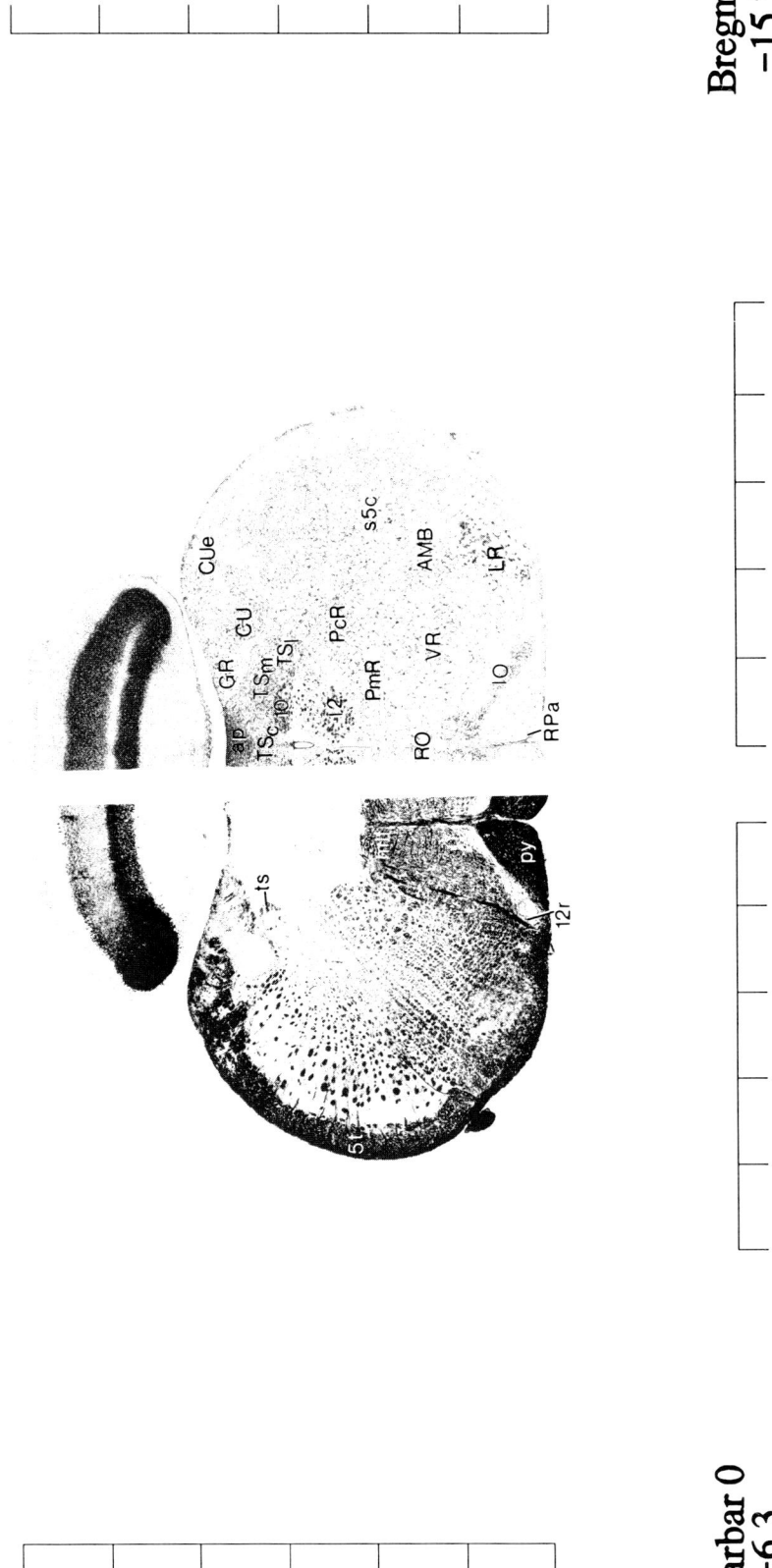

5
4
3
2
1
0
−1

5 — 4 — 3 — 2 — 1 — 0 — 1 — 2 — 3 — 4 — 5

Bregma −15.3

Earbar 0 −6.3

CUe
GR   CU
TSm   TSl
ap
TSc   io   r2
RO   PmR   PcR   s5c
VR   AMB
IO   LR
RPa
ts
py
12r
5i

Plate 60

AMB Nucleus Ambiguus
CU Cuneate Nucleus
DR Dorsal Reticular Nucleus
GR Gracile Nucleus
IO Inferior Olive
LR Lateral Reticular Nucleus
mlf Medial Longitudinal Fascicle
py Pyramidal Tract
pyx Decussation of Pyramidal Tract
s5c Spinal Trigeminal Nucleus (caudal)
TS$_m$ Nucleus of the Solitary Tract (medial)
ts Solitary Tract
VR Ventral Reticular Nucleus
5t Trigeminal Tract
10 Dorsal Motor Nucleus of Vagus
12 Hypoglossal Nucleus
12n Hypoglossal Nerve

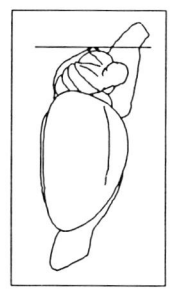

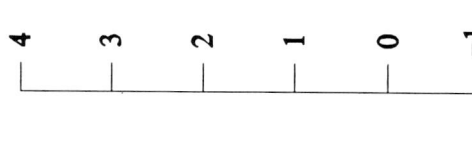

4

3

2

1

0

–1

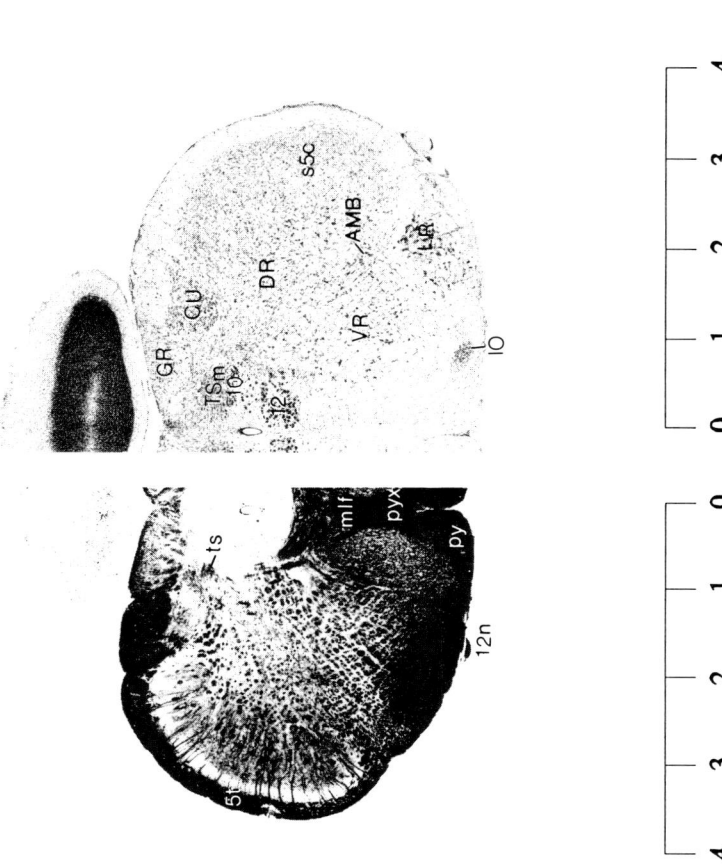

4 3 2 1 0

0 1 2 3 4

Plate 61

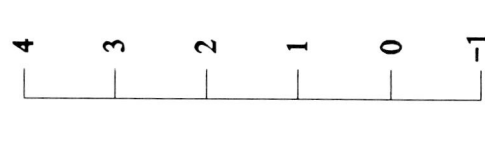

4

3

2

1

0

−1

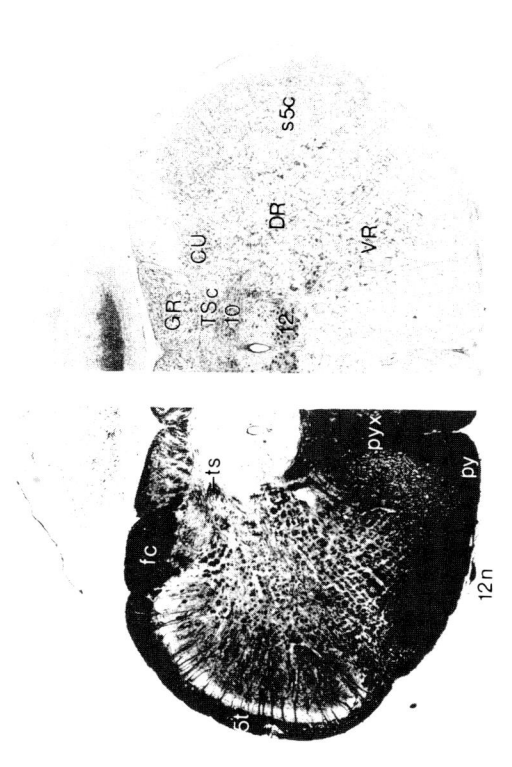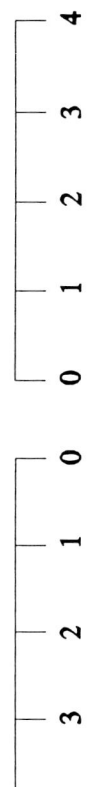

Bregma
−15.9

4   3   2   1   0

0   1   2   3   4

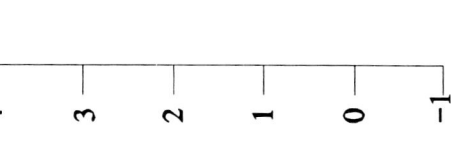

4

3

2

1

0

−1

Earbar 0
−6.9

Plate 62

CU Cuneate Nucleus
DR Dorsal Reticular Nucleus
fc Cuneate Fascicle
GR Gracile Nucleus
py Pyramidal Tract
pyx Decussation of Pyramidal Tract
s5c Spinal Trigeminal Nucleus (caudal)
sg Substantia Gelatinosa
TS$_c$ Nucleus of the Solitary Tract (commissural)
VR Ventral Reticular Nucleus
5t Trigeminal Tract
10 Dorsal Motor Nucleus of Vagus
12 Hypoglossal Nucleus
12n Hypoglossal Nerve

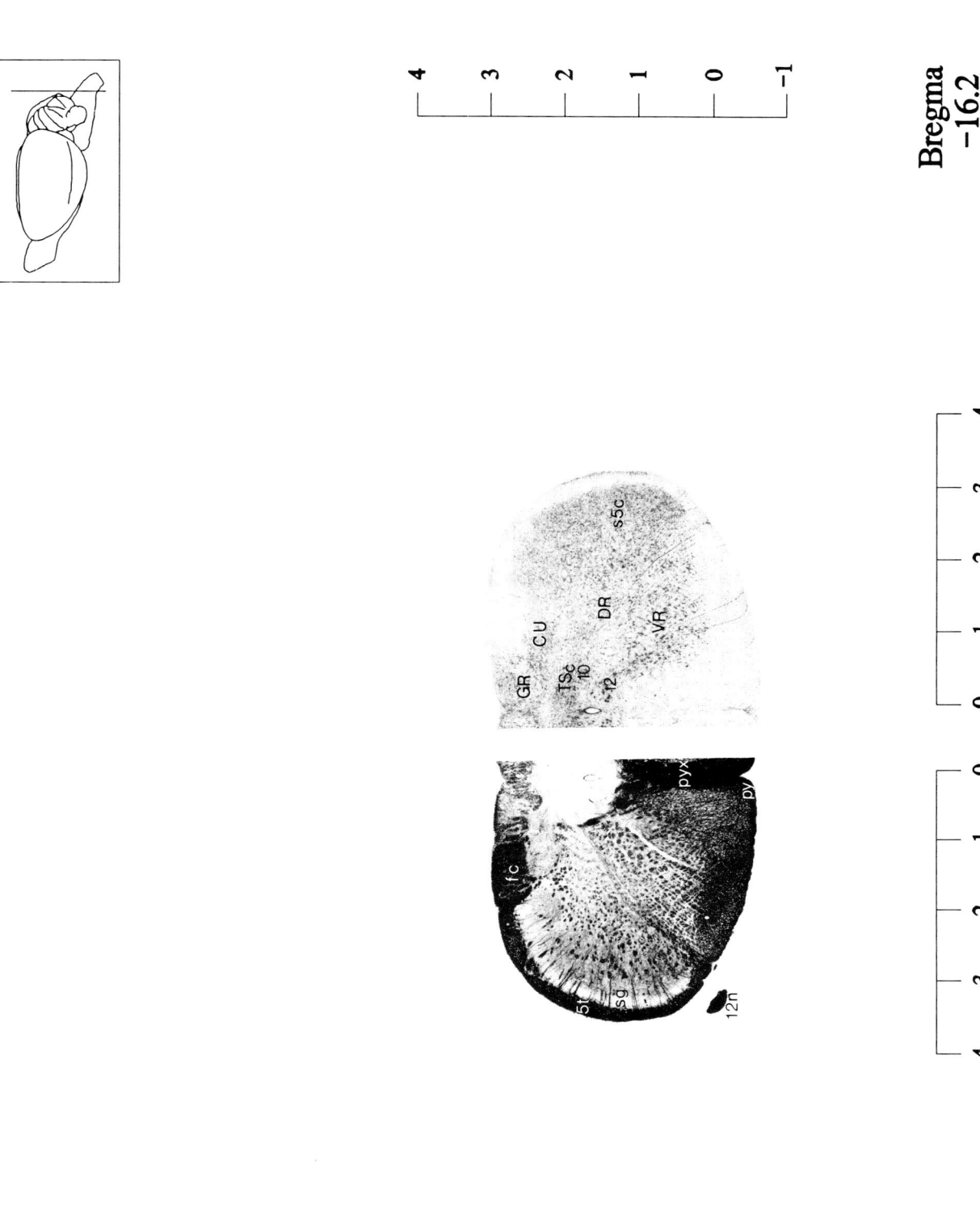

Plate 63

CG Central Gray
CU Cuneate Nucleus
DR Dorsal Reticular Nucleus
fc Cuneate Fascicle
fg Gracile Fascicle
GR Gracile Nucleus
s5c Spinal Trigeminal Nucleus (caudal)
sg Substantia Gelatinosa
TS$_c$ Nucleus of the Solitary Tract (commissural)
VR Ventral Reticular Nucleus
5t Trigeminal Tract

fc and fg: Cuneate Fascicle and Gracile Fascicle (also called, collectively, Dorsal Columns or Fasciculi and Fasciculi Cuneatus and Gracilis). The homologues in the human brain stem bear the eponyms of Burdach and Goll, respectively. These large, heavily myelinated tracts lie on the dorsal surface of the spinal cord and medulla lateral to the midline containing primarily large diameter tactile axons originating in sensory ganglion cells from lumbosacral (gracile) and cervical-thoracic (cuneate) levels. These tracts terminate in the gracile medial nucleus and the cuneate lateral nucleus of the medulla and are separable by a longitudinal fissure visible from the surface. The cuneate bundle (and nucleus) considerably exceeds the size of the gracile component.

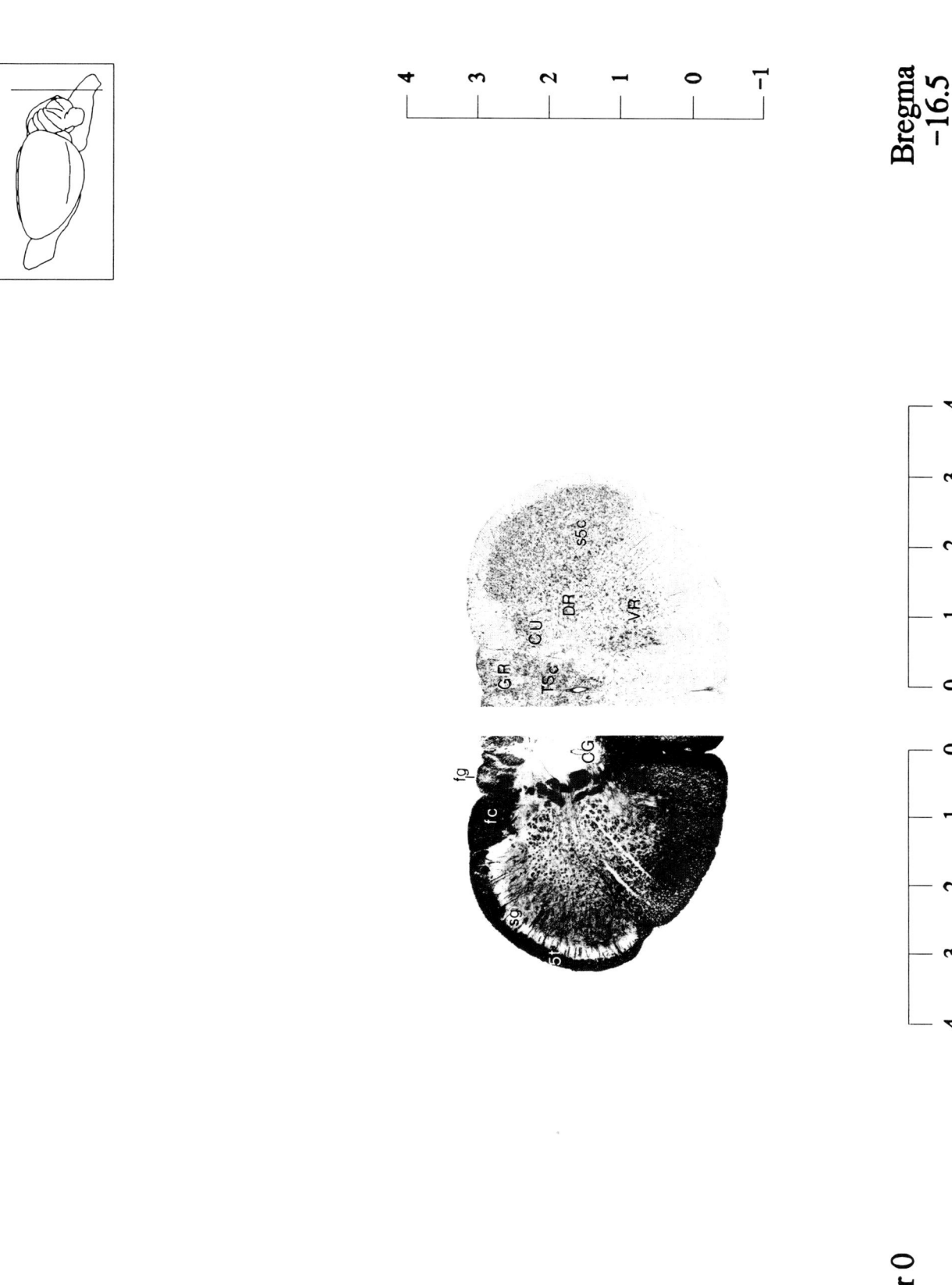

4

3

2

1

0

-1

Bregma
-16.5

4   3   2   1   0

0   1   2   3   4

4

3

2

1

0

-1

Earbar 0
-7.5

Plate 64

c Central Canal
CG Central Gray
CU Cuneate Nucleus
fc Cuneate Fascicle
fg Gracile Fascicle
GR Gracile Nucleus
s5c Spinal Trigeminal Nucleus (caudal)
sg Substantia Gelatinosa
5t Trigeminal Tract

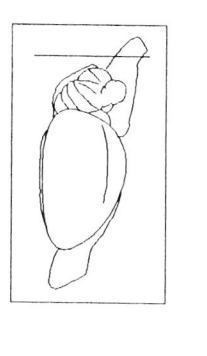

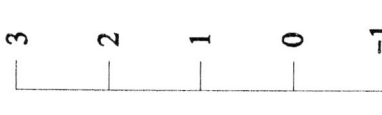

3

2

1

0

−1

Bregma
−16.8

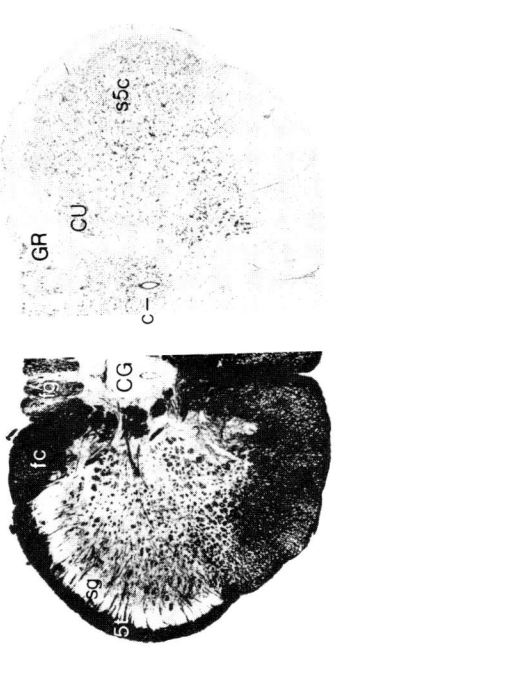

s5c

CU

GR

c

fc

CG

g

sg

3 2 1 0 1 2 3

3

2

1

0

−1

Earbar 0
−7.8

Plate 65

c   Central Canal
C_1 First Cervical Ventral Root
CG   Central Gray
fc   Cuneate Fascicle
fg   Gracile Fascicle
GR   Gracile Nucleus
s5c   Spinal Trigeminal Nucleus (caudal)
sg   Substantia Gelatinosa
Vh   Ventral Horn
5t   Trigeminal Tract

C_1: First Cervical Level of Spinal Cord. The C_1 is continuous with the caudal medulla but lacks a precise boundary. It is the only segmental spinal level in which the ventral root is usually not paired with a dorsal root.

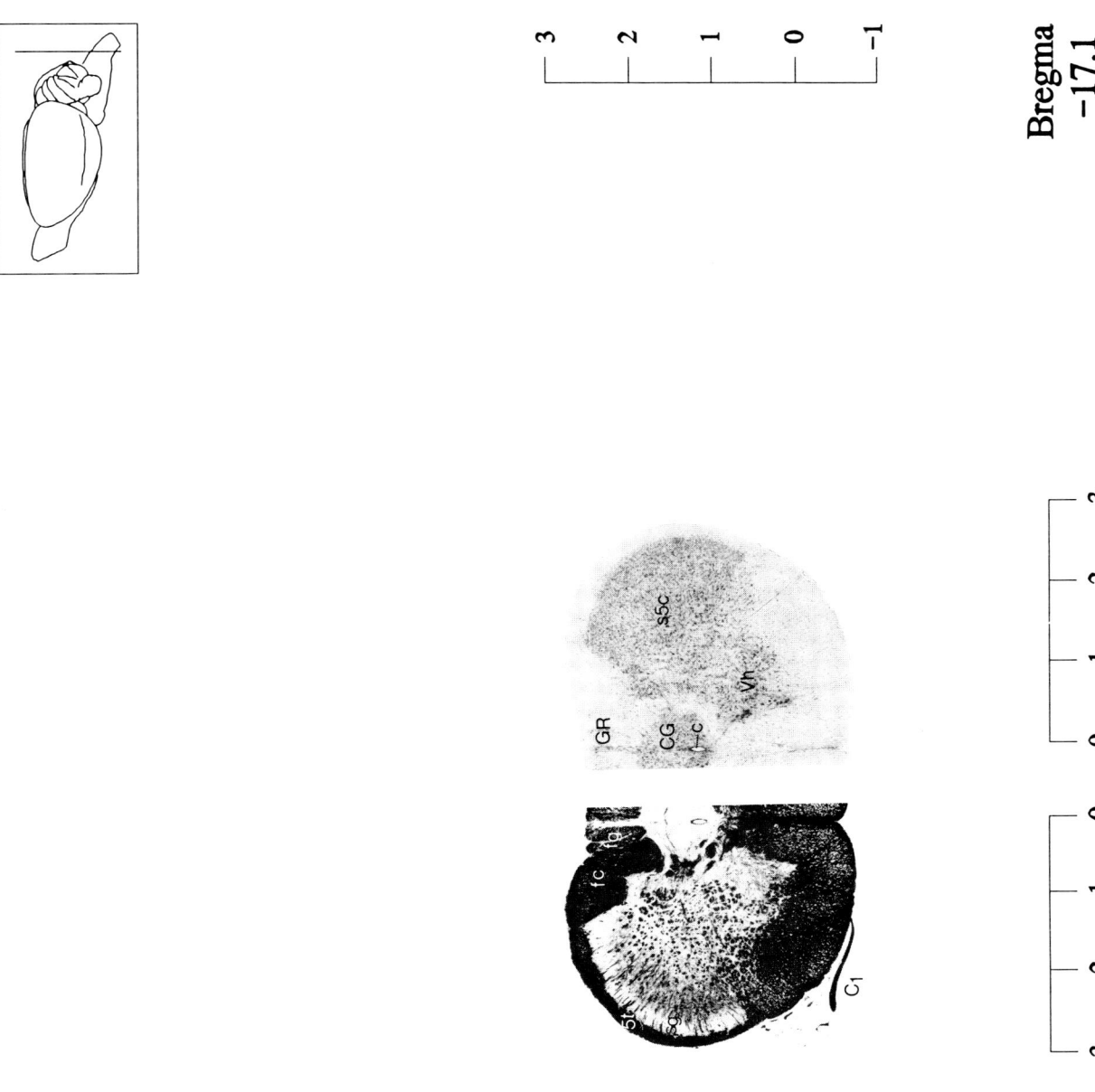

3

2

1

0

–1

3 2 1 0

0 1 2 3

3

2

1

0

–1

Plate 66

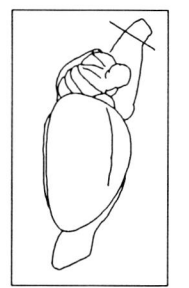

3

2

1

0

−1

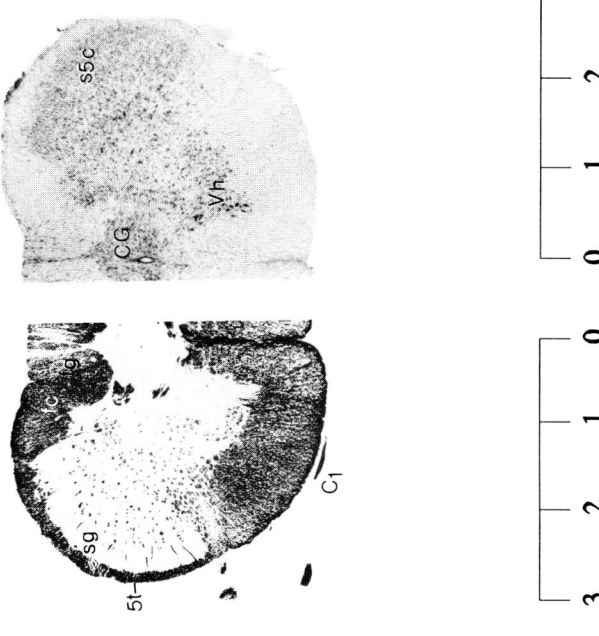

3   2   1   0   0   1   2   3

Earbar 0
−8.4

Bregma
−17.4

Plate 67

C₁ First Cervical Ventral Root
CG Central Gray
ic Cuneate Fascicle
fg Gracile Fascicle
s5c Spinal Trigeminal Nucleus (caudal)
sg Substantia Gelatinosa
Vh Ventral Horn
5t Trigeminal Tract

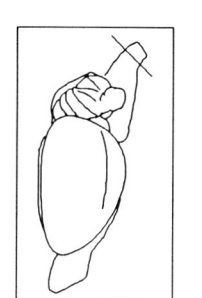

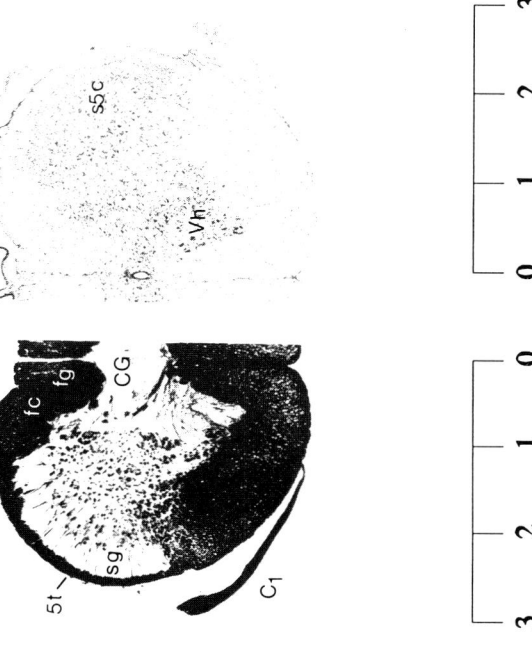

Plate 68

CeC Central Cervical Nucleus
Dh Dorsal Horn
fc Cuneate Fascicle
fg Gracile Fascicle
Lce Lateral Cervical Nucleus
s5c Spinal Trigeminal Nucleus (caudal)
sg Substantia Gelatinosa
Vh Ventral Horn

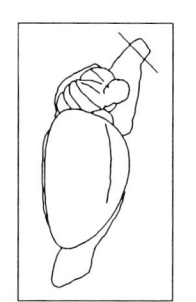

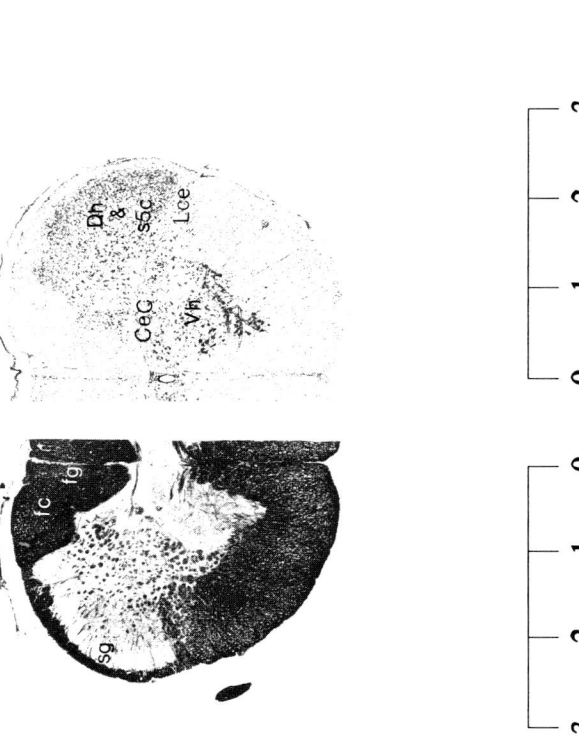

Bregma
−18.0

3 2 1 0

0 1 2 3

Earbar 0
−9.0

Plate 69

CeC Central Cervical Nucleus
Dh Dorsal Horn
fc Cuneate Fascicle
fg Gracile Fascicle
Lce Lateral Cervical Nucleus
sg Substantia Gelatinosa
Vh Ventral Horn

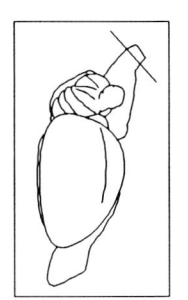

Bregma
−18.3

Earbar 0
−9.3

Plate 70

C₂ Second Cervical Ventral Root
CeC Central Cervical Nucleus
cst Corticospinal Tract
Dh Dorsal Horn
fc Cuneate Fascicle
fg Gracile Fascicle
Lce Lateral Cervical Nucleus
sg Substantia Gelatinosa
Vh Ventral Horn

Dh: Dorsal Horn. The dorsal (sensory) gray matter of the cervical spinal cord fuses with the posterior end of s5c without any apparent boundary or dorsal root entry zone comparable to lower spinal levels. The laminar pattern of the dorsal horn at the C₁ level is essentially the same as in s5c; thus the latter has been called the trigeminal dorsal horn (Gobel et al., '77).

Lce: Lateral Cervical Nucleus (also called Nucleus Cervicalis Lateralis). A distinct rostrocaudal aggregate of neurons embedded in the laterodorsal tract of the upper cervical spinal cord in humans and other primates is less distinct in smaller mammals, but the homologous somatosensory spinal relay has been identified and characterized in several species following the account of Rexed and Brodal ('51). A comparable but larger aggregate associated with the homologous region of the trigeminal nuclear complex (s5c) is more evident in the rat as the paratrigeminal nucleus (pa5). The cervical contribution is sufficiently small in rat and many other small mammals to render identification of this scattering of cells difficult in many sections, thus requiring an extensive series. See Baker and Giesler ('84) and Broman and Blomqvist ('89) for description in rat.

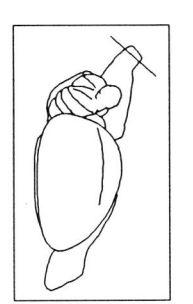

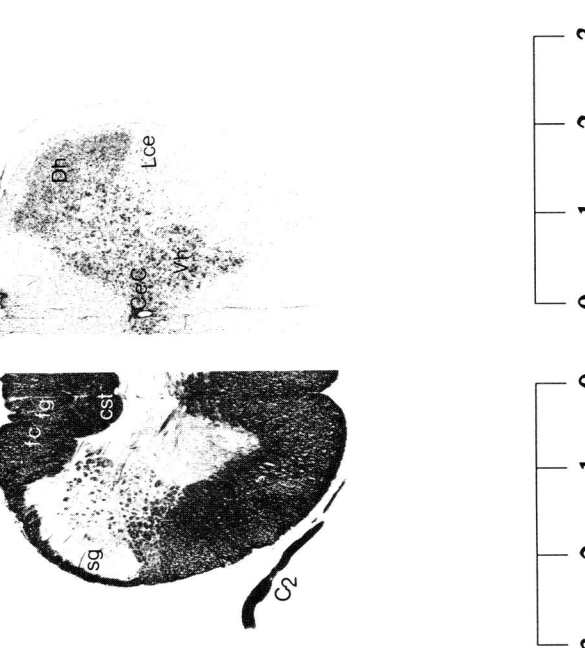

Earbar 0
−9.6

Bregma
−18.6

Plate 71

C₂ Second Cervical Ventral Root
CeC Central Cervical Nucleus
Dh Dorsal Horn
fc Cuneate Fascicle
fg Gracile Fascicle
Lce Lateral Cervical Nucleus
sg Substantia Gelatinosa
Vh Ventral Horn

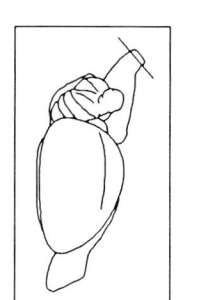

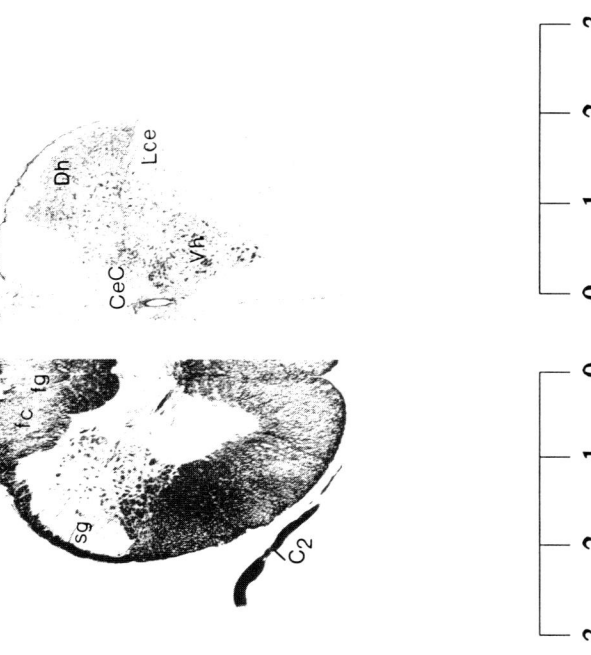

Bregma
−18.9

Earbar 0
−9.9

Plate 72

C₂ Second Cervical Roots
CeC Central Cervical Nucleus
Dh Dorsal Horn
fc Cuneate Fascicle
fg Gracile Fascicle
Lce Lateral Cervical Nucleus
sg Substantia Gelatinosa
Vh Ventral Horn

C₂: Second Cervical Level of Spinal Cord. At the C₂ level of the spinal cord the dorsal and ventral roots merge, in contrast to the C₁ level, which lacks a dorsal root (see Plates 67 and 72).

CeC: Central Cervical Nucleus. In the upper cervical spinal cord, area X of Rexed (also sometimes called *lamina or layer 10*) surrounding the central canal reveals a pair of lateral nuclear excrescences constituting the CeC (see Brichta and Grant, '85).

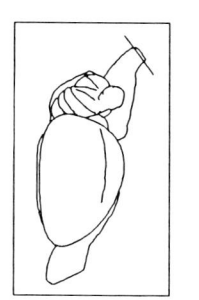

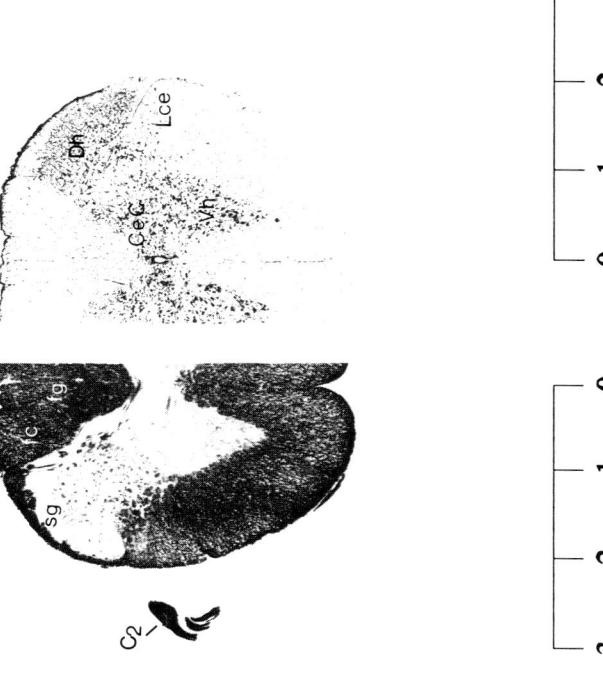

Bregma
–19.2

Earbar 0
–10.2

Plate 73

C₂ Second Cervical Roots
CeC Central Cervical Nucleus
Dh Dorsal Horn
fc Cuneate Fascicle
fg Gracile Fascicle
Lce Lateral Cervical Nucleus
sg Substantia Gelatinosa
Vh Ventral Horn

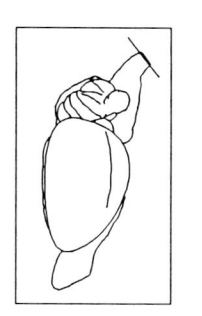

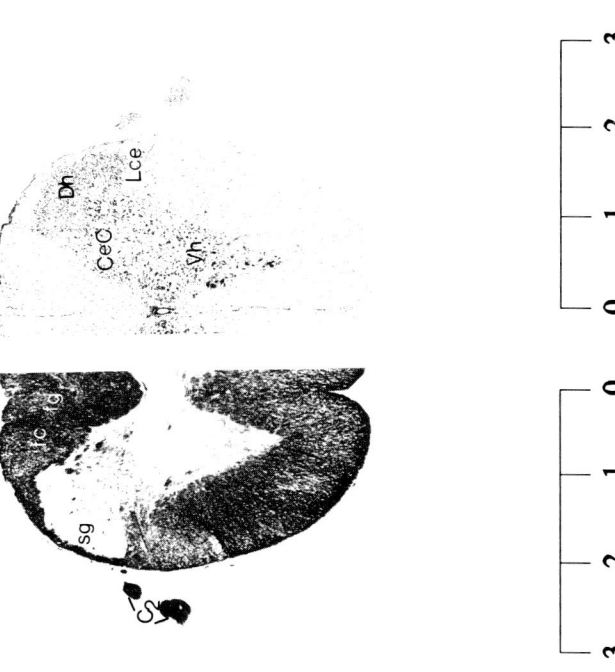

Earbar 0
−10.5

Bregma
−19.5

Plate 74

AL Anterior Limbic Area (Cortex)
Ast Area Striata (Cortex)
Pi Pineal Gland (Epiphysis)
PL Posterior Limbic Area (Cortex)
Rsp Retrosplenial Area (Cortex)
SomS Somatosensory Cortex

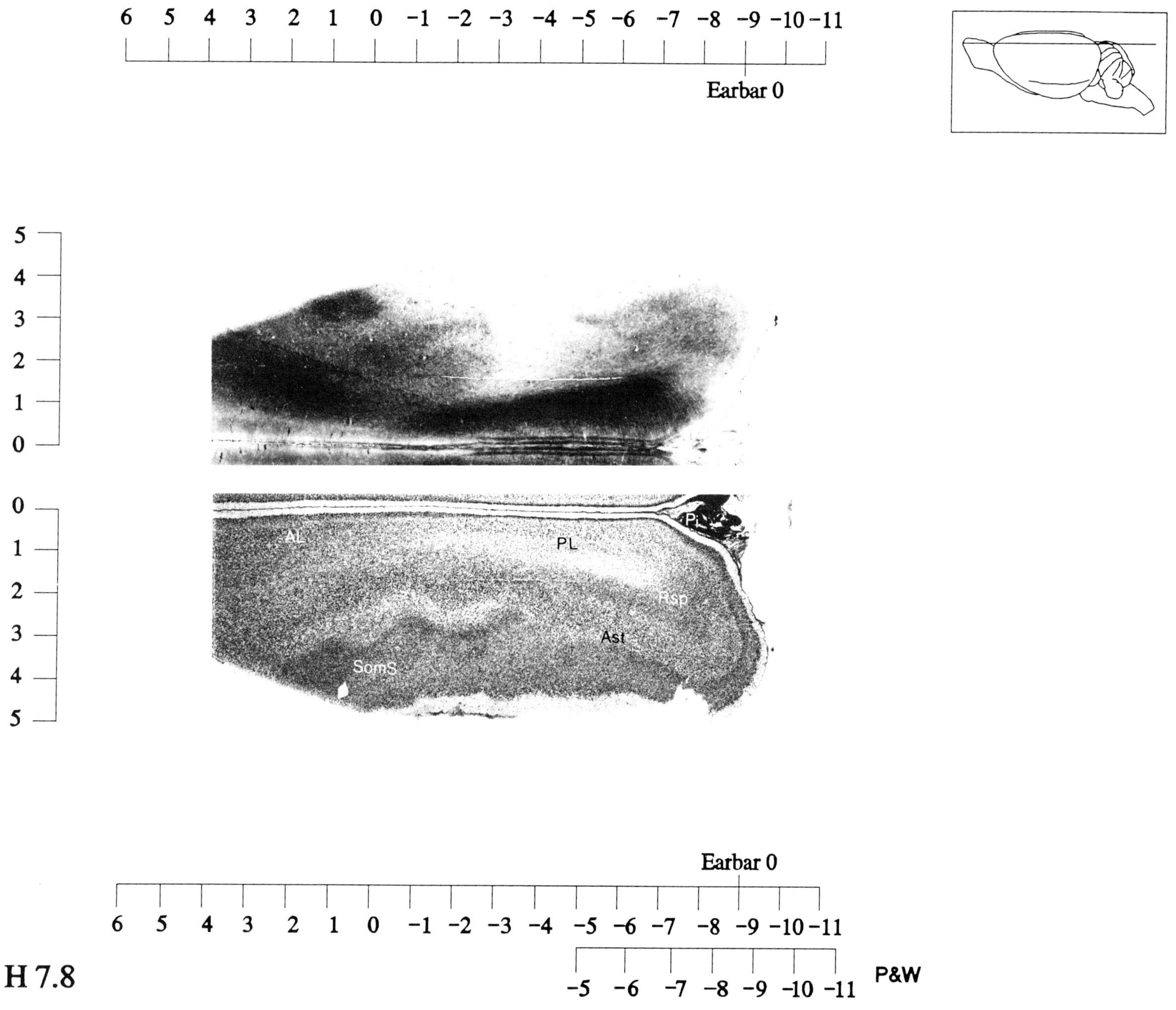

6 5 4 3 2 1 0 −1 −2 −3 −4 −5 −6 −7 −8 −9 −10 −11
Earbar 0
5 4 3 2 1 0
0 1 2 3 4 5
AL
PL
Pi
Rsp
Ast
SomS
Earbar 0
6 5 4 3 2 1 0 −1 −2 −3 −4 −5 −6 −7 −8 −9 −10 −11
−5 −6 −7 −8 −9 −10 −11
P&W

H 7.8

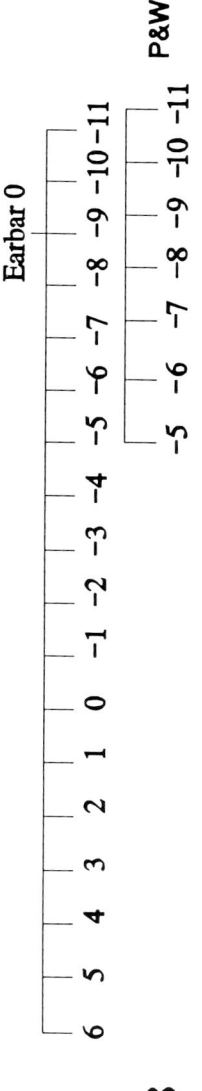

Plate 75

AL Anterior Limbic Area (Cortex)
Ast Area Striata (Cortex)
ENT Entorhinal Area
IC Inferior Colliculus
Pi Pineal Gland (Epiphysis)
Rsp Retrosplenial Area (Cortex)
SomS Somatosensory Cortex

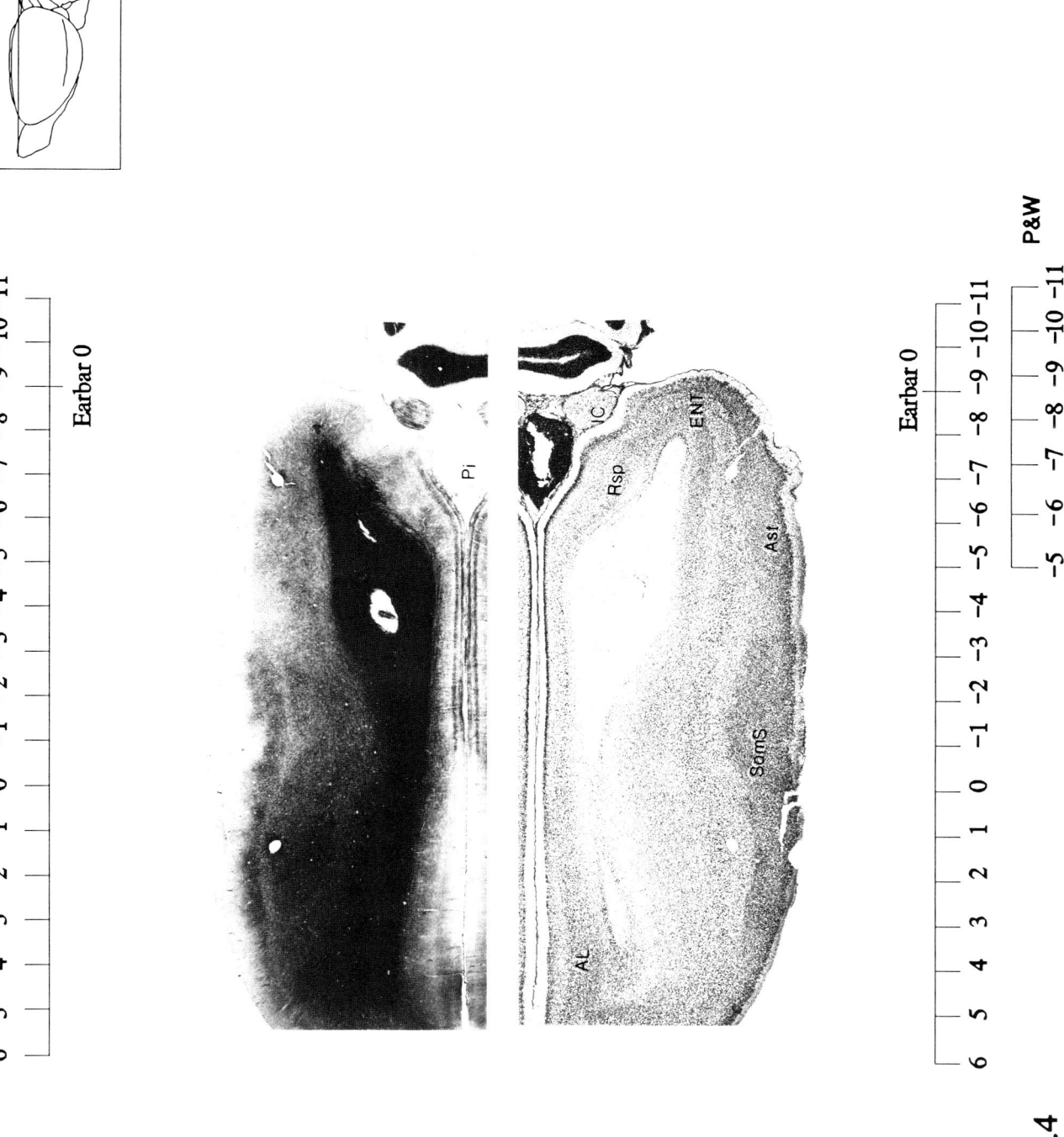

Earbar 0

6 5 4 3 2 1 0 -1 -2 -3 -4 -5 -6 -7 -8 -9 -10 -11

5
4
3
2
1
0

0
1
2
3
4
5

Pi

IC

Rsp

ENT

Ast

SomS

AL

P&W

Earbar 0

6 5 4 3 2 1 0 -1 -2 -3 -4 -5 -6 -7 -8 -9 -10 -11

-5 -6 -7 -8 -9 -10 -11

H 8.4

Plate 76

AL   Anterior Limbic Area (Cortex)
Ast   Area Striata (Cortex)
CA$_1$ Hippocampal Area CA$_1$ (Ammon's Horn)
cc   Corpus Callosum
ENT   Entorhinal Area
IC   Inferior Colliculus
Ma   Motor Agranular Cortical Area
Pi   Pineal Gland (Epiphysis)
Rsp   Retrosplenial Area (Cortex)
SC   Superior Colliculus
SomS   Somatosensory Cortex

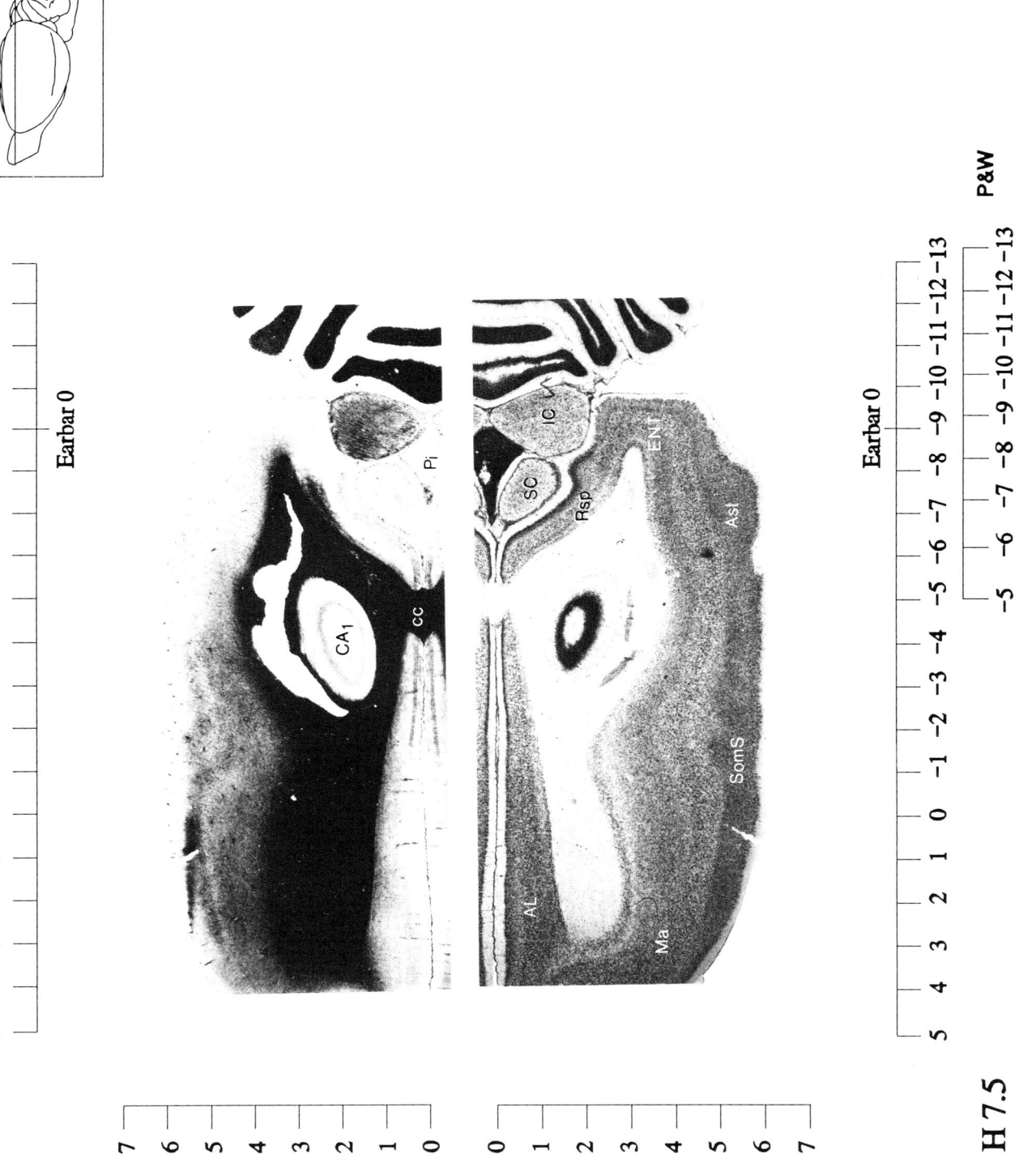

Earbar 0

5 4 3 2 1 0 -1 -2 -3 -4 -5 -6 -7 -8 -9 -10 -11 -12 -13

7 6 5 4 3 2 1 0 0 1 2 3 4 5 6 7

Earbar 0

5 4 3 2 1 0 -1 -2 -3 -4 -5 -6 -7 -8 -9 -10 -11 -12 -13

P&W

H 7.5

Plate 77

AL   Anterior Limbic Area (Cortex)
Ast   Area Striata (Cortex)
CA₁ Hippocampal Area CA₁ (Ammon's Horn)
cc   Corpus Callosum
CdP Caudoputamen
chp   Choroid Plexus
cic   Commissure of the Inferior Colliculus
DG   Dentate Gyrus
dhc   Dorsal Hippocampal Commissure
ENT   Entorhinal Area
IC   Inferior Colliculus
IG   Induseum Griseum
LV   Lateral Ventricle
Ma   Motor Agranular Cortical Area
SC   Superior Colliculus
SomS Somatosensory Cortex
Sub   Subiculum

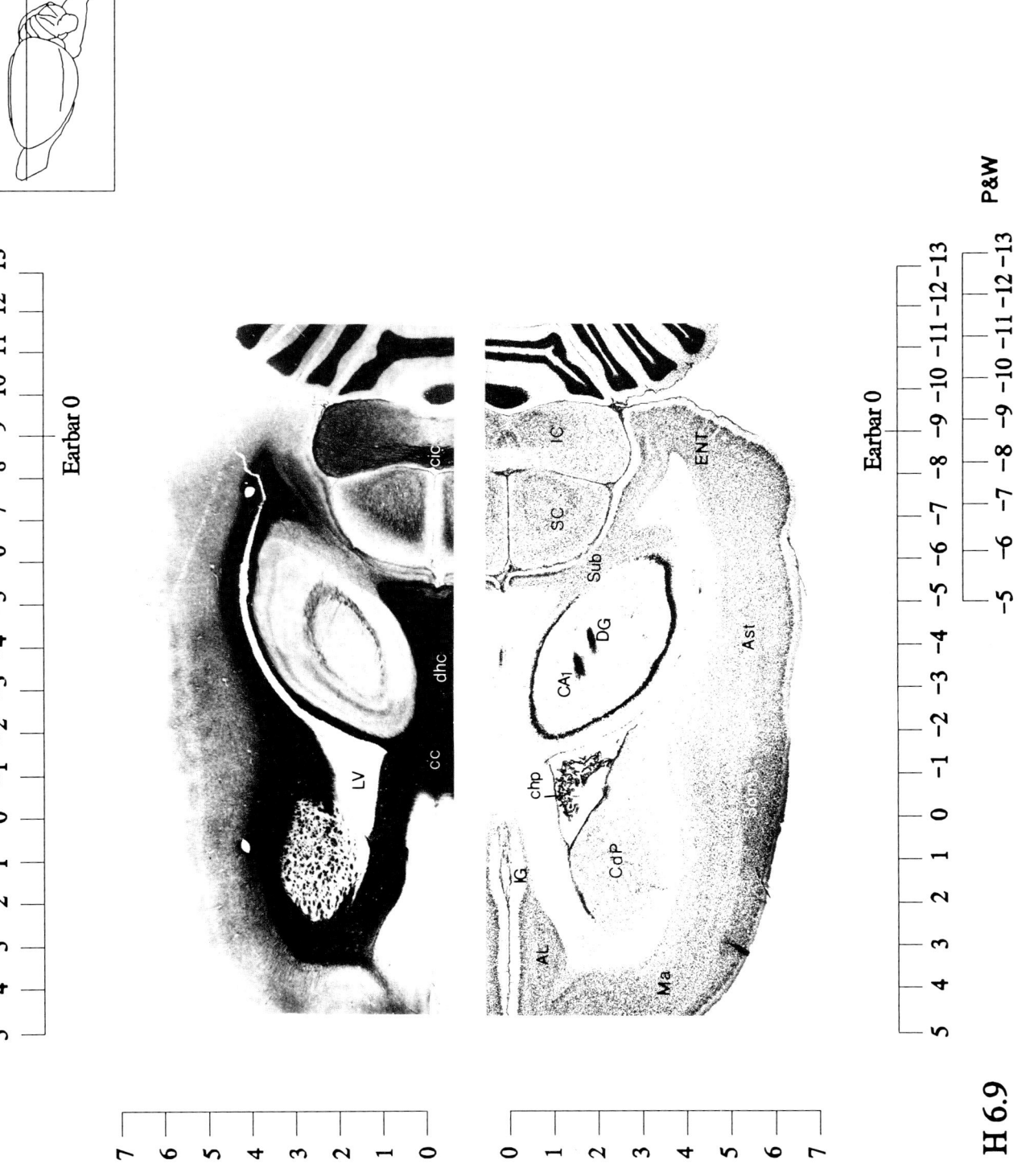

Earbar 0

5 4 3 2 1 0 −1 −2 −3 −4 −5 −6 −7 −8 −9 −10 −11 −12 −13

LV

cc

dhc

cic

7 6 5 4 3 2 1 0

CA1

DG

chp

CdP

IG

AL

Ma

Sub

SC

IC

ENT

Ast

0 1 2 3 4 5 6 7

Earbar 0

5 4 3 2 1 0 −1 −2 −3 −4 −5 −6 −7 −8 −9 −10 −11 −12 −13

P&W

H 6.9

Plate 78

alv Alveus
Ast Area Striata (Cortex)
CA₁ Hippocampal Area CA_1 (Ammon's Horn)
CA₃ Hippocampal Area CA_3 (Ammon's Horn)
cc Corpus Callosum
CdP Caudoputamen
cic Commissure of the Inferior Colliculus
DG Dentate Gyrus
ENT Entorhinal Area
FC Fasciola Cinerea
fi Fimbria
fo Fornix
IC Inferior Colliculus
IG Induseum Griseum
LV Lateral Ventricle
Ma Motor Agranular Cortical Area
SC Superior Colliculus
SomS Somatosensory Cortex
SPl_d Septal Nucleus (lateral: dorsal)
Sub Subiculum

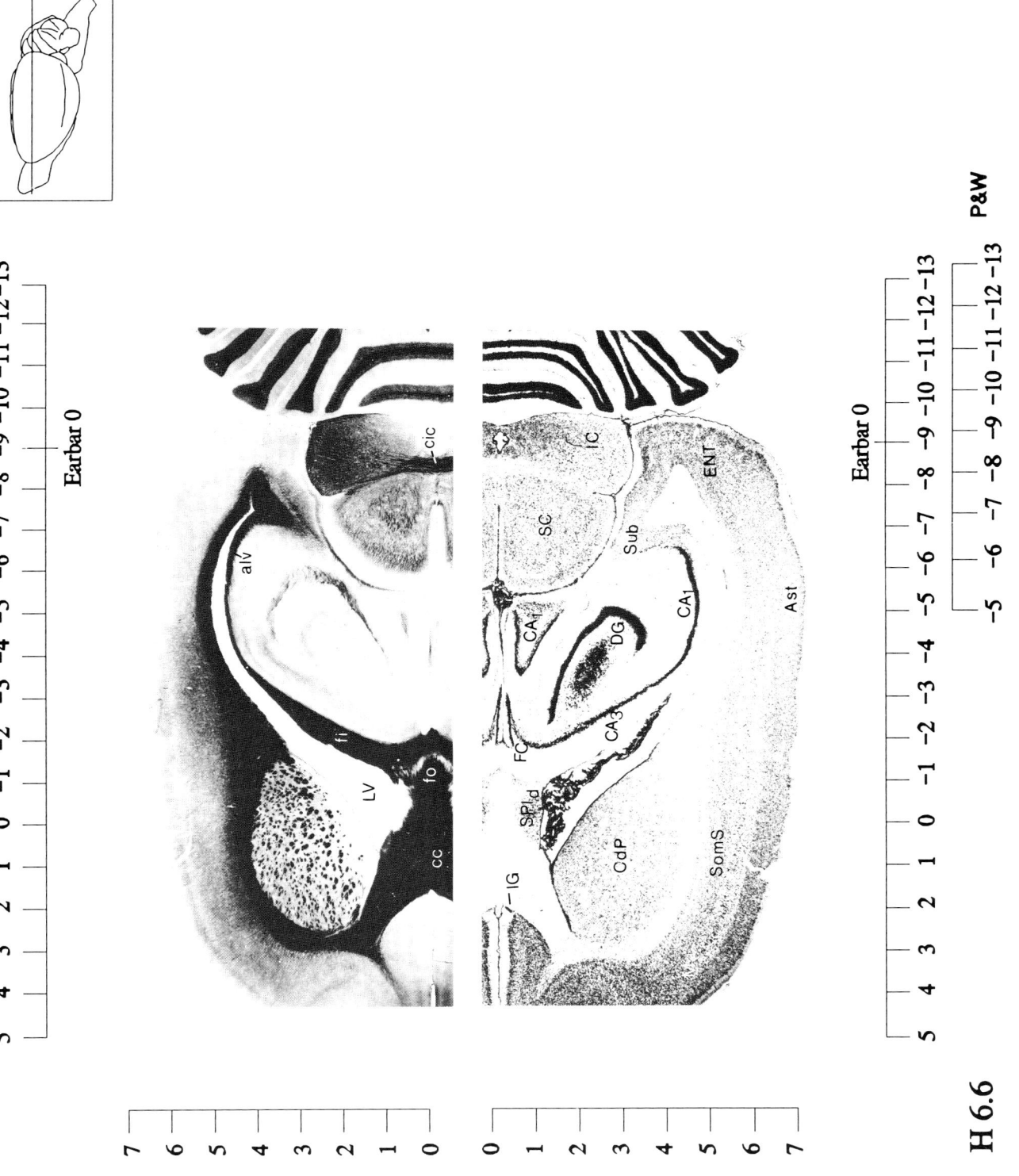

Earbar 0

5 4 3 2 1 0 −1 −2 −3 −4 −5 −6 −7 −8 −9 −10 −11 −12 −13

7 6 5 4 3 2 1 0

0 1 2 3 4 5 6 7

Earbar 0

5 4 3 2 1 0 −1 −2 −3 −4 −5 −6 −7 −8 −9 −10 −11 −12 −13

−5 −6 −7 −8 −9 −10 −11 −12 −13

P&W

H 6.6

Plate 79

AL Anterior Limbic Area (Cortex)
Ast Area Striata (Cortex)
bic Brachium of Inferior Colliculus
bsc Brachium of Superior Colliculus
CA₁ Hippocampal Area CA₁ (Ammon's Horn)
CA₃ Hippocampal Area CA₃ (Ammon's Horn)
cc Corpus Callosum
CdP Caudoputamen
chp Choroid Plexus
cic Commissure of the Inferior Colliculus
DG Dentate Gyrus
ENT Entorhinal Area
fo Fornix
IC Inferior Colliculus
IG Induseum Griseum
LD Lateral Dorsal Nucleus (Thalamus)
LV Lateral Ventricle
SC Superior Colliculus
SomS Somatosensory Cortex
SPl Septal Nucleus (lateral)
Sub Subiculum
Ts Triangular Nucleus (Septum)
vhc Ventral Hippocampal Commissure

Limbic Cortex. The nomenclature is vastly confusing because of a lack of consensus despite general acceptance of the early Brodmann ('09) numbering system. The term derives from the hilus of the hemisphere being called the *limbus*, the *lobe limbique* of Broca, but its use has been extended beyond any meaningful strict definition despite attempts at systematization (see Pribram and Kruger, '54). The terms *cingulate* and *mesocortex* have also been applied to the belt of cortical fields surrounding the corpus callosum, and they share the common feature of possessing somewhat fused supragranular layers such that it can be considered a five-layered cortex (M. Rose, '31). The anterior limbic (or cingulate) field is demarcated from the posterior limbic (or cingulate) by the appearance of a granular layer in the latter, a separation validated on the basis of its thalamic input (e.g., Rose and Woolsey, '48). The anterior and posterior limbic cortices correspond to Brodmann's ('09) areas 24 and 29, respectively, and the prelimbic and infralimbic cortices to Brodmann's areas 32 and 25, respectively. Further subdivision is quite feasible, and the distinction between posterior limbic and retrosplenial cortical fields is considered by Vogt and Peters ('81). See also Rsp: Retrosplenial Area (Plate 34).

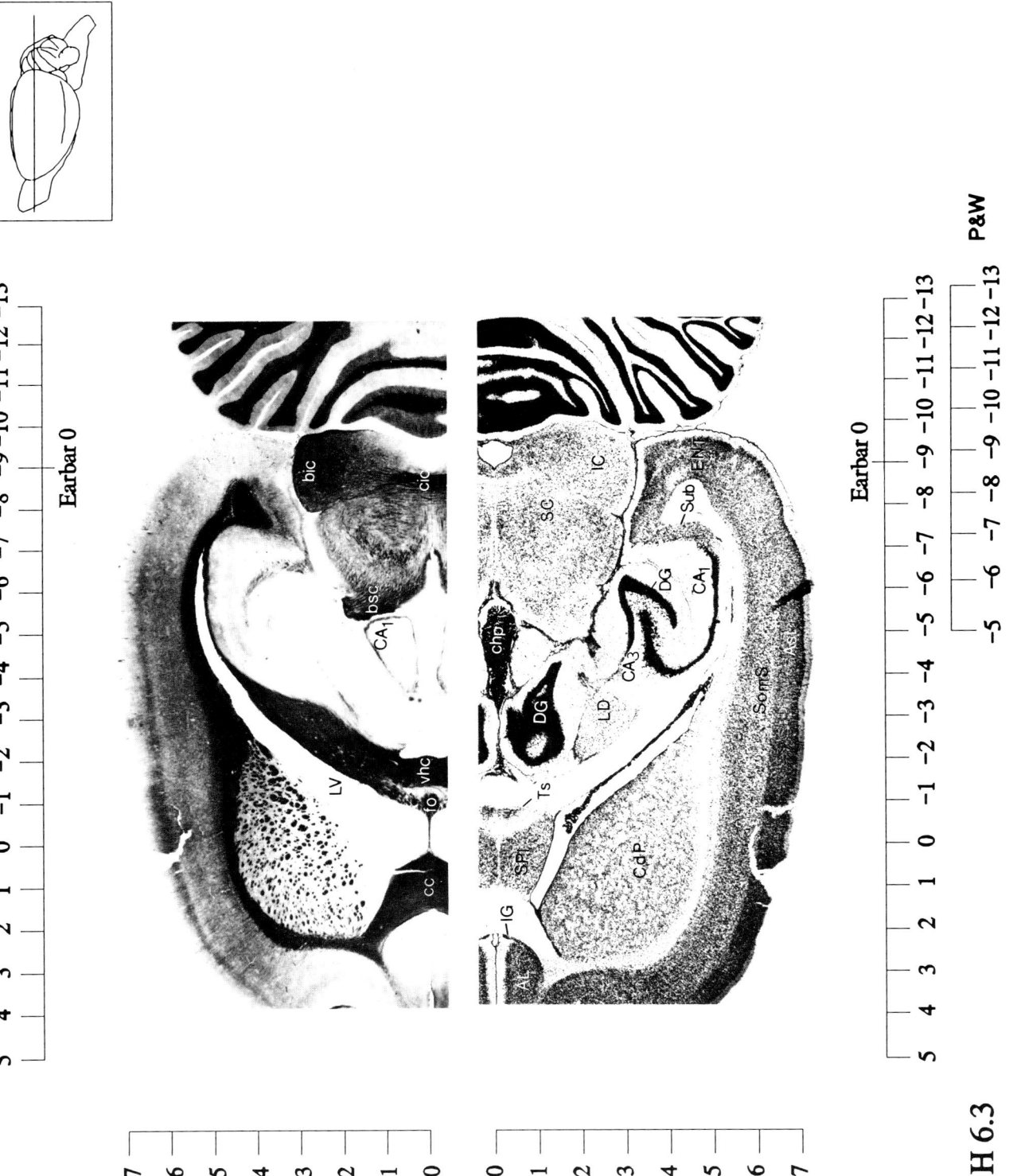

P&W

H 6.3

Plate 80

AL Anterior Limbic Area (Cortex)
Ast Area Striata (Cortex)
cc Corpus Callosum
CdP Caudoputamen
chp Choroid Plexus
DG Dentate Gyrus
ENT Entorhinal Area
fo Fornix
IC Inferior Colliculus
LD Lateral Dorsal Nucleus (Thalamus)
LGd Dorsal Lateral Geniculate Nucleus (Thalamus)
LP Lateral Posterior Nucleus (Thalamus)
SC Superior Colliculus
SomS Somatosensory Cortex
Sub Subiculum
Ts Triangular Nucleus (Septum)
tt Tenia Tecta

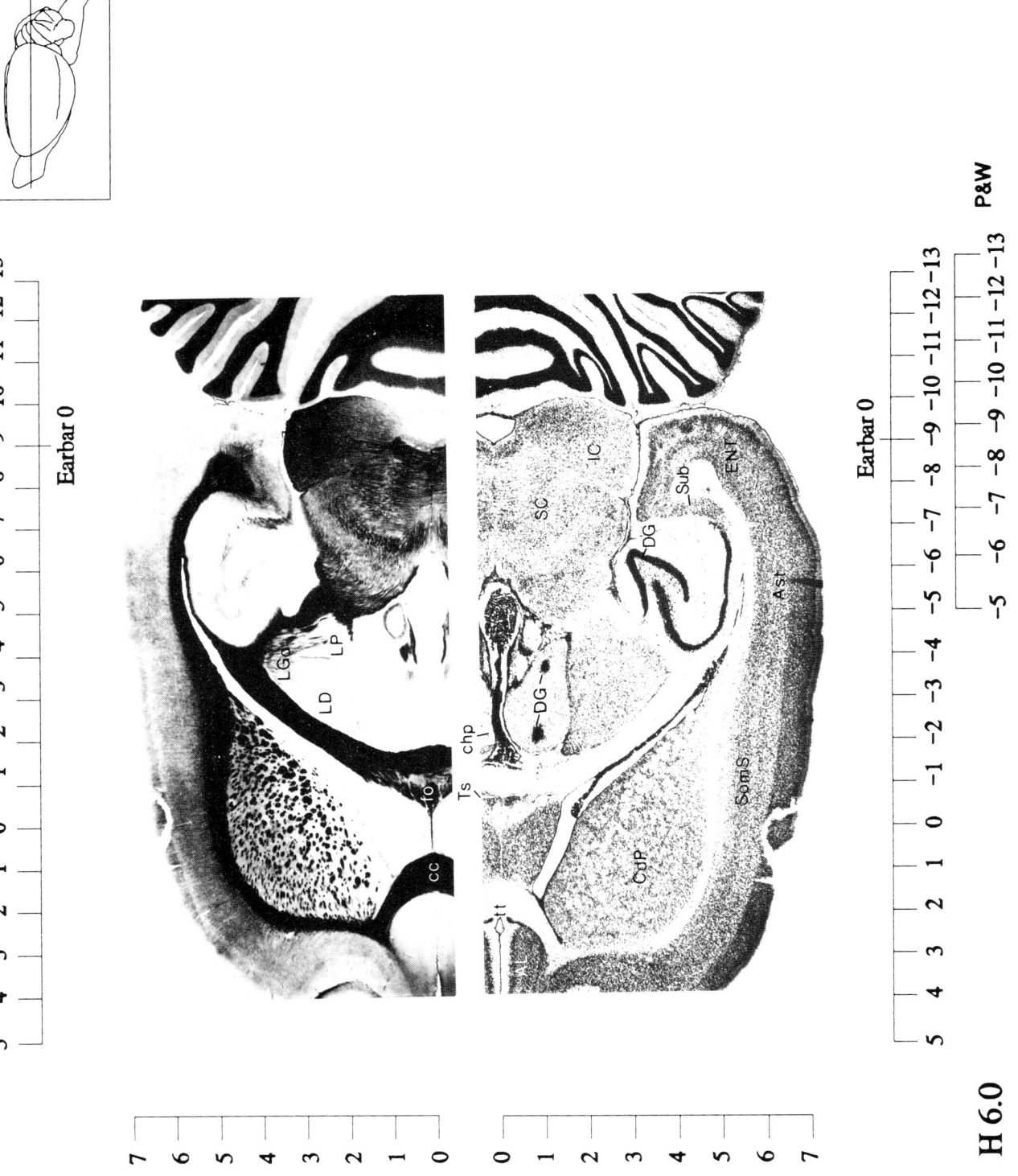

Earbar 0

5 4 3 2 1 0 -1 -2 -3 -4 -5 -6 -7 -8 -9 -10 -11 -12 -13

7 6 5 4 3 2 1 0

0 1 2 3 4 5 6 7

Earbar 0

5 4 3 2 1 0 -1 -2 -3 -4 -5 -6 -7 -8 -9 -10 -11 -12 -13

-5 -6 -7 -8 -9 -10 -11 -12 -13

P&W

H 6.0

Plate 81

AD Anterodorsal Nucleus (Thalamus)
alv Alveus
Ast Area Striata (Cortex)
AV Anteroventral Nucleus (Thalamus)
CdP Caudoputamen
CG Central Gray
cl Central Lateral Nucleus (Thalamus)
csc Commissure of the Superior Colliculus
DG Dentate Gyrus
ec External Capsule
eml External Medullary Lamina
ENT Entorhinal Area
fi Fimbria
Hb$_l$ Habenula Nucleus (lateral) (Thalamus)
Hb$_m$ Habenula Nucleus (medial) (Thalamus)
hc Habenula Commissure
IC Inferior Colliculus
IG Induseum Griseum
LD Lateral Dorsal Nucleus (Thalamus)
LGd Dorsal Lateral Geniculate Nucleus (Thalamus)
LP Lateral Posterior Nucleus (Thalamus)
LV Lateral Ventricle
pc Posterior Commissure
Prl Prelimbic Area (Cortex)
Pta Anterior Pretectal Nucleus (Thalamus)
Ptp Posterior Pretectal Nucleus (Thalamus)
SC Superior Colliculus
sm Stria Medullaris
SPl Septal Nucleus (lateral)
Sub Subiculum
Ts Triangular Nucleus (Septum)
vhc Ventral Hippocampal Commissure

cl: Central Lateral Nucleus (Thalamus) (also called Nucleus Centralis Lateralis). The laterodorsal extension of neurons embedded in the internal medullary lamina is variously outlined by different authors because it is often indistinct in Nissl-stained preparations. The location is more easily defined in fiber-stained sections as the portion of the intralaminar wing lateral and dorsal to the paracentral nucleus. The dorsal portion extending vertically is often distinct in rat and rabbit, such that some authors designate this dorsal portion of the internal medullary lamina the *nucleus angularis* (e.g., M. Rose, '35). The most medial portion of cl may be divisible into dorsal and ventral components, but this is not conventionally recognized.

LD: Lateral Dorsal Nucleus (Thalamus) (also called Laterodorsal Nucleus). The lateral nuclear group constitutes one of the remarkable expansions in the evolution of the mammalian thalamus, especially in the larger primates in which the caudal portion is usually subdivided extensively (see LP: Lateral Posterior Nucleus at Plate 115). The rostrocaudal separation from the contiguous lateral posterior (LP) nucleus is difficult, but in horizontal and sagittal sections a gradient of cell density can enable separation in Nissl-stained sections. However, the fiber architecture provides a distinctive pattern, most evident in the rostrocaudal axis, that supports recognition of separate LD and LP nuclei in the rat. Some early neuroanatomists used *lateral* for the ventral group of our designation, a confusion that has blessedly fallen into disuse.

Prl: Prelimbic Area (Cortex). See AL: Anterior Limbic Area (Plate 12).

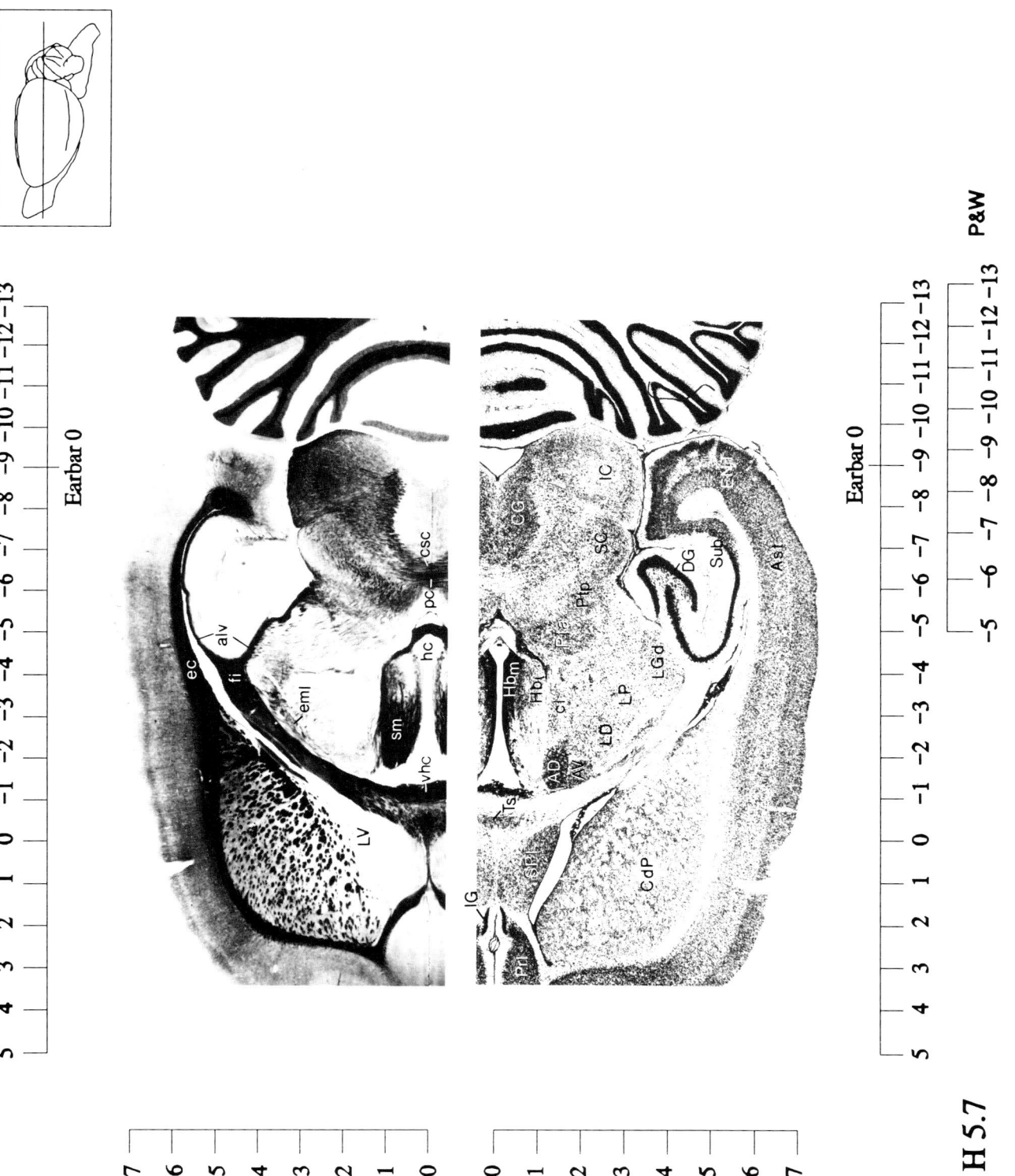

Earbar 0

−13 −12 −11 −10 −9 −8 −7 −6 −5 −4 −3 −2 −1 0 1 2 3 4 5

7 6 5 4 3 2 1 0

0 1 2 3 4 5 6 7

Earbar 0

−13 −12 −11 −10 −9 −8 −7 −6 −5

P&W

H 5.7

Plate 82

alv Alveus
Aq Cerebral Aqueduct
Ast Area Striata (Cortex)
AV Anteroventral Nucleus (Thalamus)
CdP Caudoputamen
CG Central Gray
Cl Claustrum
cl Central Lateral Nucleus (Thalamus)
DG Dentate Gyrus
ec External Capsule
eml External Medullary Lamina
ENT Entorhinal Area
ep Ependymal Layer of Ventricle
fi Fimbria
fo Fornix
fo_pr Fornix (precommissural)
fr Fasciculus Retroflexus
IC Inferior Colliculus
ic Internal Capsule
IG Induseum Griseum
In Insular Area (Cortex)
LGd Dorsal Lateral Geniculate Nucleus (Thalamus)
LP Lateral Posterior Nucleus (Thalamus)
LV Lateral Ventricle
MD Mediodorsal Nucleus (Thalamus)
ot Optic Tract
Pa_a Paraventricular Nucleus (anterior) (Thalamus)
Pa_p Paraventricular Nucleus (posterior) (Thalamus)
pc Posterior Commissure
prf Perforant Path
PrL Prelimbic Area (Cortex)
PT Paratenial Nucleus (Thalamus)
Pta Anterior Pretectal Nucleus (Thalamus)
Ptm Medial Pretectal Area (Thalamus)
Pto Olivary Pretectal Nucleus (Thalamus)
Ptp Posterior Pretectal Nucleus (Thalamus)
R Reticular Nucleus (Thalamus)
SC Superior Colliculus
sco Subcommissural Organ
sfo Subfornical Organ
Sgp Suprageniculate Nucleus of Pretectal Group
sm Stria Medullaris
SomS Somatosensory Cortex
SPl_i Septal Nucleus (lateral: intermediate)
Sub Subiculum
V3 Third Ventricle
VB Ventrobasal Nuclear Complex (Thalamus)
vhc Ventral Hippocampal Commissure
VL Ventrolateral Nucleus (Thalamus)

MD: Mediodorsal Nucleus (Thalamus) (also called Nucleus Medialis Dorsalis). The borders of this large thalamic nucleus, whose projections are often used to define the orbitofrontal cortex, are relatively clear, except laterally where it can be difficult to distinguish unequivocally from medial parts of the central lateral nucleus. MD is commonly divided into medial, central, and lateral parts (Krettek and Price, '77c), as well as a midline intermediodorsal nucleus.

PT: Paratenial Nucleus (Thalamus) (also called Nucleus Parataenialis or Parataenialis). A paired rostral nucleus lateral to the paired midline anterior paraventricular thalamic nuclei and rostral to n. medialis dorsalis. There is little disagreement in the various accounts and atlases except for Jones, '85, who places a "PT" both rostral and caudal to medialis dorsalis – a presumptive error confounded by an abbreviation list designation of "pretectum." However, this nucleus is particularly nicely (and correctly) illustrated in Jones, '83.

Pto: Olivary Pretectal Nucleus (Thalamus). This nucleus is a source of some confusion because various authors have included it with the nucleus of the optic tract (e.g., Bucher and Nauta, '54; Hayhow et al., '62) or failed to recognize it as a specific entity (Siminoff et al., '67). It is nicely described and illustrated by Scalia ('72), who provides a tabular account of the nomenclature and homologies (but see Ptm: medial Pretectal Area at Plate 25).

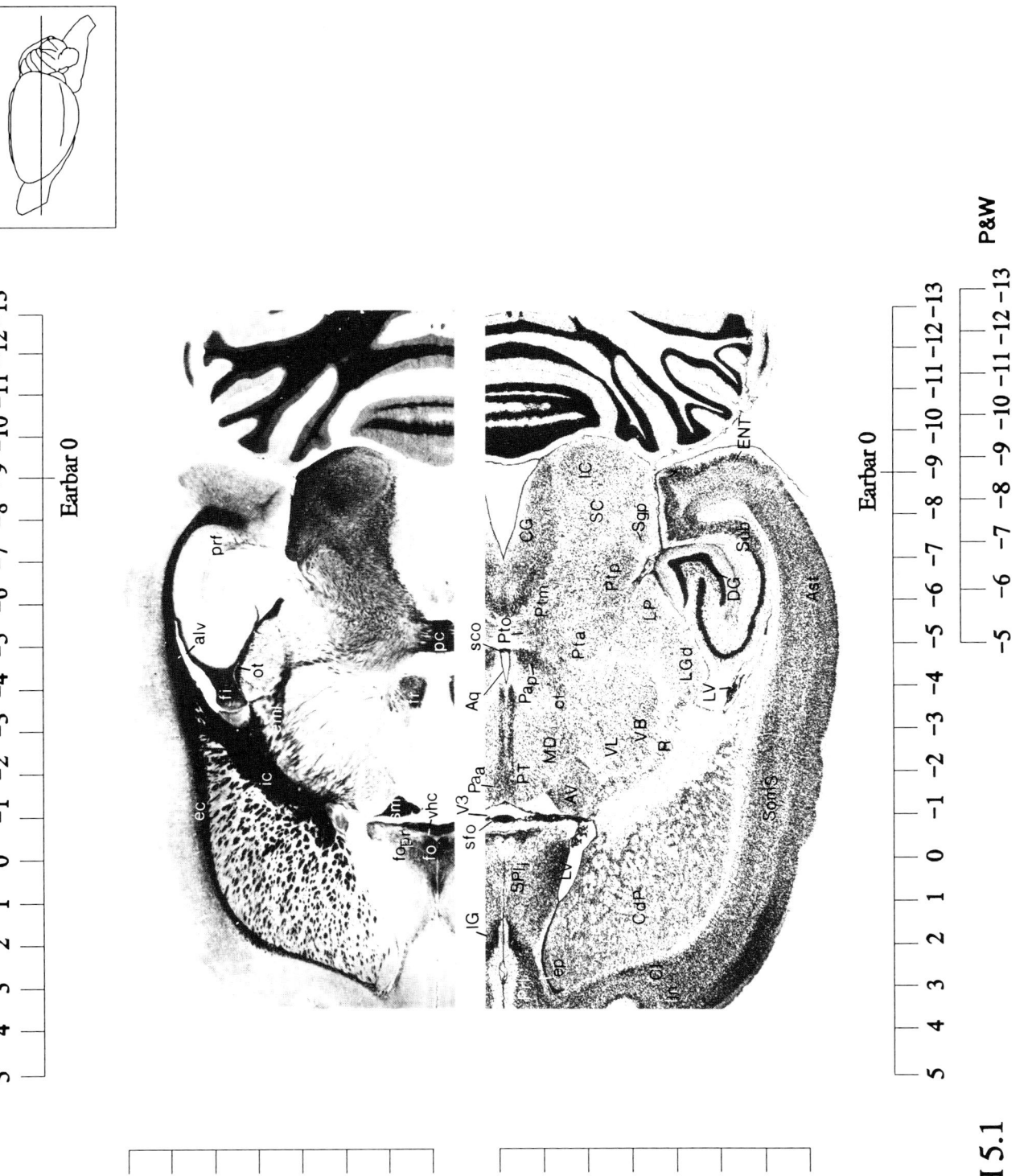

5 4 3 2 1 0 −1 −2 −3 −4 −5 −6 −7 −8 −9 −10 −11 −12 −13

ec alv prf
fi ot
rmh
ic
fo pr vhc
fo

7 6 5 4 3 2 1 0

5 4 3 2 1 0 −1 −2 −3 −4 −5 −6 −7 −8 −9 −10 −11 −12 −13

−5 −6 −7 −8 −9 −10 −11 −12 −13

cG SC IC
Aq sco Ptm Ptp Sgp ENT
Pto Pta LP Sub
Pap ot DG
V3 Paa MD VL VB LGd Agl
sfo PT R LV
IG SPf AV SomS
LV
ep CdP
in

0 1 2 3 4 5 6 7

P&W

H 5.1

Plate 83

alv Alveus
AM Anteromedial Nucleus (Thalamus)
Aq Cerebral Aqueduct
AV Anteroventral Nucleus (Thalamus)
bic Brachium of Inferior Colliculus
CdP Caudoputamen
CG Central Gray
Cl Claustrum
cl Central Lateral Nucleus (Thalamus)
DG Dentate Gyrus
dMR Deep Mesencephalic Reticular Nucleus
ec External Capsule
ENT Entorhinal Area
Fa Fastigial (medial) Cerebellar Nucleus
fi Fimbria
fo Fornix
fr Fasciculus Retroflexus
GP Globus Pallidus (external or lateral)
IC Inferior Colliculus
ic Internal Capsule
iml Internal Medullary Lamina
In Insular Area (Cortex)
Int Interpositus (Intermediate or Interposed) Cerebellar Nucleus
LGd Dorsal Lateral Geniculate Nucleus (Thalamus)
LGv Ventral Lateral Geniculate Nucleus (Thalamus)
LV Lateral Ventricle
MD Mediodorsal Nucleus (Thalamus)

MGp Principal/Parvicellular Medial Geniculate Nucleus (Thalamus)
ml Medial Lemniscus
ms5 Mesencephalic Nucleus of the Trigeminal
Pa$_a$ Paraventricular Nucleus (anterior) (Thalamus)
Pa$_p$ Paraventricular Nucleus (posterior) (Thalamus)
pc Posterior Commissure
Pf Parafascicular Nucleus (Thalamus)
Po Posterior Group (Thalamus)
prf Perforant Path
Prl Prelimbic Area (Cortex)
Pta Anterior Pretectal Nucleus (Thalamus)
Ptp Posterior Pretectal Nucleus (Thalamus)
R Reticular Nucleus (Thalamus)
rf Rhinal Fissure
Sgp Suprageniculate Nucleus of Pretectal Group
sm Stria Medullaris
SomS Somatosensory Cortex
SPl Septal Nucleus (lateral)
SPm Septal Nucleus (medial)
Sub Subiculum
tt Tenia Tecta
V4 Fourth Ventricle
VB Ventrobasal Nuclear Complex (Thalamus)
VB$_l$ Ventrobasal Nuclear Complex (lateral) (Thalamus)
VL Ventrolateral Nucleus (Thalamus)
zb Zuckerkandl Bundle

ic: Internal Capsule. In the rat, this term refers essentially to what corresponds to the posterior limb (and probably the genu) of the internal capsule in human anatomy. In the rat it separates the thalamus (diencephalon) and basal ganglia (telencephalon). It contains the subcortical (descending) projections of the isocortex, along with most of the thalamocortical projections and certain pathways associated with the basal ganglia. It continues as the cerebral peduncle, which courses along the lateral border of the hypothalamus and midbrain to become the pyramidal tract.

ms5: Mesencephalic Nucleus of the Trigeminal. Recognition of this thin rostrocaudal linear strand of large neurons surrounding the central gray as a nucleus is difficult, especially in transverse sections. But the large size and distinctive dark stain of these unusual unipolar neurons make them easy to identify at higher magnification. These neurons are unique in that they are the only sensory ganglion cells within the central nervous system. There is embryological literature arguing that they derive from the neural crest and experimental evidence that their peripheral axons terminate as "proprioceptive" sensitive mechanoreceptors in masticator muscles (Luo et al., '91) and the periodontal ligament. The ms5 is bounded by the axons of the mesencephalic root of 5 (ms5t), a rostrocaudal tract of sensory fibers most easily traced in sagittal or horizontal sections. A modern account can be found in Rokx et al. ('86).

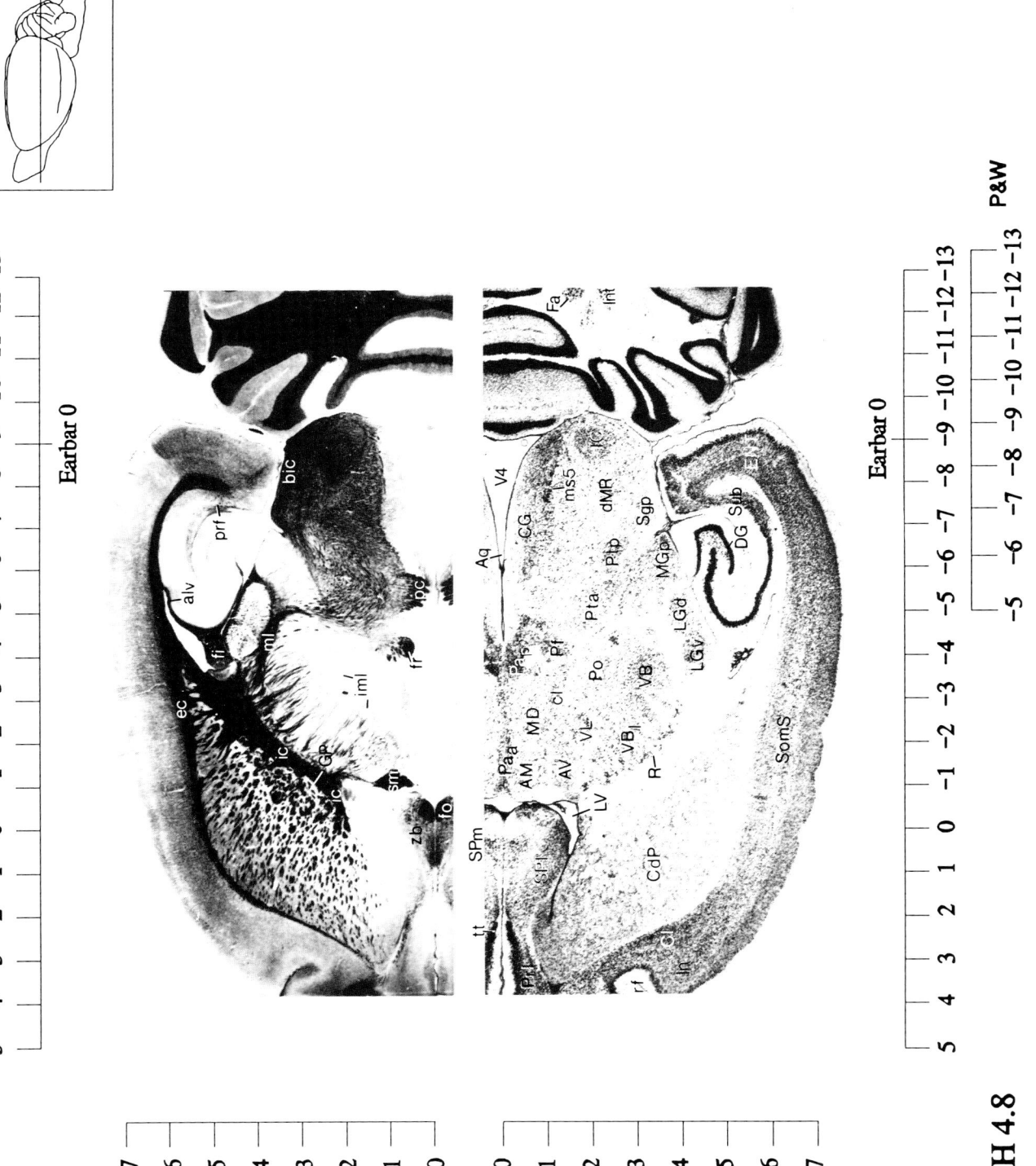

Earbar 0

-13 -12 -11 -10 -9 -8 -7 -6 -5 -4 -3 -2 -1 0 1 2 3 4 5

7 6 5 4 3 2 1 0

Earbar 0

-13 -12 -11 -10 -9 -8 -7 -6 -5 -4 -3 -2 -1 0 1 2 3 4 5

0 1 2 3 4 5 6 7

P&W

H 4.8

Plate 84

ac₀ Anterior Commissure (olfactory limb)
Acb Nucleus Accumbens
alv Alveus
AM Anteromedial Nucleus (Thalamus)
bic Brachium of Inferior Colliculus
Bst Bed Nucleus of Stria Terminalis
CA₁ Hippocampal Area CA₁ (Ammon's Horn)
CA₂ Hippocampal Area CA₂ (Ammon's Horn)
CA₃ Hippocampal Area CA₃ (Ammon's Horn)
CdP Caudoputamen
CG Central Gray
Cl Claustrum
cl Central Lateral Nucleus (Thalamus)
cm Central Medial Nucleus (Thalamus)
Cun Cuneiform Nucleus
DG Dentate Gyrus
dMR Deep Mesencephalic Reticular Nucleus
ec External Capsule
ENT Entorhinal Area
Fa Fastigial (medial) Cerebellar Nucleus
fi Fimbria
fo_p Fornix (postcommissural)
fo_pr Fornix (precommissural)
fr Fasciculus Retroflexus
GP Globus Pallidus (external or lateral)
ic Internal Capsule
In Insular Area (Cortex)
Int Interpositus (Intermediate or Interposed) Cerebellar Nucleus
LGv Ventral Lateral Geniculate Nucleus (Thalamus)

LL_d Nucleus of Lateral Lemniscus (dorsal)
ll Lateral Lemniscus
MD Mediodorsal Nucleus (Thalamus)
MGm Magnocellular Medial Geniculate Nucleus (Thalamus)
MGp Principal/Parvicellular Medial Geniculate Nucleus (Thalamus)
ms5 Mesencephalic Nucleus of the Trigeminal
Pa_a Paraventricular Nucleus (anterior) (Thalamus)
Pc Paracentral Nucleus (Thalamus)
Pf Parafascicular Nucleus (Thalamus)
Pir Piriform Area (Cortex)
Po Posterior Group (Thalamus)
prf Perforant Path
Pta Anterior Pretectal Nucleus (Thalamus)
Ptp Posterior Pretectal Nucleus (Thalamus)
R Reticular Nucleus (Thalamus)
Rd Dorsal Raphé Nucleus
rf Rhinal Fissure
sm Stria Medullaris
SomS Somatosensory Cortex
SPl Septal Nucleus (lateral)
SPm Septal Nucleus (medial)
STF Striatal Fundus
Sub Subiculum
tt Tenia Tecta
V3 Third Ventricle
V4 Fourth Ventricle
VB_l Ventrobasal Nuclear Complex (lateral) (Thalamus)
VB_m Ventrobasal Nuclear Complex (medial) (Thalamus)
VL Ventrolateral Nucleus (Thalamus)
zb Zuckerkandl Bundle

cm: Central Medial Nucleus (Thalamus) (also called Nucleus Centralis Medialis). This midline nucleus contrasts sharply by its marked cell density from the paler overlying midline *n. interanteromedialis* and *n. intermedialis dorsalis* rostrally and is easily recognized in sagittal sections. The original description by Guridian ('27) is satisfactory and is supported in an experimental study by Jones and Leavitt ('74), but some confusion has arisen from a discrepant designation in Jones ('83), which is presumably an error, differing from that of Jones ('85). In the latter, *n. rhomboidalis* is placed dorsal to centralis medialis; rostral, and farther caudal, rhomboidalis is labeled *centralis medialis* where it is bounded laterally by *n. gelatinosus*. This is perhaps a mechanical error and it differs from this and other atlases (e.g., Jones, '83; Paxinos and Watson, '86; Swanson, '92). It should be noted that centralis medialis possesses distinct dorsal and ventral sectors as well as paired lateral wings that might be interpreted as dual midline-wing systems in some species (Rose and Woolsey, '43).

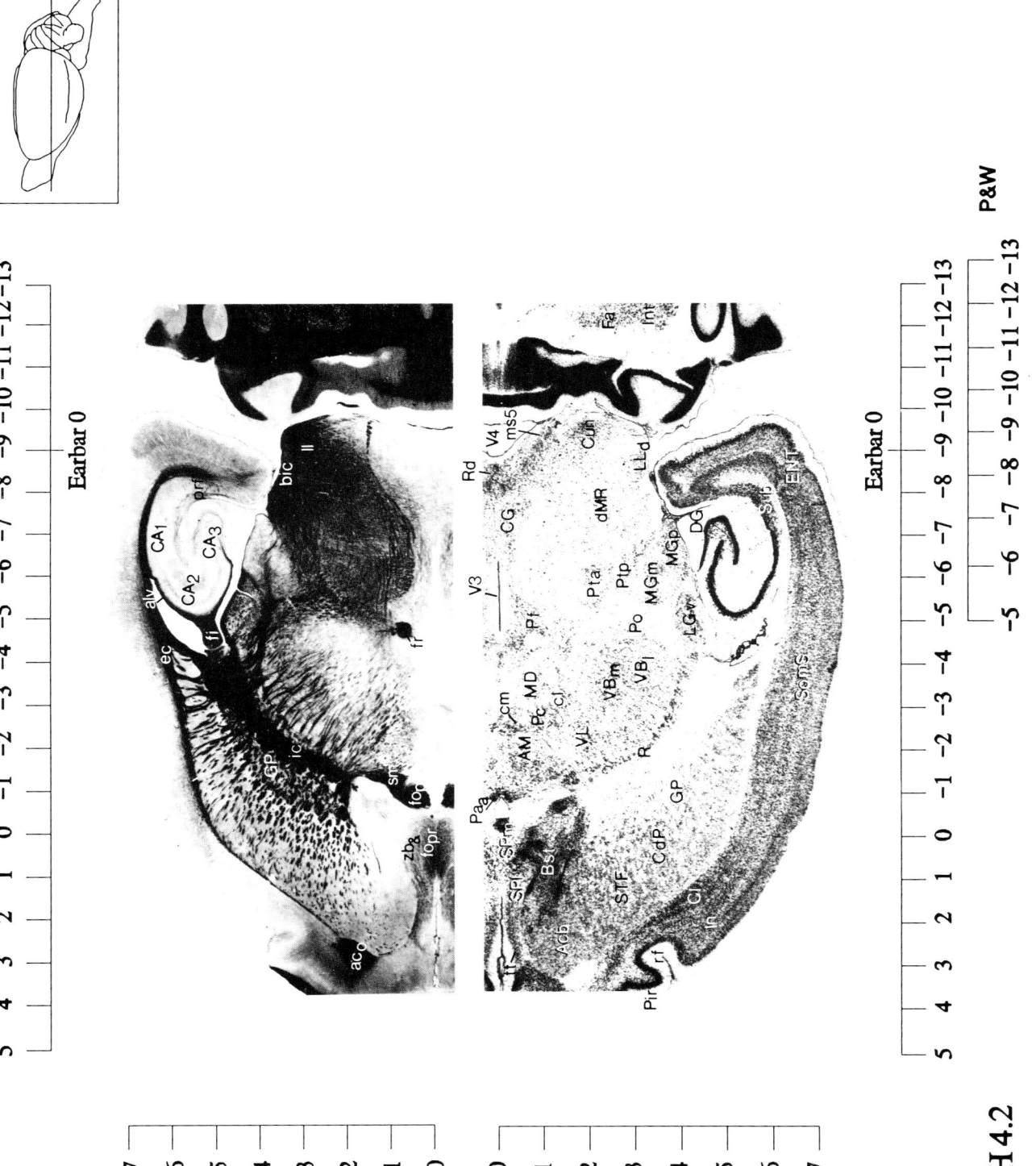

Earbar 0

−13 −12 −11 −10 −9 −8 −7 −6 −5 −4 −3 −2 −1 0 1 2 3 4 5

Earbar 0

−13 −12 −11 −10 −9 −8 −7 −6 −5 −4 −3 −2 −1 0 1 2 3 4 5

7 6 5 4 3 2 1 0

0 1 2 3 4 5 6 7

H 4.2

Plate 85

ac Anterior Commissure
ac_o Anterior Commissure (olfactory limb)
Acb Nucleus Accumbens
alv Alveus
AM Anteromedial Nucleus (Thalamus)
Aud Auditory Area (Cortex)
Bst Bed Nucleus of Stria Terminalis
CA$_1$ Hippocampal Area CA$_1$ (Ammon's Horn)
CA$_3$ Hippocampal Area CA$_3$ (Ammon's Horn)
CdP Caudoputamen
Cl Claustrum
cm Central Medial Nucleus (Thalamus)
Cun Cuneiform Nucleus
DB Nucleus of the Diagonal Band
db Diagonal Band of Broca
De Dentate (Lateral) Cerebellar Nucleus
DG Dentate Gyrus
dMR Deep Mesencephalic Reticular Nucleus
ec External Capsule
eml External Medullary Lamina
ENT Entorhinal Area
Fa Fastigial (Medial) Cerebellar Nucleus
fo$_p$ Fornix (postcommissural)
fo$_{pr}$ Fornix (precommissural)
fr Fasciculus Retroflexus
GP Globus Pallidus (external or lateral)
ic Internal Capsule
Ig Intermediate Geniculate Nucleus (leaflet) (Thalamus)
iml Internal Medullary Lamina
In Insular Area (Cortex)
Int Interpositus (Intermediate or Interposed) Cerebellar Nucleus
LDT Laterodorsal Tegmental Nucleus

LGv Ventral Lateral Geniculate Nucleus (Thalamus)
LL$_d$ Nucleus of Lateral Lemniscus (dorsal)
ll Lateral Lemniscus
LV Lateral Ventricle
MD Mediodorsal Nucleus (Thalamus)
MePO Median Preoptic Nucleus (Hypothalamus)
MGm Magnocellular Medial Geniculate Nucleus (thalamus)
MGp Principal/Parvicellular Medial Geniculate Nucleus (Thalamus)
ml Medial Lemniscus
ms5t Mesencephalic Tract (root of the Trigeminal
ot Optic Tract
Pa$_a$ Paraventricular Nucleus (anterior) (Thalamus)
PB$_l$ Parabrachial Nucleus (lateral)
Pbg Parabigeminal Nucleus
Pf Parafascicular Nucleus (Thalamus)
Pir Piriform Area (Cortex)
Po Posterior Group (Thalamus)
Pta Anterior Pretectal Nucleus (Thalamus)
R Reticular Nucleus (Thalamus)
Rd Dorsal Raphé Nucleus
rf Rhinal Fissure
sm Stria Medullaris
STF Striatal Fundus
Sub Subiculum
V4 Fourth Ventricle
VA Ventral Anterior Nucleus (Thalamus)
VB Ventrobasal Nuclear Complex (Thalamus)
VB$_l$ Ventrobasal Nuclear Complex (lateral) (Thalamus)
VL Ventrolateral Nucleus (Thalamus)
3 Oculomotor Nucleus
4 Trochlear Nucleus

Aud: Auditory Area (Cortex). The boundaries of the auditory cortical fields are difficult to determine by either cyto- or myeloarchitecture in many species, including the rat. The approximate homologous region can be inferred from a variety of studies including similar location in most mammals, electrophysiological studies, and most persuasively on anatomical studies of the medial geniculate body projection (see Winer and Larue, '87).

Bst: Bed Nucleus of the Stria Terminalis. This heterogeneous collection of neuronal aggregates is susceptible to various modes of subdivision into over a dozen subnuclei based on cytoarchitectonic criteria combined with immunohistochemical labeling. Ju and Swanson ('89) provide a detailed, illustrated account of this region that has guided the overall designation employed here, but we would not attempt subdivision without higher-magnification photomicrographs and closer spacing of sections.

CA$_{1-3}$: Hippocampal Areas (Ammon's Horn). The cortex of the hippocampal region is generally divided into three subfields of Ammon's horn (cornu ammonis) as CA$_1$, CA$_2$, and CA$_3$; the dentate gyrus (fascia dentata), and contiguous fields of the subicular and entorhinal cortical areas. The polymorph layer at the transition between CA$_3$ and the dentate gyrus is sometimes called CA$_4$ (see Plate 127), following Lorente de Nó's ('34) description in the mouse. The boundaries of CA$_2$ can be difficult to discern in Nissl-stained sections, but are more easily detected in Loyez-stained series. The distribution of CA$_3$ has been illustrated in all three planes with a specific gene marker (GAP-43 mRNA) by Kruger et al. ('92). The extensive literature has been surveyed recently by Bayer ('85) and Swanson et al. ('87) with specific reference to the rat hippocampus.

rf: Rhinal Fissure. The principal hemispheric fissure (sulcus) of the lissencephalic rat forebrain demarcates the separation of the neocortex (isocortex) from the allocortex associated with the rhinencephalon, which principally constitutes the olfactory cortex in mammals, accounting for the peculiar name of the fissure.

5 4 3 2 1 0 −1 −2 −3 −4 −5 −6 −7 −8 −9 −10 −11 −12 −13
Earbar 0
7 6 5 4 3 2 1 0
alv
ec
LV
ot
jc
ml
CP
eml
ll
aco
ic
sm
Pta
iml
ms5t
fopr
db
fop
fr
ac
Rd
Paa
MePO
DB
cm
3
4
V4
Pf
AM
MD
LDT
Acb
Bst
Pf
VA
dMR
PBl
Cun
Fa
Pta
STF
VL
VB
R
Po
Int
VBl
LLd
Pir
GP
MGm
rf
Pbg
De
MGp
LGv
Cl
CdP
In
DG
lg
CA3
ENT
CA1
Sub
Aud
0 1 2 3 4 5 6 7
Earbar 0
5 4 3 2 1 0 −1 −2 −3 −4 −5 −6 −7 −8 −9 −10 −11 −12 −13
−5 −6 −7 −8 −9 −10 −11 −12 −13
P&W

H 3.9

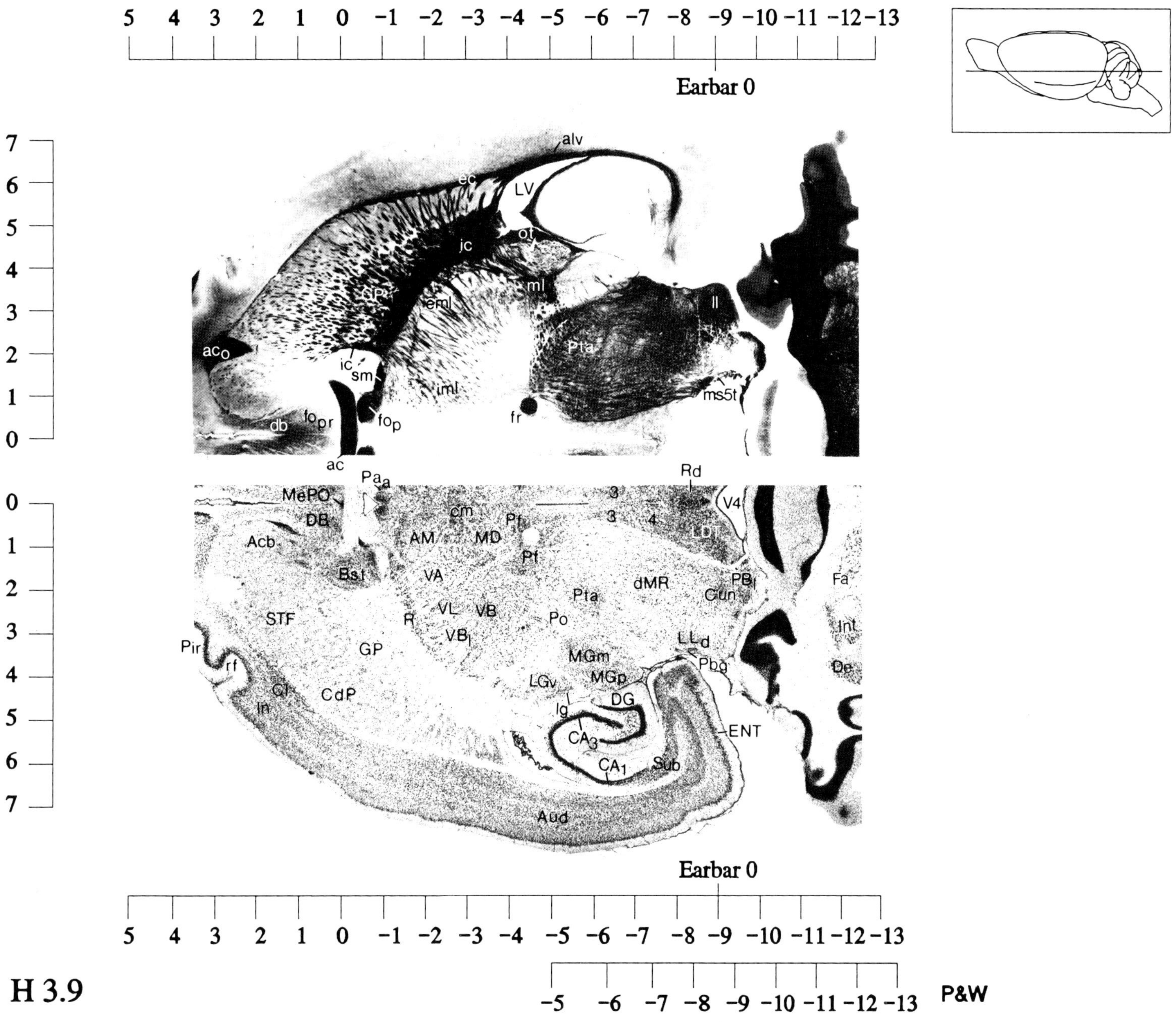

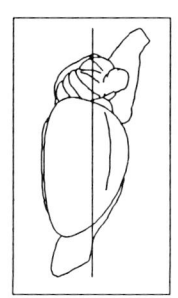

Plate 86

Aud Auditory Area (Cortex)
ac Anterior Commissure
Acb Nucleus Accumbens
alv Alveus
AM Anteromedial Nucleus (Thalamus)
BI Nucleus of the Brachium Inferior Colliculus
Bst Bed Nucleus of Stria Terminalis
CdP Caudoputamen
Cl Claustrum
CM Centromedial Nucleus (Centrum Medianum) (Thalamus)
cp Cerebral Peduncle
D Nucleus of Darkschewitsch
De Dentate (Lateral) Cerebellar Nucleus
DG Dentate Gyrus
dMR Deep Mesencephalic Reticular Nucleus
DT Dorsal Tegmental Nucleus
ec External Capsule
ENT Entorhinal Area
EW Edinger-Westphal Nucleus
Fa Fastigial (Medial) Cerebellar Nucleus
fi Fimbria
fo Fornix
fr Fasciculus Retroflexus
Ge Nucleus Gelatinosus (Nucleus Submedius) (Thalamus)
GP Globus Pallidus (external or lateral)
ic Internal Capsule
In Insular Area (Cortex)
Inc Interstitial Nucleus of Cajal
Int Interpositus (Intermediate or Interposed) Cerebellar Nucleus

LC Locus Ceruleus
LDT Laterodorsal Tegmental Nucleus
LL$_d$ Nucleus of Lateral Lemniscus (dorsal)
ll Lateral Lemniscus
MePO Median Preoptic Nucleus (Hypothalamus)
MGm Magnocellular Medial Geniculate Nucleus (Thalamus)
MGp Principal/Parvicellular Medial Geniculate Nucleus (Thalamus)
ml Medial Lemniscus
mt Mammillothalamic Tract
Pa Paraventricular Nucleus (Thalamus)
PB$_l$ Parabrachial Nucleus (lateral)
Pbg Parabigeminal Nucleus
Pf Parafascicular Nucleus (Thalamus)
Po Posterior Group (Thalamus)
pp Peripeduncular Nucleus
PpT Pedunculopontine Tegmental Nucleus
Pta Anterior Pretectal Nucleus (Thalamus)
R Reticular Nucleus (Thalamus)
Rd Dorsal Raphé Nucleus
Re Nucleus Reuniens (Thalamus)
Rh Rhomboid Nucleus (Thalamus)
scp Superior Cerebellar Peduncle
sm Stria Medullaris
SNc Substantia Nigra (compact)
Sub Subiculum
V3 Third Ventricle
V4 Fourth Ventricle
VB Ventrobasal Nuclear Complex (Thalamus)
VM Ventromedial Nucleus (Thalamus)
ZI Zona Incerta
3 Oculomotor Nucleus
4 Trochlear Nucleus

MGm: Magnocellular Medial Geniculate Nucleus (Thalamus). Other nuclei associated with the medial geniculate nucleus have been variously defined and named, and homologies with the thalamus of other mammals are frequently conjectural. In the rat thalamus, the MGm does not qualify as a distinctly magnocellular nucleus on the basis of cell size and thus it is designated as the medial portion of the medial geniculate by some workers. A "magnocellular" medial geniculate (MGm) near the ventromedial aspect of MGp is generally accepted and can be distinguished experimentally by the absence of a projection to the neuronal layers of the auditory cortex, in contrast to cells of the lateroventral and ovoid divisions of MGp (Clerici and Coleman, '90). The MGm is known to project upon layer I broadly in several neocortical fields. Jones ('85) designates an internal nucleus (MGi) and a separate MGm, but his MGi probably constitutes MGm of other atlases, including that of Jones ('83). The suprageniculate nucleus (Sg) lying dorsal to MGm also lacks a projection to conventionally recognized auditory cortex and is sometimes unrecognized or included in the posterior nuclear group (e.g., Jones, '85). See Po: Posterior Group (Plate 21) and MGp: Principal Medial Geniculate Nucleus (Plate 121) for further information.

(pp) The Peripeduncular Nucleus (Thalamus). This lies interposed between the ventrocaudal margin of the medial geniculate body and the cerebral peduncle. It has also been called the *suprapeduncular nucleus* (e.g., Winer and Larue, '87). It is continuous with the lateral portion of the "posterior intralaminar" wing, usually inaccurately designated *subparafascicular nucleus*. Some consider it to be a single nuclear entity (e.g., Kruger et al., '88b) demarcating the diencephalic-mesencephalic boundary (see Pf: Parafascicular Nucleus at Plate 23 for a more extensive discussion of this issue).

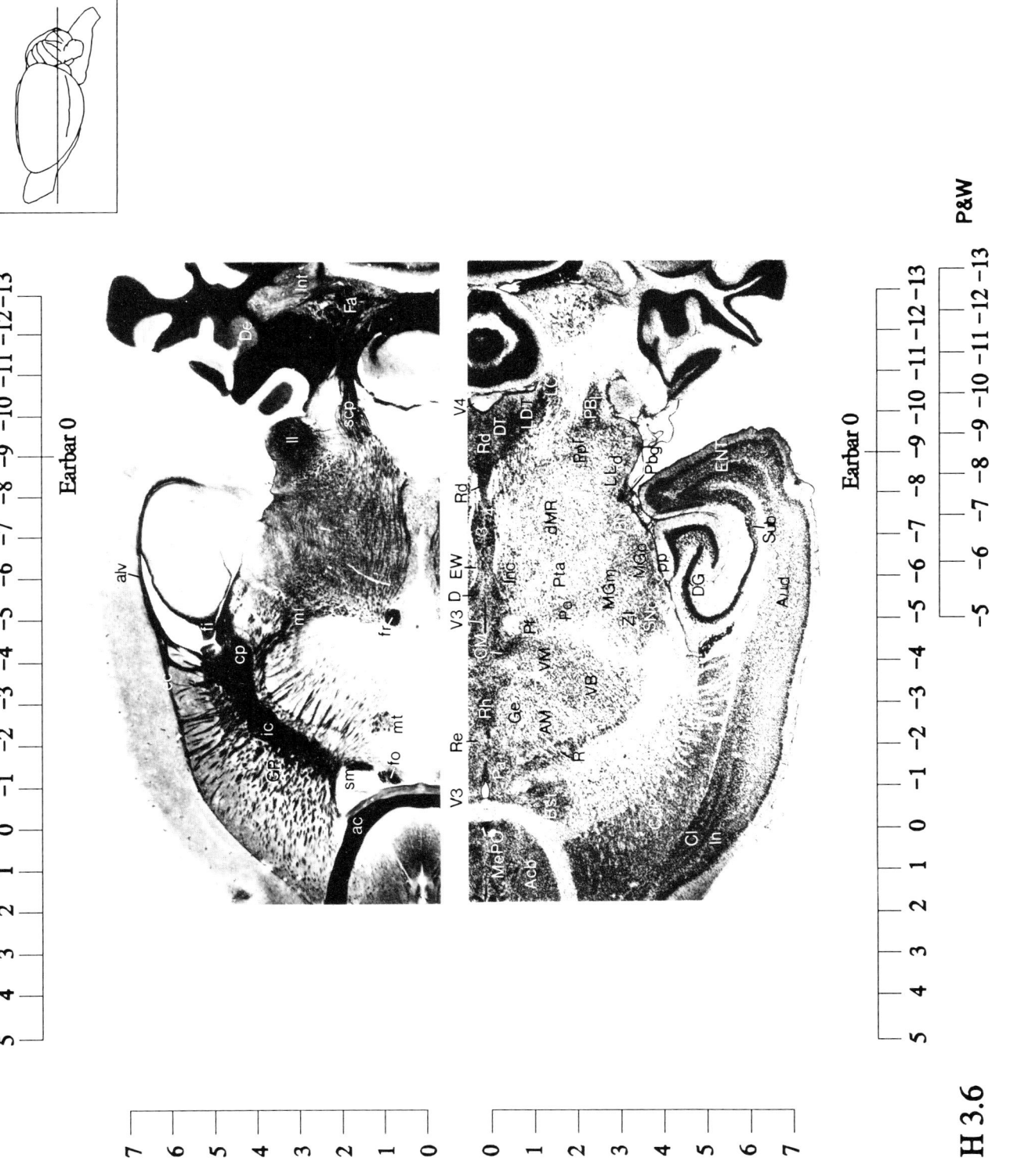

P&W

H 3.6

Plate 87

ac Anterior Commissure
ADP Anterodorsal Preoptic Nucleus (Hypothalamus)
BIs Nucleus of the Brachium Inferior Colliculus (subbrachial sector)
Bst Bed Nucleus of Stria Terminalis
CdP Caudoputamen
Cl Claustrum
cp Cerebral Peduncle
db Diagonal Band of Broca
DG Dentate Gyrus
dMR Deep Mesencephalic Reticular Nucleus
DT Dorsal Tegmental Nucleus
ec External Capsule
ENT Entorhinal Area
fo Fornix
fr Fasciculus Retroflexus
Ge Nucleus Gelatinosus (Nucleus Submedius) (Thalamus)
GP Globus Pallidus (external or lateral)
ic Internal Capsule
In Insular Area (Cortex)
LC Locus Ceruleus
LDT Laterodorsal Tegmental Nucleus
LL$_d$ Nucleus of Lateral Lemniscus (dorsal)
ll Lateral Lemniscus
IT Terminal Nuclei of Accessory Optic Root (basal root)
mcp Middle Cerebellar Peduncle
MePO Median Preoptic Nucleus (Hypothalamus)
mfb Medial Forebrain Bundle

ml Medial Lemniscus
mlf Medial Longitudinal Fascicle
ms5 Mesencephalic Nucleus of the Trigeminal
ms5t Mesencephalic Tract (root) of the Trigeminal
mt Mammillothalamic Tract
ot Optic Tract
Pa Paraventricular Nucleus (Thalamus)
PB Parabrachial Nucleus
PB$_m$ Parabrachial Nucleus (medial)
Pbg Parabigeminal Nucleus
Pf Parafascicular Nucleus (Thalamus)
Po Posterior Group (Thalamus)
pp Peripeduncular Nucleus
PpT Pedunculopontine Tegmental Nucleus
R Reticular Nucleus (Thalamus)
Rd Dorsal Raphé Nucleus
Rdl Dorsal Raphé Nucleus (lateral extension)
rf Rhinal Fissure
Rum Red Nucleus (magnocellular)
Rup Red Nucleus (parvicellular)
Sag Nucleus Sagulum
scp Superior Cerebellar Peduncle
sm Stria Medullaris
SNc Substantia Nigra (compact)
Sub Subiculum
V3 Third Ventricle
VB Ventrobasal Nuclear Complex (Thalamus)
VM Ventromedial Nucleus (Thalamus)
Vs Superior Vestibular Nucleus
vsct Ventral Spinocerebellar Tract
ZI Zona Incerta
3 Oculomotor Nucleus

IT: Lateral Terminal Nucleus, Accessory or Basal Optic Tract. A thin lateral wedge-shaped nucleus that is especially difficult to discern in transverse sections, where it appears as the caudoventral tail of the LGv and the lateral extension of the peripeduncular nucleus (pp). The separation laterally is best seen in horizontal section (Plate 87) and in the rostrocaudal axis in sagittal section (Plate 121). The discrepancy between our IT and pp designations and those of Paxinos and Watson ('86) is probably more apparent than real; the division is difficult on purely architectural grounds and is best recognized on the basis of connections in experimental material (see Giolli et al.,'85).

MePO: Median Preoptic Nucleus (Hypothalamus). This thin, vertically oriented column of densely packed neurons lies along the whole rostral end of the third ventricle along the presumed lamina terminalis (Loo, '31). In frontal sections, it begins as somewhat laterally expanded wings just dorsal to the anteroventral periventricular nucleus, and, as can be seen in midsagittal sections, it extends dorsally around the rostral aspect of the anterior commissure to end between the descending columns of the fornix dorsal to the anterior commissure. The dorsalmost part of the nucleus has sometimes been confused with the triangular nucleus of the septum (see Ts: Triangular Nucleus at Plate 10).

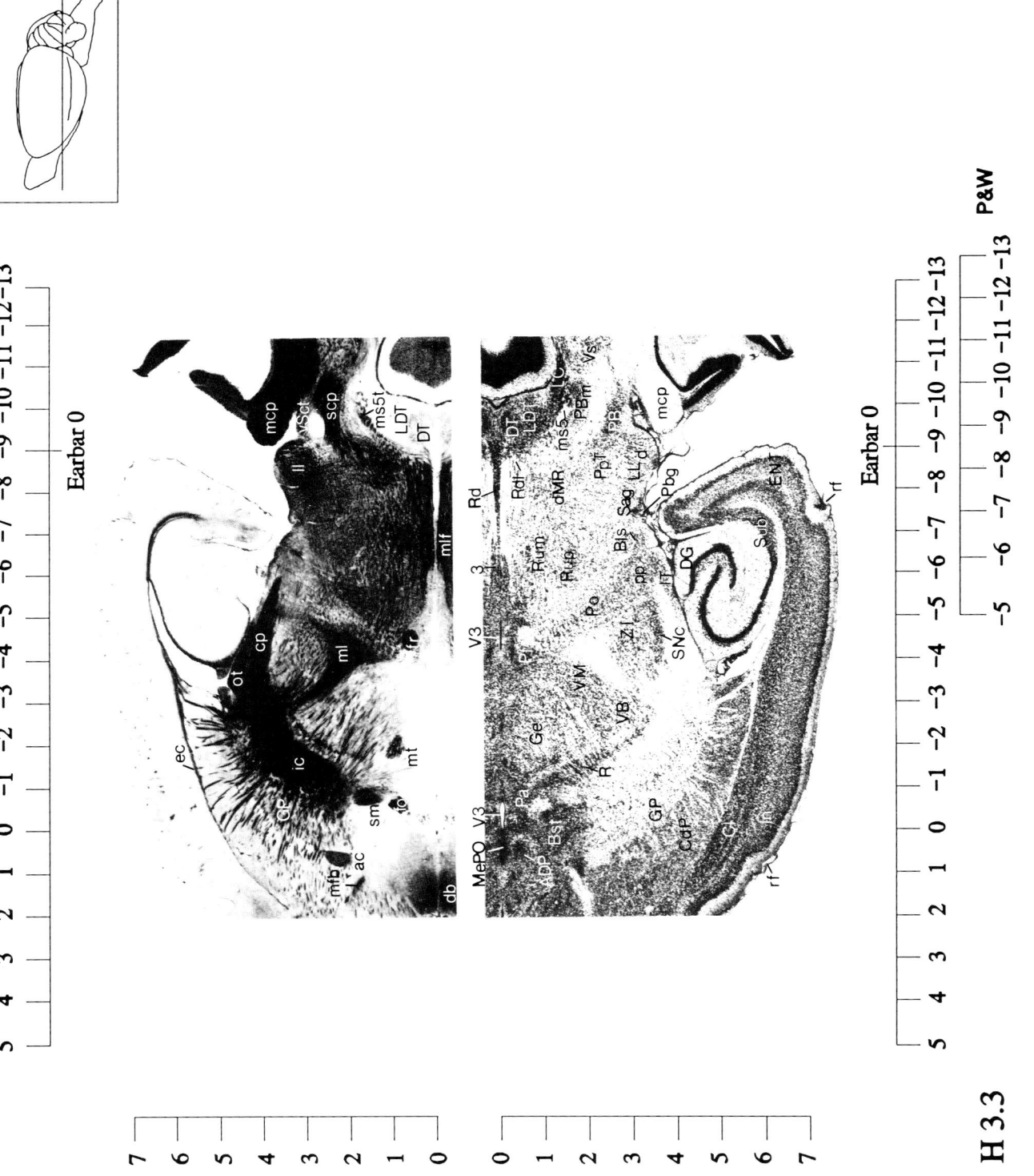

Earbar 0

5 4 3 2 1 0 −1 −2 −3 −4 −5 −6 −7 −8 −9 −10 −11 −12 −13

7 6 5 4 3 2 1 0

0 1 2 3 4 5 6 7

Earbar 0

5 4 3 2 1 0 −1 −2 −3 −4 −5 −6 −7 −8 −9 −10 −11 −12 −13

P&W

H 3.3

Plate 88

ac Anterior Commissure
ADP Anterodorsal Preoptic Nucleus (Hypothalamus)
alv Alveus
Aud Auditory Area (Cortex)
BIs Nucleus of the Brachium Inferior Colliculus (subbrachial sector)
Bst Bed Nucleus of Stria Terminalis
CA₃ Hippocampal Area CA₃ (Ammon's Horn)
CdP Caudoputamen
CG Central Gray
Cl Claustrum
cm Central Medial Nucleus (Thalamus)
CO_d Cochlear Nucleus (dorsal)
cp Cerebral Peduncle
CU Cuneate Nucleus
CUe External Cuneate Nucleus
DB Nucleus of the Diagonal Band
db Diagonal Band of Broca
dMR Deep Mesencephalic Reticular Nucleus
DT_p Dorsal Tegmental Nucleus (posterior)
dtd Dorsal Tegmental Decussation
ec External Capsule
ENT Entorhinal Area
fc Cuneate Fascicle
fg Gracile Fascicle
fo Fornix
fr Fasciculus Retroflexus
Ge Nucleus Gelatinosus (Nucleus Submedius) (Thalamus)
GP Globus Pallidus (external or lateral)
GR Gracile Nucleus
ic Internal Capsule
icp Inferior Cerebellar Peduncle
In Insular Area (Cortex)
LC Locus Ceruleus
LDT Laterodorsal Tegmental Nucleus
LL_v Nucleus of Lateral Lemniscus (ventral)
lot Lateral Olfactory Tract
mcp Middle Cerebellar Peduncle
MePO Median Preoptic Nucleus (Hypothalamus)

mfb Medial Forebrain Bundle
ml Medial Lemniscus
mlf Medial Longitudinal Fascicle
ms5 Mesencephalic Nucleus of the Trigeminal
mt Mammillothalamic Tract
ot Optic Tract
pp Peripeduncular Nucleus
on Optic Nerve
Pbg Parabigeminal Nucleus
PB_l Parabrachial Nucleus (lateral)
PB_m Parabrachial Nucleus (medial)
Pf Parafascicular Nucleus (Thalamus)
Pir Piriform Area (Cortex)
Po Posterior Group (Thalamus)
PpT Pedunculopontine Tegmental Nucleus
prf Perforant Path
PVh Paraventricular Nucleus (Hypothalamus)
R Reticular Nucleus (Thalamus)
Rd Dorsal Raphé Nucleus
Rdl Dorsal Raphé Nucleus (lateral extension)
Re Nucleus Reuniens (Thalamus)
rf Rhinal Fissure
RLc Nucleus Raphé Linearis (caudal)
Rum Red Nucleus (Magnocellular)
Rup Red Nucleus (Parvicellular)
Sag Nucleus Sagulum
scp Superior Cerebellar Peduncle
scpx Decussation of the Superior Cerebellar Peduncle
sm Stria Medullaris
SN Substantia Nigra
Sub Subiculum
Tu Olfactory Tubercle
V3 Third Ventricle
VA Ventral Anterior Nucleus (Thalamus)
VB Ventrobasal Nuclear Complex (Thalamus)
Vl Lateral Vestibular Nucleus
VM Ventromedial Nucleus (Thalamus)
Vm Medial Vestibular Nucleus
Vs Superior Vestibular Nucleus
Y Nucleus Y
ZI Zona Incerta

GR and Cu: Gracile Nucleus and Cuneate Nucleus (also called Gracilis and Cuneatus Nuclei, Dorsal Column Nuclei, and Nuclei of Goll and Burdach, respectively). These longitudinally elongate nuclei lying near the dorsomedial surface of the medulla constitute the principal recipient of dorsal column (funiculus) somatosensory axons and in turn project upon the contralateral ventrobasal thalamus. A midline condensation constituting the tactile representation of the tail is sometimes called the *median* or *accessory gracile nucleus* (and in other species, the *tail nucleus of Bischoff*). The separation of gracile and cuneate tracts and their terminal nuclei in the rat is less pronounced than in most larger mammals, where further subdivision is usually feasible. The two nuclei are bridged ventrally but form two distinct dorsal clusters. There is considerable ambiguity in determining the rostral limit of the gracile nucleus and thus there are significant discrepancies among several atlases. In cat and monkey, the cuneate nucleus extends farther rostral than the gracile component and this may have swayed attempts to achieve conformity in the rat. However, retrograde horseradish peroxidase labeling following large injections into the thalamus clearly reveals that the gracile nucleus extends nearly as far forward as the cuneate nucleus; this is readily evident in horizontal sections showing that the nucleus extends above the obex (Feldman and Kruger, '80), and in studies of the afferent hindlimb projection (LaMotte et al., '91; Maslany et al., '91). The experimental findings testify to the difficulty of employing cytoarchitectonic criteria for establishing boundaries in this region and especially of some of the associated small nuclei (discussed on other pages).

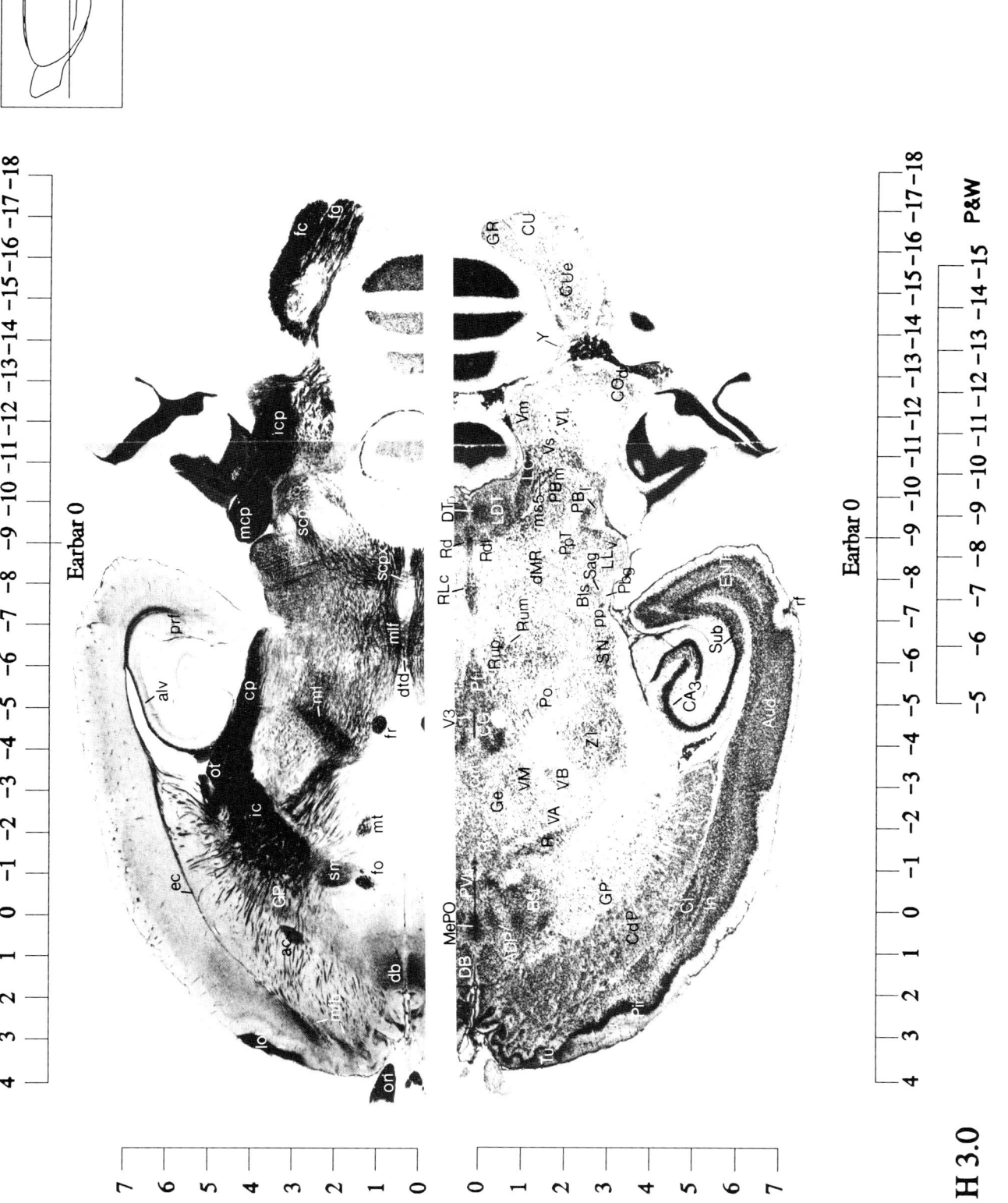

Earbar 0

Earbar 0

P&W

H 3.0

Plate 89

LL: Nuclei of the Lateral Lemniscus. We have followed tradition in dividing these neuronal aggregates embedded with the heavily myelinated axons of the lateral lemniscus into recognizable dorsal and ventral components, partly justified by the segregated projection of commissural fibers between the dorsal nuclei of both sides in other species (Goldberg and Moore, '67). Some authors, however, subdivide further: Harrison and Warr ('62) recognize a medial sector and Paxinos and Watson ('86) designate dorsal, intermediate, and ventral subnuclei, the latter extending ventrally as a sliver lateral to the lateral lemniscus that they call the *paralemniscal nucleus* and which we have included in our *external periolivary nucleus*. The transition between the lateral-caudal limit of the deep pontine nuclei (Pd) (see sagittal Plate 111) is contiguous with the dense ventral sector of the ventral nucleus of the lateral lemniscus illustrated by Paxinos and Watson ('86) but not separated by these authors.

PH: Posterior Hypothalamic Nucleus. This poorly understood cell group, which is said to reach its highest development in the human brain (Nauta and Haymaker, '69), lies between the periaqueductal part of the central gray dorsally and the supramammillary nucleus ventrally. The rostral extension of the nucleus, as defined by Swanson ('92), includes what some have referred to as the *dorsal hypothalamic area* (Bleier et al., '79; Paxinos and Watson, '86), although no criteria for distinguishing the two have been established.

Re: Nucleus Reuniens (Thalamus) (also called Nucleus Medialis Ventralis). This nucleus constitutes the ventral limit of the midline tier of nuclei and is outlined by a dorsolateral fibrous capsule. There is little discrepancy in the literature on designation of the rostral portion, but the caudal limit differs in various accounts. A central linear condensation can be recognized in this as well as in several of the more dorsal midline nuclei; this is given separate status as the *xiphoid nucleus* by Paxinos and Watson ('86), although Gurdjian ('27) earlier called it the *median part of the nucleus reuniens*. The region between the subparafascicular nuclei remains ambiguous and is unlabeled by Paxinos and Watson ('86) and called VMp (*principal ventral medial nucleus*) by Jones ('85).

ac Anterior Commissure	**mfb** Medial Forebrain Bundle	**SNc** Substantia Nigra (compact)
alv Alveus	**MP** Medial Preoptic Nucleus (Hypothalamus)	**SNl** Substantia Nigra (lateral)
ap Area Postrema	**mt** Mammillothalamic Tract	**SNr** Substantia Nigra (reticular)
BIs Nucleus of the Brachium Inferior Colliculus (subbrachial sector)	**on** Optic Nerve	**ST** Subthalamic Nucleus
BLA Basolateral Nucleus of the Amygdala	**ot** Optic Tract	**st** Stria Terminalis
CA₁ Hippocampal Area CA₁ (Ammon's Horn)	**PH** Posterior Hypothalamic Nucleus	**STF** Striatal Fundus
CA₃ Hippocampal Area CA₃ (Ammon's Horn)	**Ph** Perihypoglossal Nucleus	**Sub** Subiculum
Cbl Cerebellum	**pp** Peripeduncular Nucleus	**TSₗ** Nucleus of the Solitary Tract (lateral)
CeA Central Nucleus of the Amygdala	**PRₒ** Pontine Reticular Nucleus (oral)	**TSₘ** Nucleus of the Solitary Tract (medial)
CG Central Gray	**PVh** Paraventricular Nucleus (Hypothalamus)	**ts** Solitary Tract
chp Choroid Plexus	**Re** Nucleus Reuniens (Thalamus)	**Tu** Olfactory Tubercle
Cl Claustrum	**RL** Nucleus Raphé Linearis (rostral)	**V4** Fourth Ventricle
CO_d Cochlear Nucleus (dorsal)	**RLc** Nucleus Raphé Linearis (central)	**Vl** Lateral Vestibular Nucleus
CO_p Cochlear Nucleus (ventral posterior)	**Rm** Median Raphé Nucleus	**Vm** Medial Vestibular Nucleus
cp Cerebral Peduncle	**Rum** Red Nucleus (magnocellular)	**Vsp** Spinal Vestibular Nucleus
CU Cuneate Nucleus	**Rup** Red Nucleus (parvicellular)	**VT** Ventral Tegmental Nucleus
CUe External Cuneate Nucleus	**S5** Supratrigeminal Nucleus	**Z** Nucleus Z
DB Nucleus of the Diagonal Band	**s5c** Spinal Trigeminal Nucleus (caudal)	**ZI** Zona Incerta
db Diagonal Band of Broca	**Sag** Nucleus Sagulum	**5t** Trigeminal Tract
dtd Dorsal Tegmental Decussation	**scpx** Decussation of the Superior Cerebellar Peduncle	**7g** Genu of Facial Nerve
ec External Capsule	**sg** Substantia Gelatinosa	
ENTₗ Entorhinal Area (lateral)	**Sge** Supragenual Nucleus	
ENTₘ Entorhinal Area (medial)	**SI** Substantia Innominata	
fo Fornix		
fr Fasciculus Retroflexus		
IA Intercalated Mass of the Amygdala		
ic Internal Capsule		
icp Inferior Cerebellar Peduncle		
In Insular Area (Cortex)		
KF Kölliker-Fuse Nucleus		
LA Lateral Nucleus of the Amygdala		
LL_v Nucleus of Lateral Lemniscus (ventral)		
ll Lateral Lemniscus		
lot Lateral Olfactory Tract		
LPO Lateral Preoptic Area (Hypothalamus)		
LV Lateral Ventricle		
m5 Motor Trigeminal Nucleus		
MaPO Magnocellular Preoptic Nucleus (Hypothalamus)		
mcp Middle Cerebellar Peduncle		

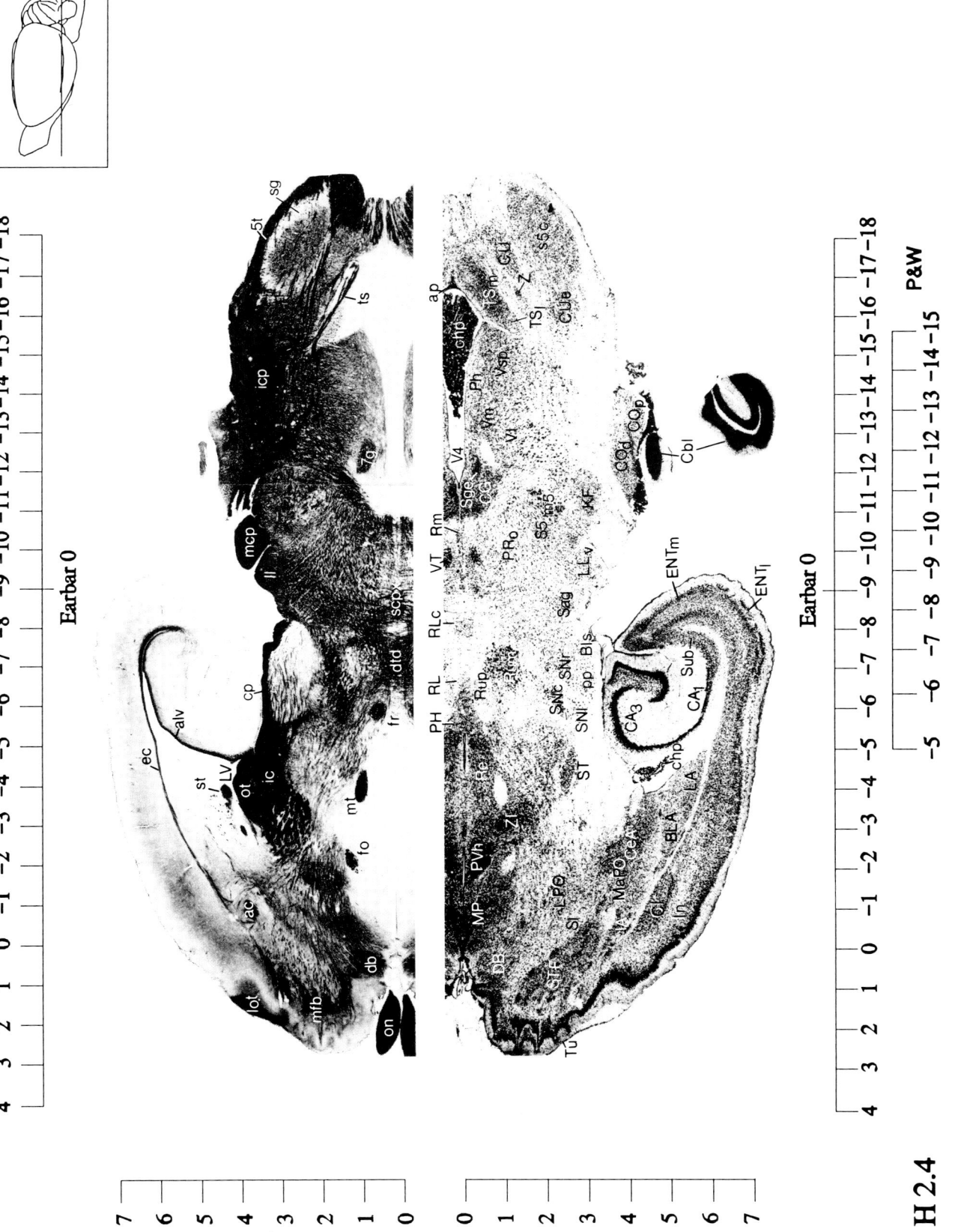

Earbar 0

−18 −17 −16 −15 −14 −13 −12 −11 −10 −9 −8 −7 −6 −5 −4 −3 −2 −1 0 1 2 3 4

7 6 5 4 3 2 1 0

P&W

−18 −17 −16 −15 −14 −13 −12 −11 −10 −9 −8 −7 −6 −5 −4 −3 −2 −1 0 1 2 3 4

Earbar 0

0 1 2 3 4 5 6 7

H 2.4

Plate 90

BLA Basolateral Nucleus of the Amygdala
CA₁ Hippocampal Area CA₁ (Ammon's Horn)
CA₃ Hippocampal Area CA₃ (Ammon's Horn)
Cbl Cerebellum
CeA Central Nucleus of the Amygdala
CO_p Cochlear Nucleus (ventral posterior)
CO_v Cochlear Nucleus (ventral anterior)
cp Cerebral Peduncle
CU Cuneate Nucleus
DG Dentate Gyrus
DMh Dorsomedial Nucleus (Hypothalamus)
ENT_l Entorhinal Area (lateral)
ENT_m Entorhinal Area (medial)
EP Endopiriform Nucleus
fo Fornix
fr Fasciculus Retroflexus
IA Intercalated Mass of the Amygdala
Ic Intercalated Nucleus (Medulla)
ic Internal Capsule
icp Inferior Cerebellar Peduncle
LA Lateral Nucleus of the Amygdala
LHA Lateral Hypothalamic Area
LL_v Nucleus of Lateral Lemniscus (ventral)
ll Lateral Lemniscus
lot Lateral Olfactory Tract
LPO Lateral Preoptic Area (Hypothalamus)
m5 Motor Trigeminal Nucleus
MeA Medial Nucleus of the Amygdala
mfb Medial Forebrain Bundle
MP Medial Preoptic Nucleus (Hypothalamus)
mt Mammillothalamic Tract
on Optic Nerve
ot Optic Tract

Pam Periamygdaloid Cortex
PcR Parvicellular Reticular Nucleus
Pe_av Periventricular Nucleus (anteroventral) (Hypothalamus)
PH Posterior Hypothalamic Nucleus
Ph Perihypoglossal Nucleus
Pir Piriform Area (Cortex)
Pr5 Principal Sensory Trigeminal Nucleus
PR_o Pontine Reticular Nucleus (oral)
PVh Paraventricular Nucleus (Hypothalamus)
RL Nucleus Raphé Linearis (rostral)
RLc Nucleus Raphé Linearis (central)
Rm Median Raphé Nucleus
RP Nucleus Raphé Pontis
s5c Spinal Trigeminal Nucleus (caudal)
s5i Spinal Trigeminal Nucleus (interpolar)
s5o Spinal Trigeminal Nucleus (oral)
sg Substantia Gelatinosa
SI Substantia Innominata
SNc Substantia Nigra (compact)
SNr Substantia Nigra (reticular)
ST Subthalamic Nucleus
st Stria Terminalis
STa Subthalamic Nucleus (accessory)
STF Striatal Fundus
Sub Subiculum
TS_l Nucleus of the Solitary Tract (lateral)
TS_m Nucleus of the Solitary Tract (medial)
Tu Olfactory Tubercle
VTA Ventral Tegmental Area (of Tsai)
ZI Zona Incerta
5m Trigeminal Nerve Root (motor or minor division)
5s Trigeminal Nerve Root (sensory or major division)
5t Trigeminal Tract
10 Dorsal Motor Nucleus of Vagus
12 Hypoglossal Nucleus

ENT_l,m: Entorhinal Area (also called Entorhinal Cortical Field and Area Entorhinalis). The caudal portion of the hippocampal gyrus in the human brain is an enormous, generally subdivided cortical field. It occupies the same position in the rat brain bordered dorsolaterally by visual isocortex and medially by the subicular fields. It is customary to distinguish a medial (ENT_m) portion with a single layer II from a lateral portion (ENT_l) in which layer II is split.

LA: Lateral Nucleus of the Amygdala. A relatively homogeneous nucleus sitting astride the basolateral nucleus expanding toward the caudal pole of the amygdala. Subdivision is most evident in fiber architecture but not readily reconciled with cytoarchitectonic divisions (see Krettek and Price, '78).

Pe_av: Periventricular Nucleus (anteroventral) (Hypothalamus). Bleier et al. ('79) identified this sexually dimorphic nucleus (which is larger in females) and called it, along with the suprachiasmatic preoptic nucleus, the *medial preoptic nucleus*, referring to what Gurdjian ('27) had called the *medial preoptic nucleus* as the *anterior hypothalamic nucleus*. The Pe_av is an oval cell group lying just caudal to the vascular organ of the lamina terminalis and is best seen in horizontal sections, where a cell-sparse capsular region is obvious (see Simerly and Swanson, '87).

TS_m,l,c: Nucleus of the Solitary Tract (also called Solitary Nucleus and Nucleus Solitarius). Strict usage would require calling this aggregate of nuclei a nuclear complex joined by the feature of constituting the principal and primary relay for general visceral and gustatory inputs derived from V, VII, IX and X cranial nerves. We have divided it only into medial, lateral and commissural subdivisions, but more detailed nomenclature is frequently employed. A useful description is found in Kalia and Sullivan ('82) for the rat, and the detailed account in hamster by Whitehead ('88) includes a tabular presentation of nomenclature and a good discussion of grounds for functional subdivision.

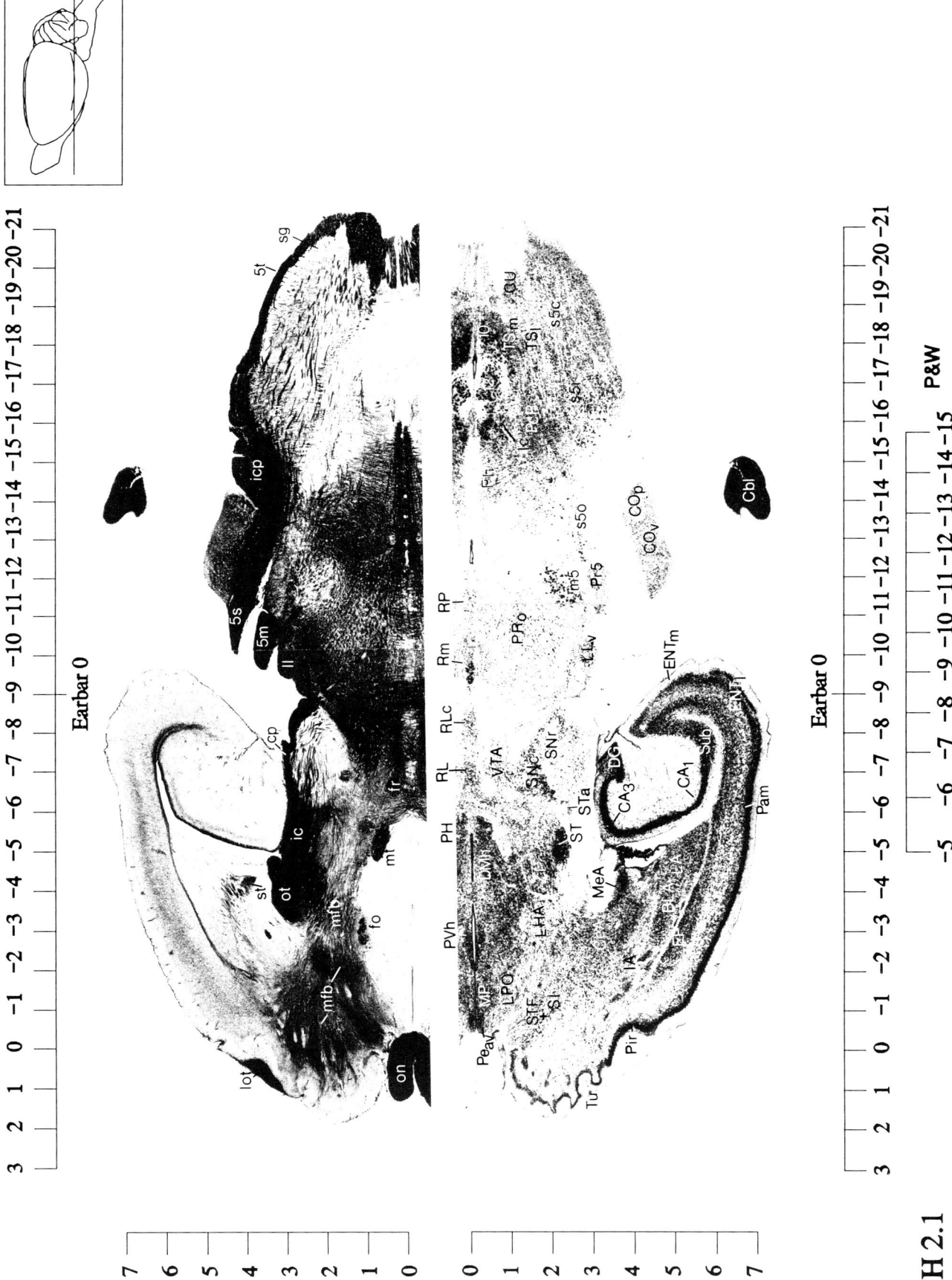

H 2.1

CeA: Central Nucleus of the Amygdala. The dorsal region in transition with the overlying caudoputamen and ventrolaterally bordered by the intercalated amygdaloid masses is easily recognized as the CeA of most descriptions. Some authors divide it into medial and lateral divisions, best seen in the richer fiber density of the medial sector, and more extensive subdivision can be found in the original literature.

MeA: Medial Nucleus of the Amygdala. This most medial component of the "periamygdaloid cortex" of earlier usage is discussed at CoA: Cortical Nucleus of the Amygdala (Plate 13). It is variously subdivided by most authors. Price ('81) presents a useful delineation on cytoarchitectonic grounds and others lean toward connectional criteria (see Krettek and Price, '78; De Olmos et al., '85; Bayer, '80; Paxinos and Watson, '86; Swanson '92). We indicate only anterior and posterior sectors, but recognize that further subdivision is feasible.

Pe: Periventricular Nucleus (Hypothalamus). This is a narrow zone of vertically oriented fusiform neurons surrounding hypothalamic parts of the third ventricle (except in ventral midline regions associated with the median eminence where a few neurons may nevertheless be observed; Ishikawa et al., '92). It has been somewhat arbitrarily divided into preoptic, anterior, tuberal, and posterior levels (Gurdjian, '27; Swanson, '92), which cannot be distinguished readily as sectors in our plates, but it displays several relatively clear differentiations, including the median preoptic nucleus (MePO), anteroventral periventricular nucleus (Pe_{av}), paraventricular nucleus of the hypothalamus (PVh), and arcuate nucleus (Arc) (see individual notes at Plates 87, 90, 14, and 17, respectively). The periventricular nucleus or zone as a whole is characterized by the presence of pools of neuroendocrine neurons controlling the secretion of anterior pituitary hormones (see Everitt et al., '86; Swanson, '87). For further discussion see PVh: Paraventricular Nucleus (Plate 14).

sumx: Supramammillary Decussation. Most of the fibers in this poorly understood fiber system appear to cross the midline between the posterior hypothalamic and supramammillary nuclei (Gurdjian, '27; Nauta and Haymaker, '69).

Plate 91

Ah   Anterior Hypothalamic Nucleus
AHZ   Amygdalo-Hippocampal Area
BLA   Basolateral Nucleus of the Amygdala
BMA_p Basomedial Nucleus of the Amygdala (posterior)
CA_3 Hippocampal Area CA_3 (Ammon's Horn)
CeA   Central Nucleus of the Amygdala
CO_p Cochlear Nucleus (ventral posterior)
CO_v Cochlear Nucleus (ventral anterior)
cp   Cerebral Peduncle
DG   Dentate Gyrus
DMh_a Dorsomedial Nucleus (anterior) (Hypothalamus)
DMh_p Dorsomedial Nucleus (posterior) (Hypothalamus)
ENT   Entorhinal Area
EP   Endopiriform Nucleus
fo   Fornix
fr   Fasciculus Retroflexus
IA   Intercalated Mass of the Amygdala
ic   Internal Capsule
IP   Interpeduncular Nucleus
LA   Lateral Nucleus of the Amygdala
LHA   Lateral Hypothalamic Area
LL_v Nucleus of Lateral Lemniscus (ventral)
lot   Lateral Olfactory Tract
LPO   Lateral Preoptic Area (Hypothalamus)
m5   Motor Trigeminal Nucleus
MeA_a Medial Nucleus of the Amygdala (anterior)
MeA_p Medial Nucleus of the Amygdala (posterior)
mfb   Medial Forebrain Bundle
MP   Medial Preoptic Nucleus (Hypothalamus)
mt   Mammillothalamic Tract
oc   Optic Chiasm
ot   Optic Tract

Pam   Periamygdaloid Cortex
PcR   Parvicellular Reticular Nucleus
Pe_a Periventricular Nucleus (anterior) (Hypothalamus)
PgR   Paragigantocellular Reticular Nucleus
PH   Posterior Hypothalamic Nucleus
Pir   Piriform Area (Cortex)
Pr5   Principal Sensory Trigeminal Nucleus
PR_c Pontine Reticular Nucleus (caudal)
PR_o Pontine Reticular Nucleus (oral)
RL   Nucleus Raphé Linearis (rostral)
RLc   Nucleus Raphé Linearis (central)
Rm   Median Raphé Nucleus
Rol   Nucleus of Roller
RP   Nucleus Raphé Pontis
s5c   Spinal Trigeminal Nucleus (caudal)
s5i   Spinal Trigeminal Nucleus (interpolar)
s5o   Spinal Trigeminal Nucleus (oral)
sg   Substantia Gelatinosa
SI   Substantia Innominata
SNc   Substantia Nigra (compact)
SNr   Substantia Nigra (reticular)
ST   Subthalamic Nucleus
st   Stria Terminalis
STF   Striatal Fundus
Sub   Subiculum
sumx   Supramammillary Decussation
TS_l Nucleus of the Solitary Tract (lateral)
TS_m Nucleus of the Solitary Tract (medial)
Tu   Olfactory Tubercle
VTA   Ventral Tegmental Area (of Tsai)
vtd   Ventral Tegmental Decussation
5m   Trigeminal Nerve Root (motor or minor division)
5s   Trigeminal Nerve Root (sensory or major division)
5t   Trigeminal Tract
6   Abducens Nucleus
10   Dorsal Motor Nucleus of Vagus
12   Hypoglossal Nucleus
12n   Hypoglossal Nerve

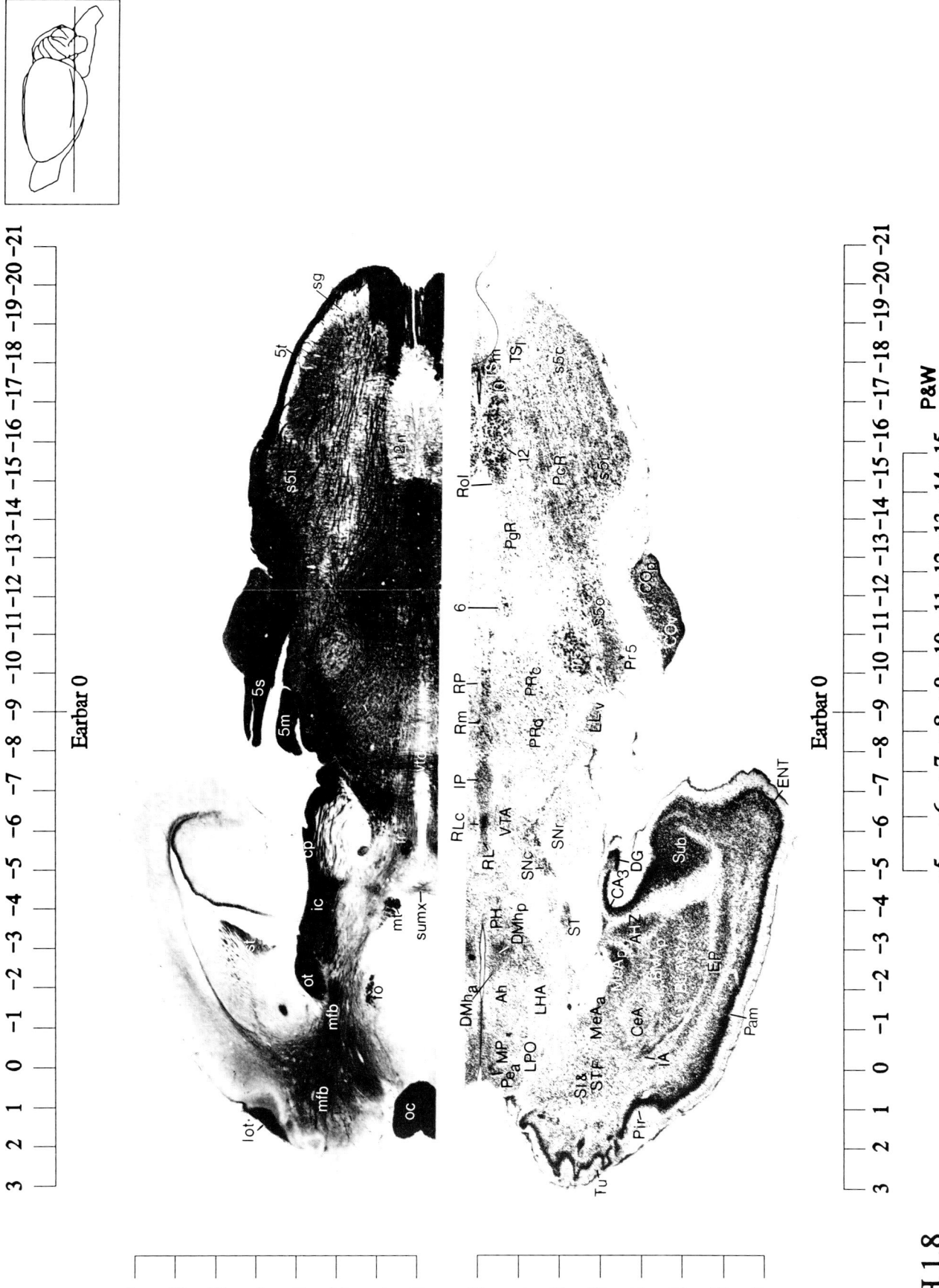

Earbar 0

3 2 1 0 -1 -2 -3 -4 -5 -6 -7 -8 -9 -10 -11 -12 -13 -14 -15 -16 -17 -18 -19 -20 -21

P&W

Earbar 0

3 2 1 0 -1 -2 -3 -4 -5 -6 -7 -8 -9 -10 -11 -12 -13 -14 -15 -16 -17 -18 -19 -20 -21

H 1.8

Plate 92

AE Amygdalo-Entorhinal Transition Area
Ah$_a$ Anterior Hypothalamic Nucleus (anterior)
AHZ Amygdalo-Hippocampal Area
BLA$_p$ Basolateral Nucleus of the Amygdala (posterior)
BMA$_p$ Basomedial Nucleus of the Amygdala (posterior)
CA$_3$ Hippocampal Area CA$_3$ (Ammon's Horn)
CeA Central Nucleus of the Amygdala
CO$_p$ Cochlear Nucleus (ventral posterior)
CO$_v$ Cochlear Nucleus (ventral anterior)
cp Cerebral Peduncle
CU Cuneate Nucleus
dc Dorsal Column
DG Dentate Gyrus
DMh$_p$ Dorsomedial Nucleus (posterior) (Hypothalamus)
DMh$_v$ Dorsomedial Nucleus (ventral) (Hypothalamus)
EP Endopiriform Nucleus
fo Fornix
GcR Gigantocellular Reticular Nucleus
GR Gracile Nucleus
IA Intercalated Mass of the Amygdala
ic Internal Capsule
IP Interpeduncular Nucleus
IPf Interpeduncular Fossa
LHA Lateral Hypothalamic Area
LL$_v$ Nucleus of Lateral Lemniscus (ventral)
lot Lateral Olfactory Tract
m5 Motor Trigeminal Nucleus
MeA Medial Nucleus of the Amygdala
mfb Medial Forebrain Bundle
ml Medial Lemniscus
mlf Medial Longitudinal Fascicle

mp Mammillary Peduncle
mT Medial Terminal Nucleus of Accessory Optic Tract
mt Mammillothalamic Tract
oc Optic Chiasm
ot Optic Tract
PcR Parvicellular Reticular Nucleus
Pir Piriform Area (Cortex)
PM$_d$ Premammillary Nucleus (dorsal) (Hypothalamus)
PmR Paramedian Reticular Nucleus
Pr5 Principal Sensory Trigeminal Nucleus
PR$_c$ Pontine Reticular Nucleus (caudal)
PR$_o$ Pontine Reticular Nucleus (oral)
Rm Median Raphé Nucleus
Rol Nucleus of Roller
RP Nucleus Raphé Pontis
s5c Spinal Trigeminal Nucleus (caudal)
s5i Spinal Trigeminal Nucleus (interpolar)
s5o Spinal Trigeminal Nucleus (oral)
SCh Suprachiasmatic Nucleus (Hypothalamus)
sg Substantia Gelatinosa
SI Substantia Innominata
SNc Substantia Nigra (compact)
STF Striatal Fundus
Sub Subiculum
SUM$_m$ Supramammillary Nucleus (medial)
TS$_l$ Nucleus of the Solitary Tract (lateral)
TS$_m$ Nucleus of the Solitary Tract (medial)
ts Solitary Tract
Tu Olfactory Tubercle
VMh Ventromedial Nucleus (Hypothalamus)
5m Trigeminal Nerve Root (motor or minor division)
5s Trigeminal Nerve Root (sensory or major division)
5t Trigeminal Tract
12 Hypoglossal Nucleus

BLA: Basolateral Nucleus of the Amygdala. This heterogeneous nuclear region or "division" is usually subdivided, largely because of its more extensive differentiation in larger primates compared with the rat. A posterior amygdaloid nucleus is recognized by some workers but we have designated this a posterior sector of the BLA for the rat following Krettek and Price ('78). Homology with the primate *basal* nucleus of the amygdala has led to a suggested change in terminology (Price et al., '87). The basolateral *division* sometimes includes the central and intercalated amygdaloid nuclei but we follow modern usage in giving these distinct aggregates separate nuclear status (see Krettek and Price, '78, and Millhouse, '86, for useful illustrated descriptions).

mt: Mammillothalamic Tract. The principal mammillary tract ascends from the medial and lateral mammillary nuclei and can be seen to emerge in fascicles (most evident in the horizontal plane, see Plate 92). Just dorsal to the dorsal premammillary nucleus, it bifurcates into the ascending mammillothalamic tract (which ends in the anterior thalamic group) and the descending mammillotegmental tract (which can be traced for some distance just medial to the fasciculus retroflexus), before merging with other fibers, seen in silver-stained preparations (see Gurdjian, '27; Fry and Cowan, '72; Allen and Hopkins, '90). Both projections (thalamic and tegmental) arise from the same neurons, with their myelinated axons bifurcating in the supramammillary region. Some of the thalamic axons divide, with one branch forming one of the myelinated midline commissures enveloping the thalamic anteromedial nucleus.

Rol: Nucleus of Roller. This nucleus is generally classified with the perihypoglossal region but is distinct from the perihypoglossal nucleus. It derives its eponymic label from human neuroanatomy and is known to project to the deep cerebellar nuclei in rat (Roste, '89), but there appears to be some uncertainty about its exact position in some descriptions. In rat as in other mammals, it lies lateral to the dorsal edge of the medial longitudinal fascicle and ventral to the rostral portion of the hypoglossal nucleus. The position in Roste's experimental study and that indicated by Paxinos and Watson ('86) are not fully reconcilable but this is illustrated with some clarity in Swanson ('92).

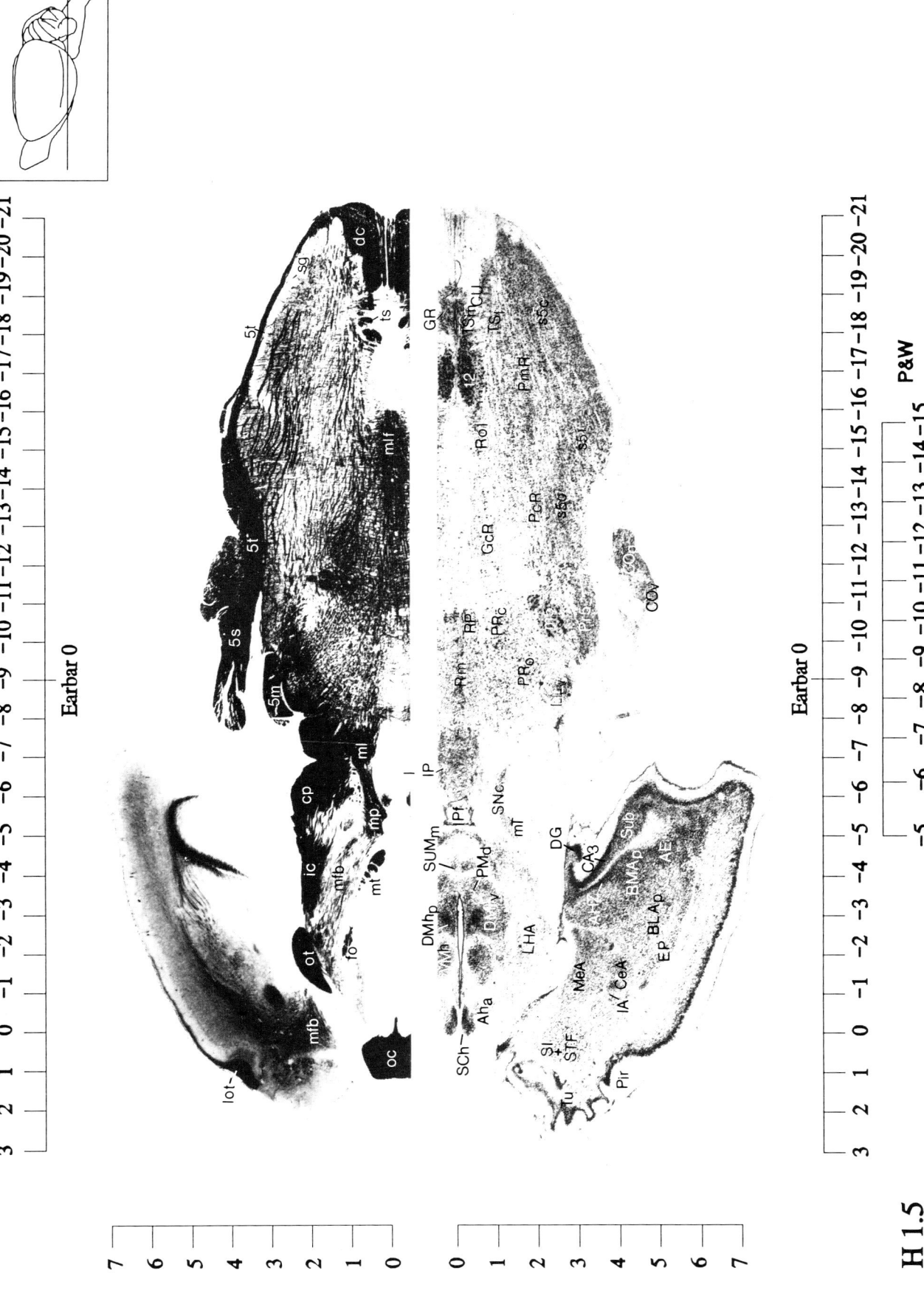

Earbar 0

3 2 1 0 −1 −2 −3 −4 −5 −6 −7 −8 −9 −10 −11 −12 −13 −14 −15 −16 −17 −18 −19 −20 −21

P&W

Earbar 0

3 2 1 0 −1 −2 −3 −4 −5 −6 −7 −8 −9 −10 −11 −12 −13 −14 −15 −16 −17 −18 −19 −20 −21

H 1.5

Plate 93

AE   Amygdalo-Entorhinal Transition Area
AHZ   Amygdalo-Hippocampal Area
Arc   Arcuate Nucleus (Hypothalamus)
BLA$_p$ Basolateral Nucleus of the Amygdala (posterior)
BMA   Basomedial Nucleus of the Amygdala
c   Central Canal
CO$_p$ Cochlear Nucleus (ventral posterior)
dc   Dorsal Column
DMh   Dorsomedial Nucleus (Hypothalamus)
EP   Endopiriform Nucleus
fo   Fornix
GcR   Gigantocellular Reticular Nucleus
IP   Interpeduncular Nucleus
LL   Nucleus of Lateral Lemniscus
Lot   Nucleus of Lateral Olfactory Tract
lot   Lateral Olfactory Tract
MeA$_a$ Medial Nucleus of the Amygdala (anterior)
MeA$_p$ Medial Nucleus of the Amygdala (posterior)
mlf   Medial Longitudinal Fascicle

MM   Medial Mammillary Nucleus (Hypothalamus)
ot   Optic Tract
Pam   Periamygdaloid Cortex
PcR   Parvicellular Reticular Nucleus
Pe$_p$ Periventricular Nucleus (posterior)
Pir   Piriform Area (Cortex)
PM$_d$ Premammillary Nucleus (dorsal)
PmR   Paramedian Reticular Nucleus
Pr5   Principal Sensory Trigeminal Nucleus
PR$_c$ Pontine Reticular Nucleus (caudal)
PR$_o$ Pontine Reticular Nucleus (oral)
Rm   Median Raphé Nucleus
RP   Nucleus Raphé Pontis
s5c   Spinal Trigeminal Nucleus (caudal)
s5i   Spinal Trigeminal Nucleus (interpolar)
s5o   Spinal Trigeminal Nucleus (oral)
sg   Substantia Gelatinosa
Tu   Olfactory Tubercle
V3   Third Ventricle
VMh   Ventromedial Nucleus (Hypothalamus)
3   Oculomotor Nucleus
4   Trochlear Nucleus
5n   Trigeminal Nerve

BMA: Basomedial Nucleus of the Amygdala. This term has been employed differently in several accounts, as summarized by De Olmos et al. ('85). The area is frequently divided into an ovoid anterior portion and a flattened posterior sector that is more difficult to delimit from surrounding nuclei in Nissl preparations but possesses a richer fiber network than the anterior sector. We lacked compelling arguments for indicating subdivision.

Pam: Periamygdaloid Cortex. The ambiguity of usage in designating the three-layered primary olfactory cortex overlying the amygdaloid complex is discussed at Pir: Piriform Area (Plate 6).

s5o: Spinal Trigeminal Nucleus (oral) (also called Pars Oralis). The initial (rostral) portion of the spinal V nuclear complex fuses with the caudal portion of the principal V nucleus and swings ventrally without sharp separation. It was suggested by Kruger ('79) that this nucleus may not deserve separate status on cytoarchitectonic grounds, but this view can be discarded on the basis of connectional data demonstrating that the s5o is not retrogradely labeled after large thalamic injections of tracer (Kruger et al., '88a). It is most easily discerned in horizontal and sagittal sections and is best illustrated and discussed in rat by Jacquin and Rhoades ('90) and in a series of papers by Falls and others (Falls et al., '85; Falls and Albin, '86; Falls et al., '90).

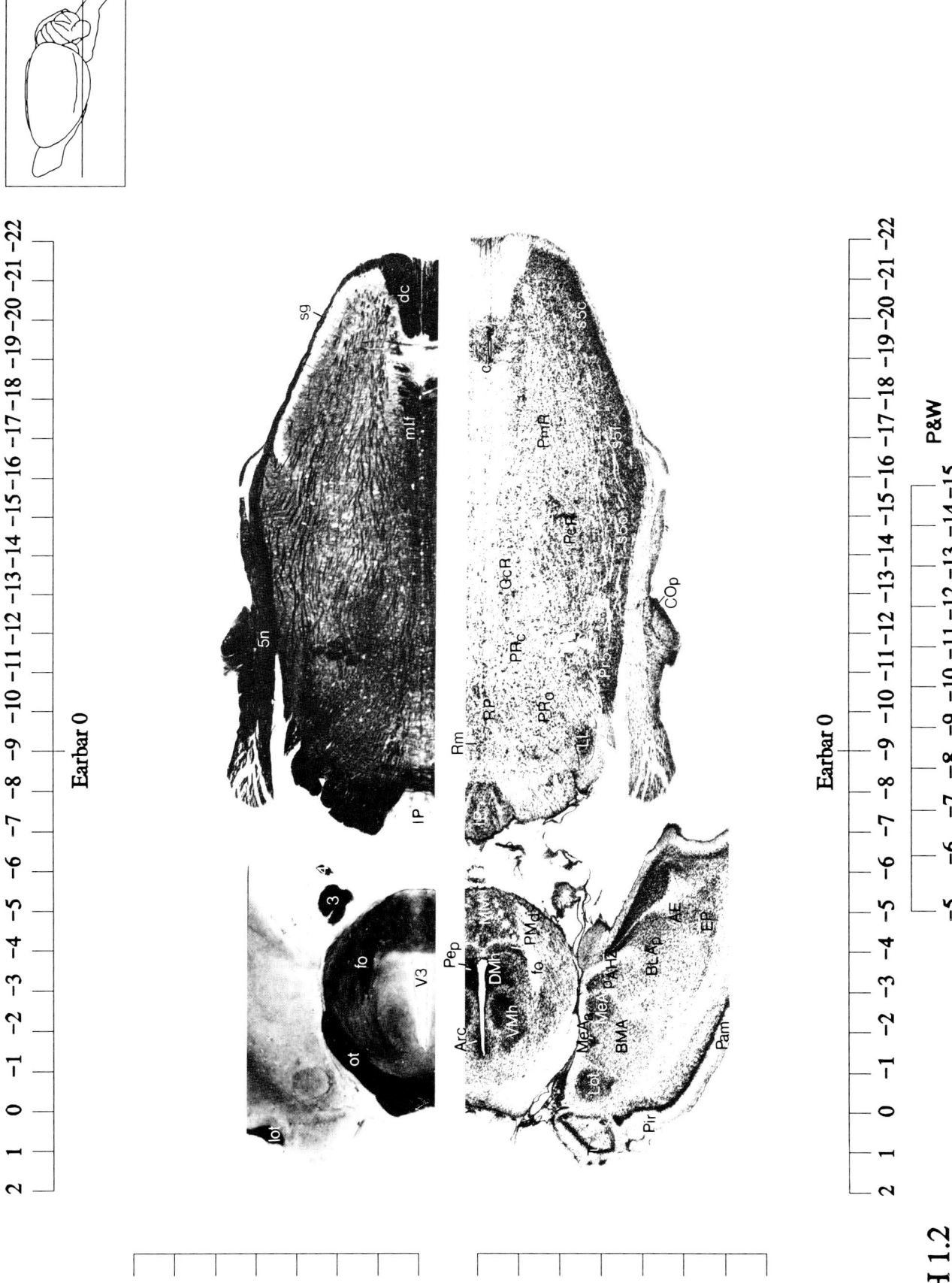

Earbar 0

−22 −21 −20 −19 −18 −17 −16 −15 −14 −13 −12 −11 −10 −9 −8 −7 −6 −5 −4 −3 −2 −1 0 1 2

P&W

Earbar 0

−22 −21 −20 −19 −18 −17 −16 −15 −14 −13 −12 −11 −10 −9 −8 −7 −6 −5 −4 −3 −2 −1 0 1 2

−15 −14 −13 −12 −11 −10 −9 −8 −7 −6 −5

7 6 5 4 3 2 1 0   0 1 2 3 4 5 6 7

H 1.2

Plate 94

AMB Nucleus Ambiguus
Arc Arcuate Nucleus (Hypothalamus)
GcR Gigantocellular Reticular Nucleus
LL Nucleus of Lateral Lemniscus
LM Lateral Mammillary Nucleus
lot Lateral Olfactory Tract
ml Medial Lemniscus
MM Medial Mammillary Nucleus (Hypothalamus)
ot Optic Tract
PcR Parvicellular Reticular Nucleus
Pd Pontine Gray (Deep)
Pe$_p$ Periventricular Nucleus (posterior)
PM$_v$ Premammillary Nucleus (ventral)
PmR Paramedian Reticular Nucleus
PR$_c$ Pontine Reticular Nucleus (caudal)
PR$_o$ Pontine Reticular Nucleus (oral)
RO Nucleus Raphé Obscurus
RP Nucleus Raphé Pontis
s5c Spinal Trigeminal Nucleus (caudal)
sg Substantia Gelatinosa
TB Nucleus of the Trapezoid Body
TM Tuberomammillary Nucleus (Hypothalamus)
VMh Ventromedial Nucleus (Hypothalamus)
5n Trigeminal Nerve
7n Facial Nerve

MM and LM: Medial and (LM) Lateral Mammillary Nuclei (Hypothalamus). These two cell groups are easily defined, although the medial nucleus is extensively subdivided by some authors (see Gurdjian, '27; Krieg, '32; Allen and Hopkins, '88). Neurons in the lateral nucleus are typically larger and more darkly staining than those in the medial nucleus. J. E. Rose ('39) provides a nicely illustrated account in a variety of mammals together with volumetric measurements. The medial nucleus is traversed by numerous myelinated axons, extending across the midline, and thus is most striking in the fiber-stained sections.

s5c: Spinal Trigeminal Nucleus (caudal) (also called Pars Caudalis). This is the largest component of the spinal V complex extending from interpolaris to the upper cervical dorsal horn; it is sometimes called the *medullary dorsal horn*. The laminar organization of the outer portion of this nucleus, with a distinct substantia gelatinosa, renders it easily separable from the interpolar subnucleus, which also contains larger, more dispersed neurons. The marginal and gelatinosa layers fuse imperceptibly with the dorsal horn, but the homologies of the deeper laminae may be difficult to establish, with current practice relying largely on the analysis of Gobel et al. ('77) in the cat. The portion below the gelatinosa, constituting the nucleus proprius, is dominated by vibrissal input and has been extensively analyzed in the rat (Ma and Woolsey, '84), including detailed electrophysiological mapping (Nord, '67; Renehan et al., '86).

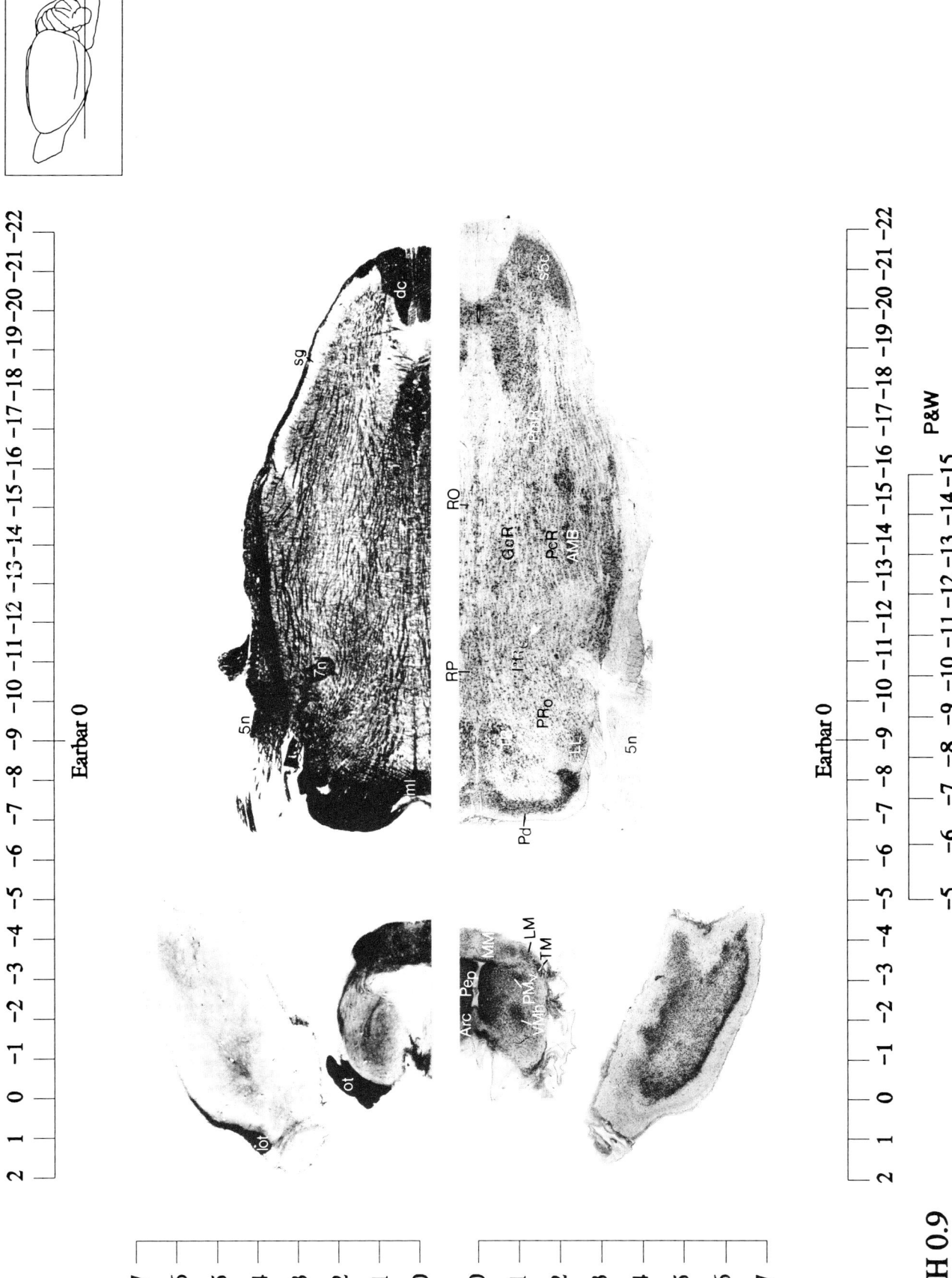

Earbar 0

2 1 0 −1 −2 −3 −4 −5 −6 −7 −8 −9 −10 −11 −12 −13 −14 −15 −16 −17 −18 −19 −20 −21 −22

P&W

Earbar 0

2 1 0 −1 −2 −3 −4 −5 −6 −7 −8 −9 −10 −11 −12 −13 −14 −15 −16 −17 −18 −19 −20 −21 −22

H 0.9

Plate 95

cst Corticospinal Tract

dc Dorsal Column

GcR Gigantocellular Reticular Nucleus

LR$_l$ Lateral Reticular Nucleus (lateral)

LR$_m$ Lateral Reticular Nucleus (medial)

mcp Middle Cerebellar Peduncle

ml Medial Lemniscus

Pd Pontine Gray (Deep)

PmR Paramedian Reticular Nucleus

PR$_c$ Pontine Reticular Nucleus (caudal)

PR$_o$ Pontine Reticular Nucleus (oral)

RM Nucleus Raphé Magnus

RO Nucleus Raphé Obscurus

RP Nucleus Raphé Pontis

TB Nucleus of the Trapezoid Body

5n Trigeminal Nerve

7 Facial Nucleus

7n Facial Nerve

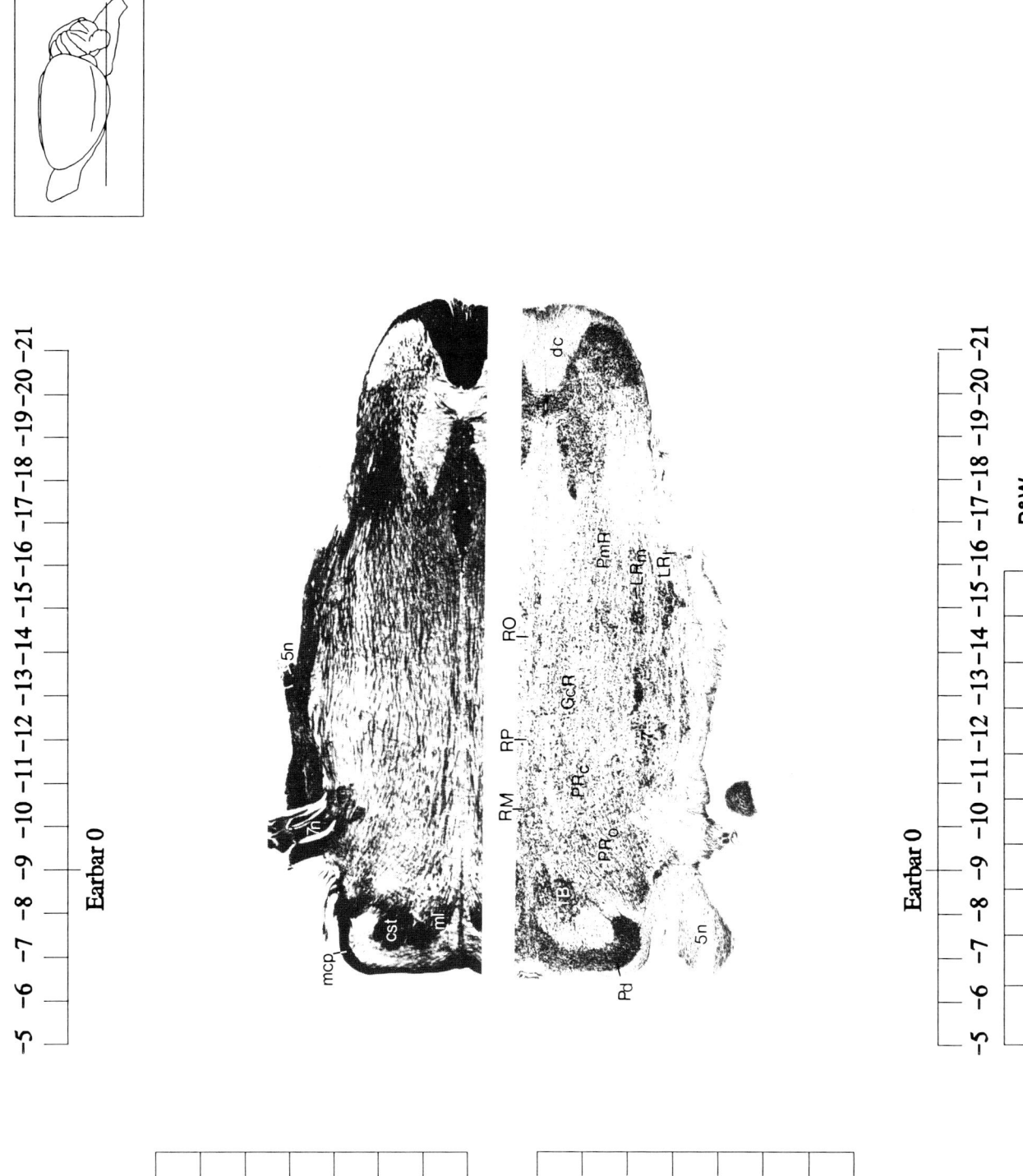

Earbar 0

-5 -6 -7 -8 -9 -10 -11 -12 -13 -14 -15 -16 -17 -18 -19 -20 -21

7 6 5 4 3 2 1 0 0 1 2 3 4 5 6 7

Earbar 0

-5 -6 -7 -8 -9 -10 -11 -12 -13 -14 -15 -16 -17 -18 -19 -20 -21

P&W

-5 -6 -7 -8 -9 -10 -11 -12 -13 -14 -15

H 0.6

Plate 96

cst Corticospinal Tract
dc Dorsal Column
Dh Dorsal Horn
IO Inferior Olive
LR$_p$ Lateral Reticular Nucleus (posterior)
mcp Middle Cerebellar Peduncle
McR Magnocellular Reticular Nucleus
ml Medial Lemniscus
mlf Medial Longitudinal Fascicle
Olep External Periolivary Nucleus
Pd Pontine Gray (Deep)
PgR Paragigantocellular Reticular Nucleus
PR$_v$ Pontine Reticular Nucleus (ventral)
RM Nucleus Raphé Magnus
RO Nucleus Raphé Obscurus
SO$_m$ Superior Olivary Complex (medial)
TB Nucleus of the Trapezoid Body
Vh Ventral Horn
5n Trigeminal Nerve
7 Facial Nucleus
7n Facial Nerve

mcp: Middle Cerebellar Peduncle (also called Brachium Pontis). This massive tract sweeps as transverse myelinated axons derived largely from pontine nuclei, crossing to the opposite side and coursing lateral and dorsal to the inferior cerebellar peduncle to enter the white matter core of the cerebellum. Despite its size, its boundaries are not readily discerned throughout much of its extent.

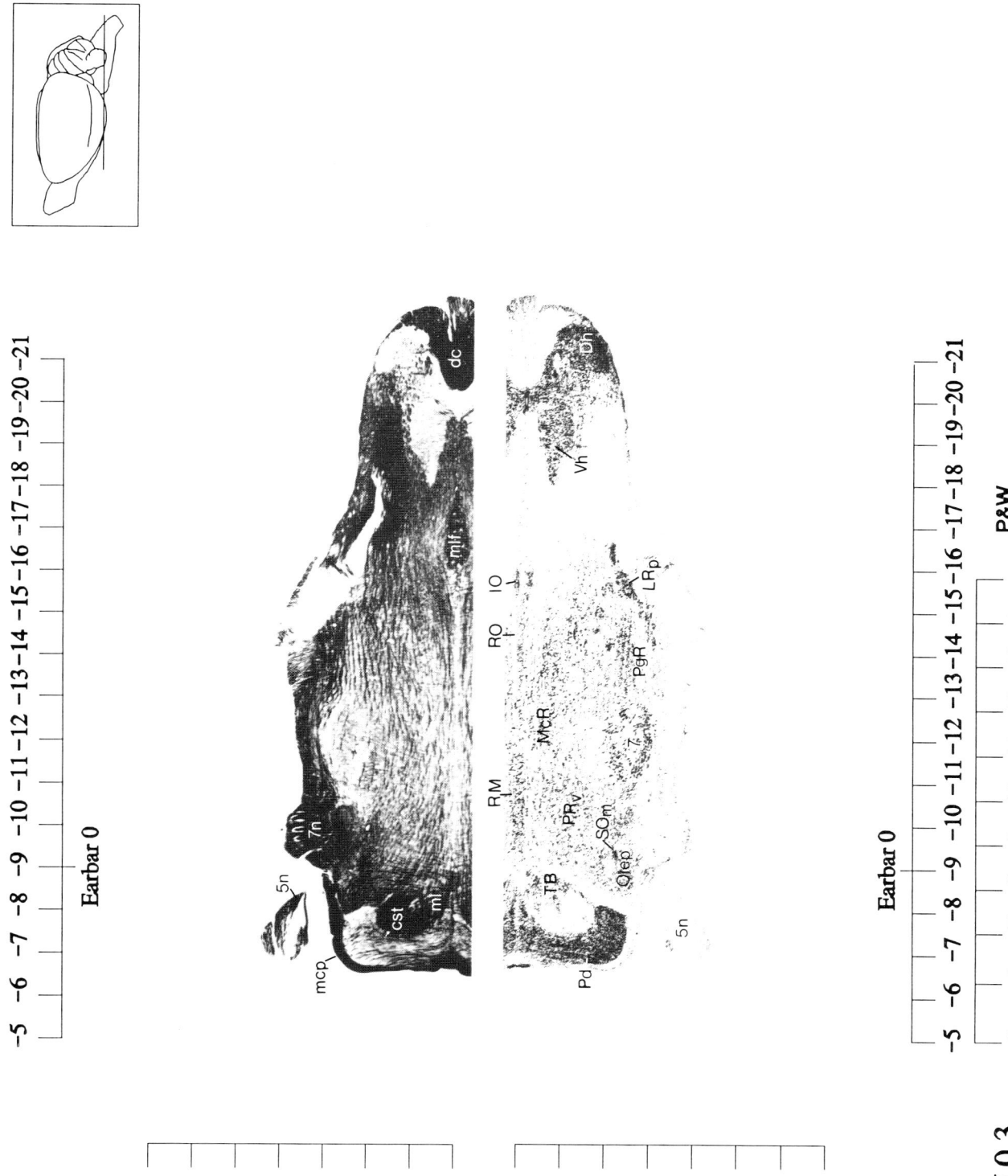

Earbar 0

−5 −6 −7 −8 −9 −10 −11 −12 −13 −14 −15 −16 −17 −18 −19 −20 −21

Earbar 0

−5 −6 −7 −8 −9 −10 −11 −12 −13 −14 −15 −16 −17 −18 −19 −20 −21

−5 −6 −7 −8 −9 −10 −11 −12 −13 −14 −15

P&W

H 0.3

Plate 97

dc   Dorsal Column
Dh   Dorsal Horn
IO   Inferior Olive
LR$_p$ Lateral Reticular Nucleus (posterior)
mcp   Middle Cerebellar Peduncle
McR   Magnocellular Reticular Nucleus
ml   Medial Lemniscus
mlf   Medial Longitudinal Fascicle
Olep   External Periolivary Nucleus
Olsp   Superior Paraolivary Nucleus
Pd   Pontine Gray (Deep)
PgR   Paragigantocellular Reticular Nucleus
PR$_v$ Pontine Reticular Nucleus (ventral)
RM   Nucleus Raphé Magnus
RO   Nucleus Raphé Obscurus
SO$_l$ Superior Olivary Complex (lateral)
SO$_m$ Superior Olivary Complex (medial)
TB   Nucleus of the Trapezoid Body
Vh   Ventral Horn
7   Facial Nucleus

Olsp: Superior Paraolivary Nucleus. This nucleus is fairly distinct in outline, lying dorsolateral to the nucleus of the trapezoid body (tb) and medial to the lateral superior olive, which it caps and with which it merges (Plate 43). The lateral portion, designated the *dorsal periolivary region* by Paxinos and Watson ('86), can indeed be seen as a separate caudal nucleus in the horizontal plane (Plate 97).

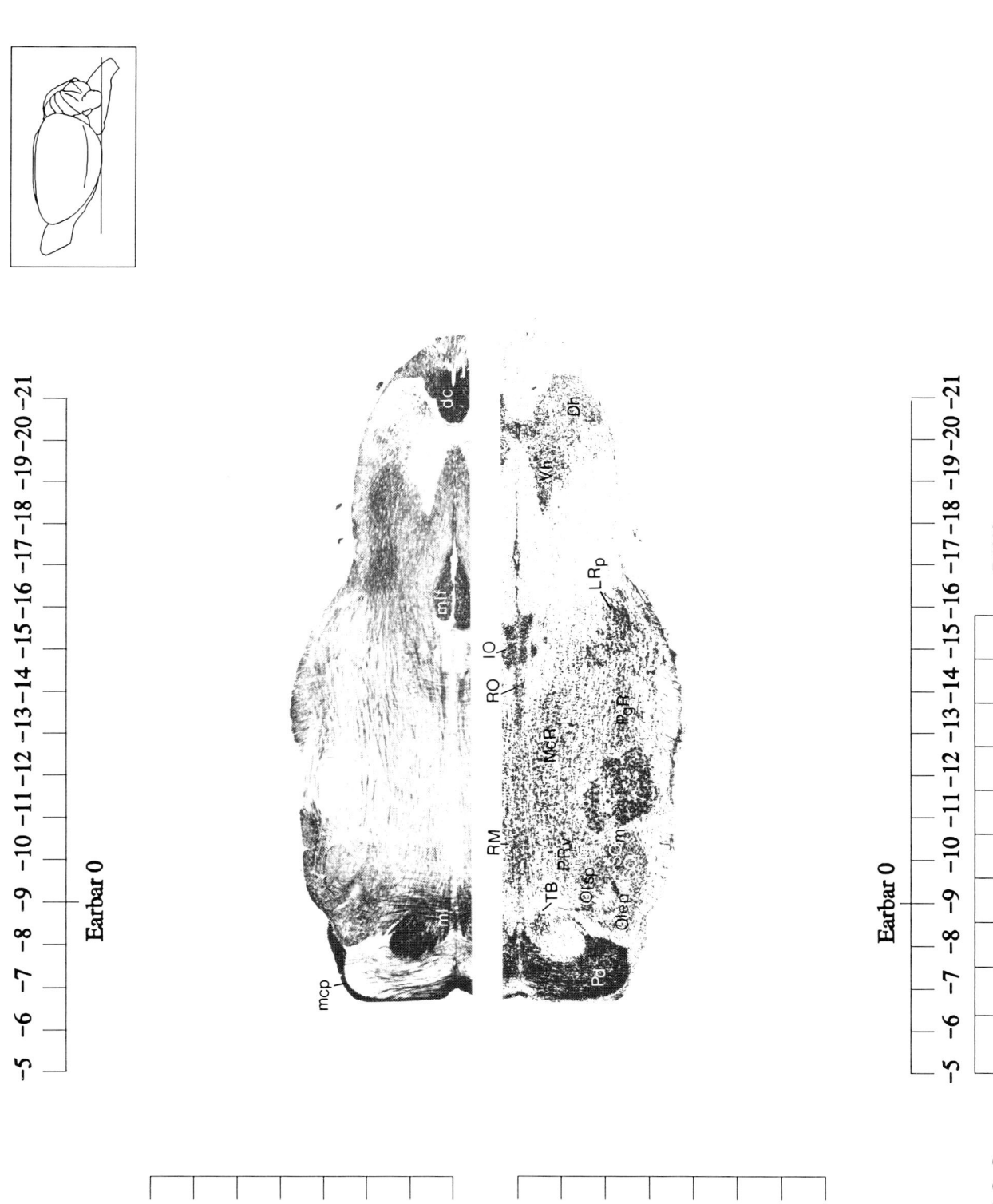

Earbar 0

−5 −6 −7 −8 −9 −10 −11 −12 −13 −14 −15 −16 −17 −18 −19 −20 −21

P&W

Earbar 0

−5 −6 −7 −8 −9 −10 −11 −12 −13 −14 −15 −16 −17 −18 −19 −20 −21

−5 −6 −7 −8 −9 −10 −11 −12 −13 −14 −15

7 6 5 4 3 2 1 0

0 1 2 3 4 5 6 7

H 0.0

Plate 98

C₁   First Cervical Level of Spinal Cord
cst   Corticospinal Tract
Dh   Dorsal Horn
dsct   Dorsal Spinocerebellar Tract
IO   Inferior Olive
LR$_p$ Lateral Reticular Nucleus (posterior)
mcp   Middle Cerebellar Peduncle
Olep   External Periolivary Nucleus
Olsp   Superior Paraolivary Nucleus
Pd   Pontine Gray (Deep)
py   Pyramidal Tract
RM   Nucleus Raphé Magnus
RPa   Nucleus Raphé Pallidus
SO$_l$ Superior Olivary Complex (lateral)
SO$_m$ Superior Olivary Complex (medial)
TB   Nucleus of the Trapezoid Body
Vh   Ventral Horn
7   Facial Nucleus

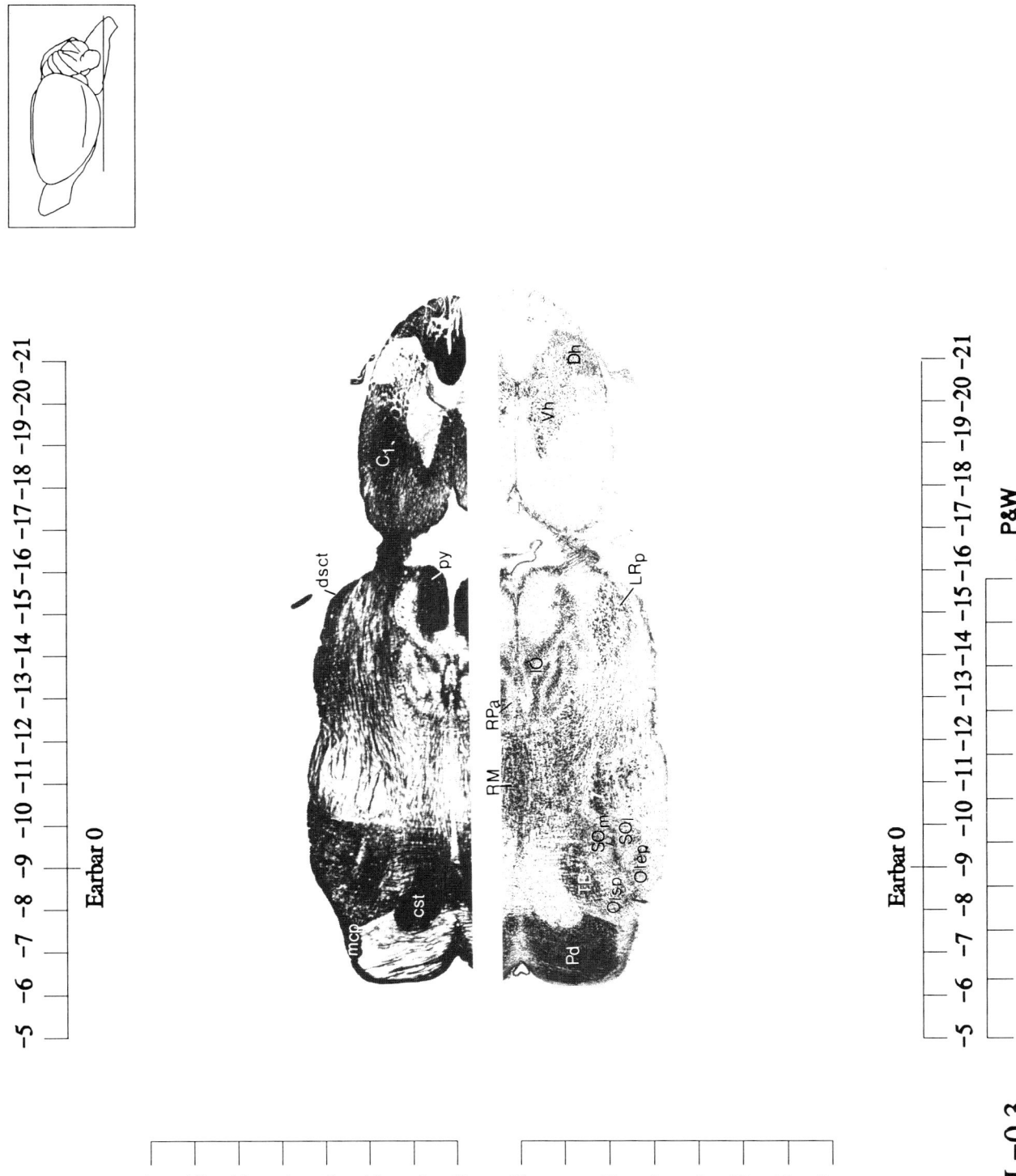

Earbar 0

-5 -6 -7 -8 -9 -10 -11 -12 -13 -14 -15 -16 -17 -18 -19 -20 -21

Earbar 0

-5 -6 -7 -8 -9 -10 -11 -12 -13 -14 -15 -16 -17 -18 -19 -20 -21

P&W

H −0.3

Plate 99

C₂ Second Cervical Level of Spinal Cord
cst Corticospinal Tract
Dh Dorsal Horn
IO Inferior Olive
mcp Middle Cerebellar Peduncle
Pd Pontine Gray (Deep)
py Pyramidal Tract
RPa Nucleus Raphé Pallidus
SOₘ Superior Olivary Complex (medial)
TB Nucleus of the Trapezoid Body
Vh Ventral Horn

cst and py: Corticospinal Tract and Pyramidal Tract. The main descending cerebral pathway to the spinal cord is largely intermingled with other axonal bundles except at the base of the medulla, where in the human brain its decussation was described as pyramid-shaped in early gross descriptions. Although axons can be traced from cerebral motor cortex (largely arising from layer V *pyramidal* cells, confusing the issue further), it is not wholly cortical in origin and only the portion decussating at the base of the medullary pyramid is designated as the pyramidal tract in a strict sense. In the rat, unlike in human and many other larger brains, the axons ascend from the pyramid to enter the base of the dorsal columns before descending into the spinal cord, a condition that differs markedly from the lateral corticospinal (pyramidal) tract of primates. Although boundaries are not evident, except in experimental tracing studies, the difference in fiber density and glial nuclei makes it possible to distinguish this descending tract from the overlying cuneate and gracile fascicles.

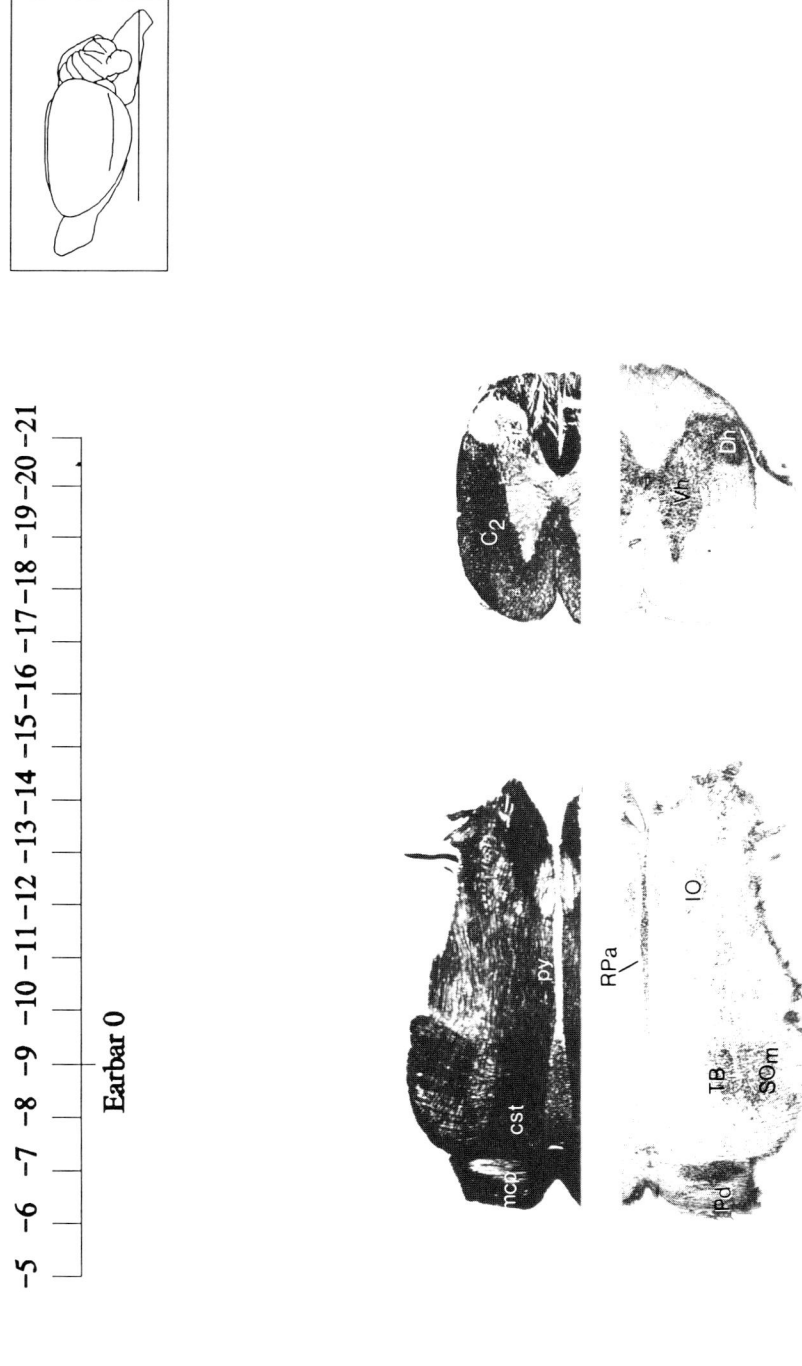

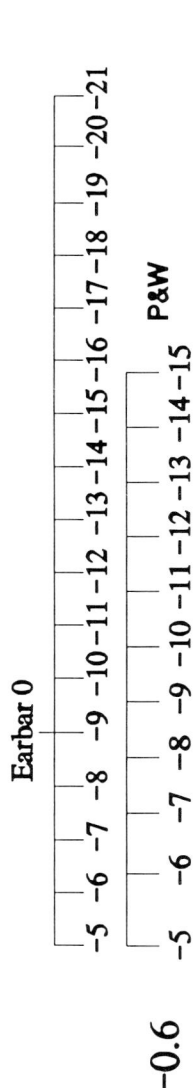

Earbar 0

-5 -6 -7 -8 -9 -10 -11 -12 -13 -14 -15 -16 -17 -18 -19 -20 -21

5 4 3 2 1 0

0 1 2 3 4 5

Earbar 0

-5 -6 -7 -8 -9 -10 -11 -12 -13 -14 -15 -16 -17 -18 -19 -20 -21

-5 -6 -7 -8 -9 -10 -11 -12 -13 -14 -15

P&W

H −0.6

Plate 100

Note. The sagittal series of sections are difficult to interpret, especially near the midline, because the plane of section diverges from the true midline in both the dorsoventral and rostrocaudal axes. The problem arises mainly from the failure to achieve perfect knife alignment at the midline from dorsal to ventral (the direction of cutting) and buckling of the brain in the long axis during embedment. One consequence of this skewing is that Plate 100, which passes through the third ventricle in the hypothalamus, is more lateral to the midline in the thalamus than in Plate 102 where the thalamic commissures of the midline are most evident. Also note that the central canal in the caudal medulla moves rostral from Plate 100 to Plate 102.

(12) Hypoglossal Nucleus. In the sagittal plane the heterogeneity of this large aggregate of motor neurons is most apparent, but we have refrained from distinguishing the more densely packed caudoventral component with a separate name and from describing a subnucleus that can easily be recognized on strictly cytoarchitectural grounds. Nor have we indulged in labeling the accessory hypoglossal nucleus lying near the ventrolateral edge in transverse sections. The differences relate to the separate skeletal muscle groups innervated, as revealed by application of experimental tracing techniques.

ac   Anterior Commissure
ADP   Anterodorsal Preoptic Nucleus (Hypothalamus)
AMi   Interanteromedial Nucleus (Thalamus)
ap   Area Postrema
Arc   Arcuate Nucleus (Hypothalamus)
cc   Corpus Callosum
CG   Central Gray
chp   Choroid Plexus
cm   Central Medial Nucleus (Thalamus)
D   Nucleus of Darkschewitsch
DG   Dentate Gyrus
dhc   Dorsal Hippocampal Commissure
DMh   Dorsomedial Nucleus (Hypothalamus)
DT   Dorsal Tegmental Nucleus
DT$_p$ Dorsal Tegmental Nucleus (posterior)
ep   Ependymal Layer of Ventricle
Fa   Fastigial (Medial) Cerebellar Nucleus
fo   Fornix
GcR   Gigantocellular Reticular Nucleus
GR   Gracile Nucleus
Hb$_m$ Habenula Nucleus (medial) (Thalamus)
IC   Inferior Colliculus
Ic   Intercalated Nucleus (Medulla)
IF   Interfascicular Nucleus
IO   Inferior Olive
IP   Interpeduncular Nucleus
MDi   Intermediodorsal Nucleus (Thalamus)
MM   Medial Mammillary Nucleus
MP   Medial Preoptic Nucleus (Hypothalamus)
mr   Mammillary Recess
oc   Optic Chiasm
Pa$_a$ Paraventricular Nucleus (anterior) (Thalamus)
Pa$_p$ Paraventricular Nucleus (posterior) (Thalamus)
Pd   Pontine Gray (Deep)
Pe$_{av}$ Periventricular Nucleus (anteroventral) (Hypothalamus)
Pe$_p$ Periventricular Nucleus (posterior) (Hypothalamus)
Pf   Parafascicular Nucleus (Thalamus)

PH   Posterior Hypothalamic Nucleus
Ph   Perihypoglossal Nucleus
Pi   Pineal Gland (Epiphysis)
PT   Paratenial Nucleus (Thalamus)
Ptm   Medial Pretectal Area (Thalamus)
PVh   Paraventricular Nucleus (Hypothalamus)
Rd   Dorsal Raphé Nucleus
Re   Nucleus Reuniens (Thalamus)
Rh   Rhomboid Nucleus (Thalamus)
RL   Nucleus Raphé Linearis (rostral)
RLc   Nucleus Raphé Linearis (central)
RM   Nucleus Raphé Magnus
Rm   Median Raphé Nucleus
RO   Nucleus Raphé Obscurus
Rol   Nucleus of Roller
RP   Nucleus Raphé Pontis
RPa   Nucleus Raphé Pallidus
Rsp   Retrosplenial Area (Cortex)
s5c   Spinal Trigeminal Nucleus (caudal)
SC   Superior Colliculus
SCh   Suprachiasmatic Nucleus (Hypothalamus)
SF   Septofimbrial Nucleus
sfo   Subfornical Organ
sgd   Deep Gray Layer of Superior Colliculus
Sge   Supragenual Nucleus
sgi   Intermediate Gray Layer of Superior Colliculus
sgs   Superficial Gray Layer of Superior Colliculus
sPf   Subparafascicular Nucleus (Thalamus)
SPm   Septal Nucleus (medial)
Sub   Subiculum
SUM$_m$ Supramammillary Nucleus (medial)
TS$_c$ Nucleus of the Solitary Tract (Commissural)
V3   Third Ventricle
vhc   Ventral Hippocampal Commissure
VMh   Ventromedial Nucleus (Hypothalamus)
VT   Ventral Tegmental Nucleus
ZI   Zona Incerta
3   Oculomotor Nucleus
4   Trochlear Nucleus
10   Dorsal Motor Nucleus of Vagus
12   Hypoglossal Nucleus

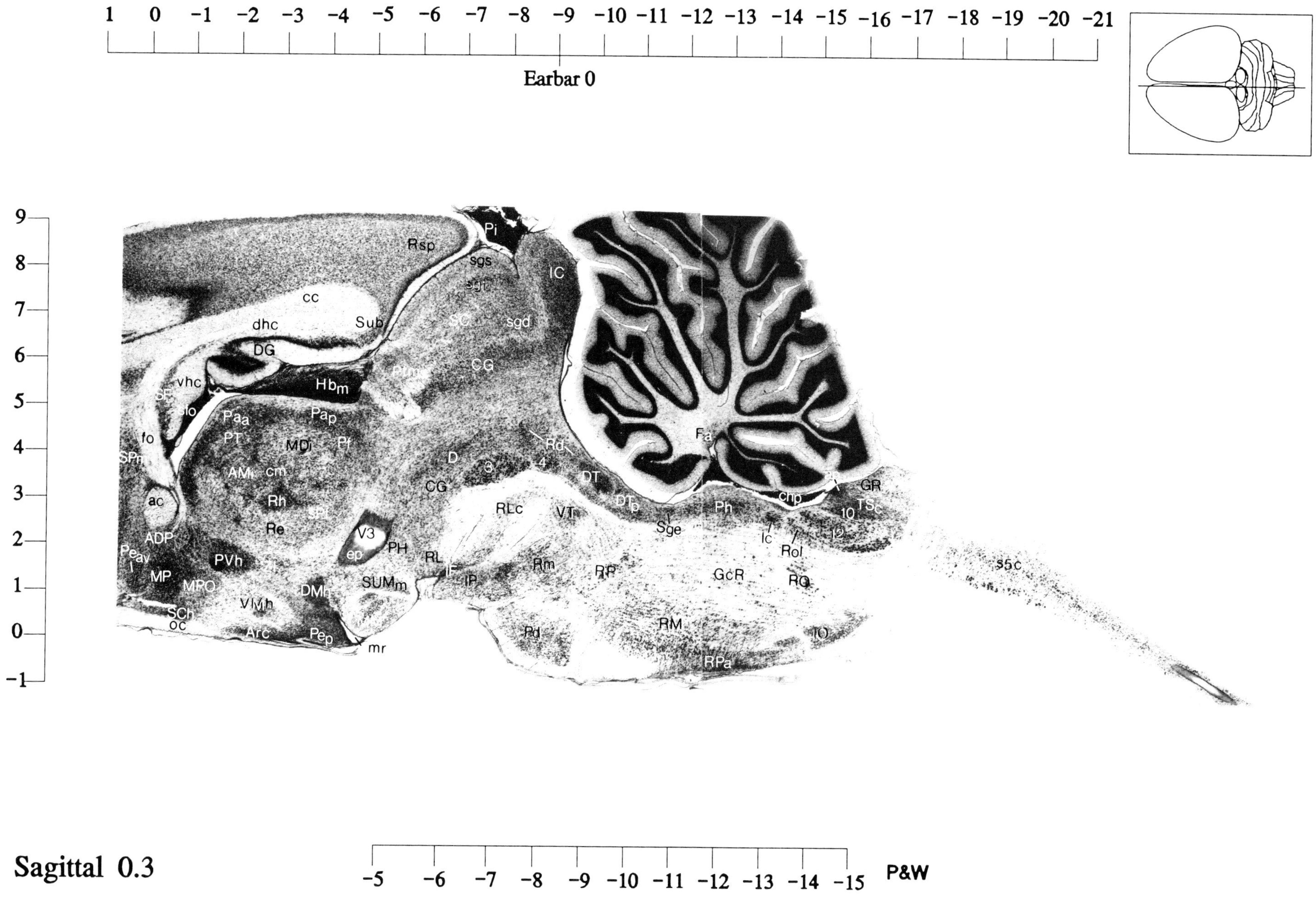

Earbar 0

1 0 -1 -2 -3 -4 -5 -6 -7 -8 -9 -10 -11 -12 -13 -14 -15 -16 -17 -18 -19 -20 -21

9
8
7
6
5
4
3
2
1
0
-1

Sagittal 0.3

P&W

-5 -6 -7 -8 -9 -10 -11 -12 -13 -14 -15

Plate 101

ac   Anterior Commissure
cc   Corpus Callosum
csc   Commissure of the Superior Colliculus
dc   Dorsal Column
fo   Fornix
fo$_p$ Fornix (postcommissural)
Hb$_m$ Habenula Nucleus (medial) (Thalamus)
hc   Habenula Commissure
IC   Inferior Colliculus
IO   Inferior Olive
IPf   Interpeduncular Fossa
mcp   Middle Cerebellar Peduncle
mlf   Medial Longitudinal Fascicle
mr   Mammillary Recess
oc   Optic Chiasm
Pa   Paraventricular Nucleus (Thalamus)
pc   Posterior Commissure
Pi   Pineal Gland (Epiphysis)
pv   Periventricular Fiber System
py   Pyramidal Tract
pyx   Decussation of Pyramidal Tract
SC   Superior Colliculus
scpx   Decussation of the Superior Cerebellar Peduncle
sm   Stria Medullaris
so   Optic Layer of Superior Colliculus
sumx   Supramammillary Decussation
swd   Deep White Layer of Superior Colliculus
swi   Intermediate White Layer of Superior Colliculus
tb   Trapezoid Body
tc   Thalamic Commissures
V3   Third Ventricle
vhc   Ventral Hippocampal Commissure
zo   Zonal Lamina of Superior Colliculus

Note. Numerous apparent fiber bundles are evident on this and subsequent sagittal plates but remain unlabeled. Examples include the medial or internal arcuate fiber bundles (above the pyramidal tract decussation), which course in an arc from the dorsal column nuclei to enter and cross the medial lemniscus. Thus, the arcuate fibers constitute the initial portion of the axons forming this large ascending tract (ml) and do not warrant separate status despite the distinctive topography. In the caudal hypothalamus we do not attempt to distinguish axons emerging from the mammillary nuclei until they separate into the mammillothalamic and mammillotegmental tracts, nor do we label the commissure between the medial mammillary nuclei, nor the ependymal layer of the third ventricle and rostral to the mammillary recess. Extensive discrete bundles, such as the conspicuous longitudinal periventricular axons at the base of the hypothalamic midline, deserve distinct status and a name, but this is rarely recognized as a long tract and is generally referred to simply on topographic grounds as "periventricular fibers" or ignored. Some fiber bundles are labeled in conformity with common usage; for example, the trapezoid body fibers associated with these nuclei appear to merit the name derived from gross anatomical description that can apply to either the fibers or cells in this region.

Note. A broad **Y**-shaped system of thin fibers at the caudal end of the diencephalon and rostral end of the midbrain central gray is obvious in this plate. The two arms of the **Y** are associated with the periventricular system of the thalamus and hypothalamus, while the stem lies in the central gray; the system as a whole harbors a complex mixture of both ascending and descending fibers (see Pe: Periventricular Nucleus at Plate 91). The most condensed part of this system (especially in humans) in the midbrain is known as the *dorsal longitudinal fascicle (of Schütz)*, which lies ventrolateral to the cerebral aqueduct. Cajal indicated that fibers of the system surround the aqueduct and suggested that the more general term *periependymal longitudinal fascicle* would be more appropriate. Relatively little is known about the precise origins and terminations of fibers in this system.

pv: Periventricular Fiber System. Longitudinal myelinated axons are associated throughout the length of the thalamic paraventricular nuclei (Pa) and surrounding the third ventricle associated with the hypothalamic periventricular nuclei (Pe) (see Sutin, '66).

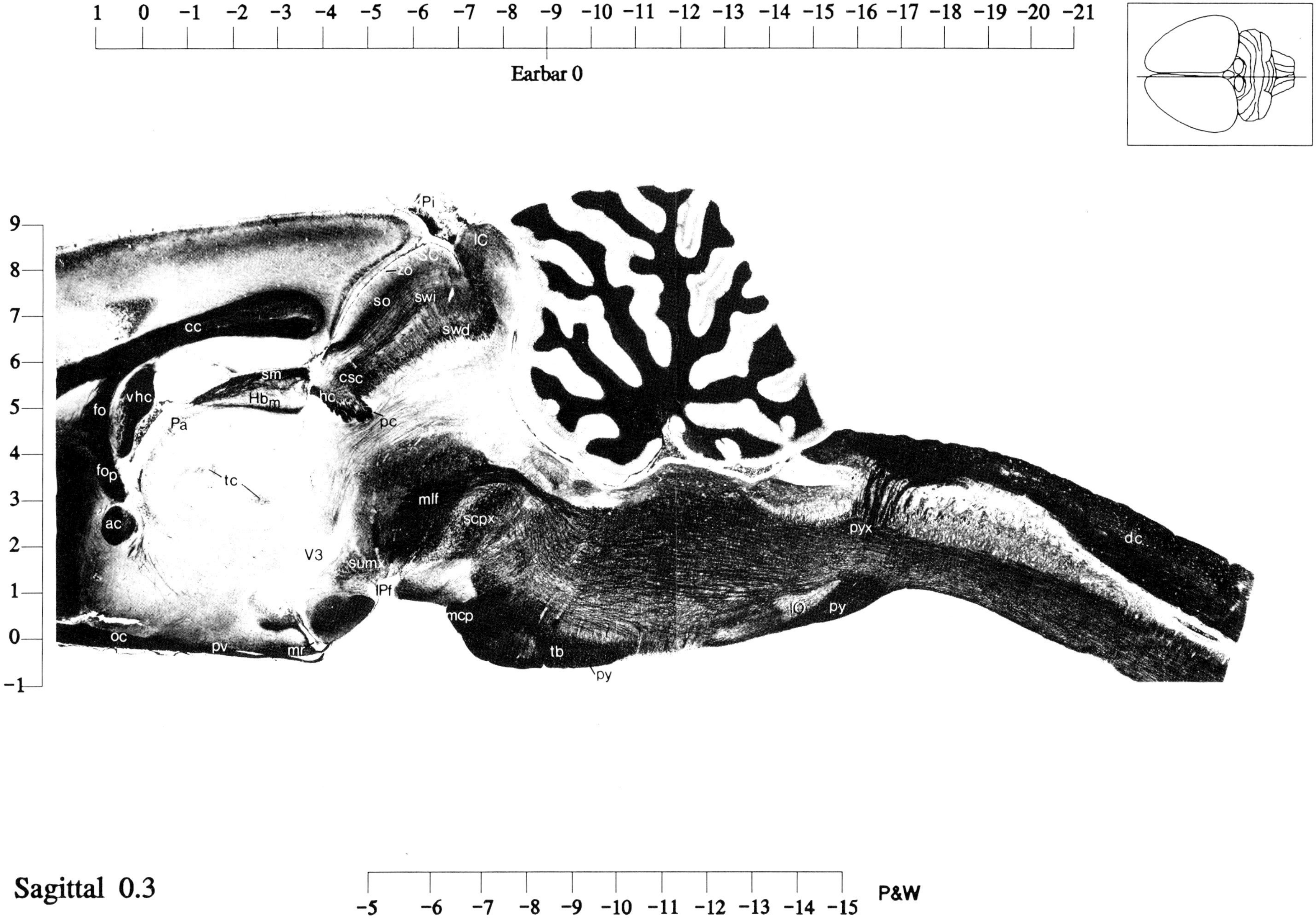

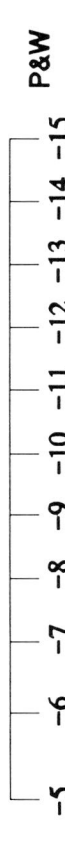

Earbar 0

1 0 −1 −2 −3 −4 −5 −6 −7 −8 −9 −10 −11 −12 −13 −14 −15 −16 −17 −18 −19 −20 −21

9 8 7 6 5 4 3 2 1 0 −1

Sagittal 0.3

P&W

−5 −6 −7 −8 −9 −10 −11 −12 −13 −14 −15

Plate 102

Note. The slight deviation from the true parasagittal axis in the entire sagittal series renders the thalamic portion of the diencephalon slightly closer to the midline than the first section (Plate 100) in which the hypothalamus is closer to the midline. In addition to the slight dorsoventral tilt there is some curvature in the antero-posterior axis, accounting for deviation of the central canal out of the section.

c: Central Canal. The midline, central channel formed by closure of the embryonic neural tube persists in the caudal medulla and spinal cord as a tiny canal surrounded by a single layer of ependymal epithelium.

Pa: Paraventricular Nuclei (Thalamus). The paired midline nuclei underlying the third ventricle constitute a nuclear complex capping the top of the thalamus across its entire extent, including its rostral pole (best seen in sagittal section). It is usually divided into anterior and posterior paraventricular nuclei, evident in horizontal and sagittal sections. There is little controversy concerning nomenclature although a central, paler core is evident but usually not labeled separately, and the exact border between the anterior and posterior parts is difficult to see.

sfo: Subfornical Organ. A "glandlike," darkly stained protuberance below the hippocampal commissure (ventral psalterium), lying in the third ventricle above the interventricular foramen, described in mammals by Pines (see Akert et al., '61), containing neurons connected with the medial septum and portions of the hypothalamus (Lind et al., '82). It lies outside the blood–brain barrier and contains a rich plexus of fenestrated capillaries.

SUM: Supramammillary Nucleus. A thin, flat, distinctive nucleus forming a cap overlying the medial and lateral mammillary nuclei. SUM can be divided into an unpaired midline part containing relatively small neurons, and a lateral part with larger neurons extending to the medial edge of the cerebral peduncle (Swanson, '82). It is separated from the overlying posterior hypothalamic nucleus by fibers of the supramammillary decussation.

ac Anterior Commissure
Ah Anterior Hypothalamic Nucleus
Ah$_a$ Anterior Hypothalamic Nucleus (anterior)
AM Anteromedial Nucleus (Thalamus)
ap Area Postrema
Aq Cerebral Aqueduct
Arc Arcuate Nucleus (Hypothalamus)
Bst Bed Nucleus of Stria Terminalis
c Central Canal
cc Corpus Callosum
CG Central Gray
cm Central Medial Nucleus (Thalamus)
D Nucleus of Darkschewitsch
DB Nucleus of the Diagonal Band
DG Dentate Gyrus
DMh Dorsomedial Nucleus (Hypothalamus)
DT Dorsal Tegmental Nucleus
DT$_p$ Dorsal Tegmental Nucleus (posterior)
ep Ependymal Layer of Ventricle
EW Edinger-Westphal Nucleus
fo Fornix
GcR Gigantocellular Reticular Nucleus
Ge Nucleus Gelatinosus (Nucleus Submedius) (Thalamus)
GR Gracile Nucleus
Hb$_m$ Habenula Nucleus (medial) (Thalamus)
IC Inferior Colliculus
IG Induseum Griseum
IO Inferior Olive
LDT Laterodorsal Tegmental Nucleus
MD Mediodorsal Nucleus (Thalamus)
MM Medial Mammillary Nucleus
MP Medial Preoptic Nucleus (Hypothalamus)
oc Optic Chiasm
Pa$_p$ Paraventricular Nucleus (posterior) (Thalamus)

pc Posterior Commissure
pcn Nucleus of Posterior Commissure
Pd Pontine Gray (Deep)
Pe$_p$ Periventricular Nucleus (posterior)
Pf Parafascicular Nucleus (Thalamus)
PH Posterior Hypothalamic Nucleus
Pi Pineal Gland (Epiphysis)
PM$_d$ Premammillary Nucleus (dorsal) (Hypothalamus)
PmR Paramedian Reticular Nucleus
PT Paratenial Nucleus (Thalamus)
PVh Paraventricular Nucleus (Hypothalamus)
R Reticular Nucleus (Thalamus)
Rd Dorsal Raphé Nucleus
Re Nucleus Reuniens (Thalamus)
Rh Rhomboid Nucleus (Thalamus)
RL Nucleus Raphé Linearis (rostral)
RLc Nucleus Raphé Linearis (central)
RM Nucleus Raphé Magnus
RO Nucleus Raphé Obscurus
RP Nucleus Raphé Pontis
Rpm Paramedian Raphé Nucleus
Rsp Retrosplenial Area (Cortex)
SC Superior Colliculus
sfo Subfornical Organ
sgd Deep Gray Layer of Superior Colliculus
sgi Intermediate Gray Layer of Superior Colliculus
sgs Superficial Gray Layer of Superior Colliculus
SUM$_m$ Supramammillary Nucleus (medial)
TR Tegmental Reticular Nucleus
Ts Triangular Nucleus (Septum)
Vh Ventral Horn
VMh Ventromedial Nucleus (Hypothalamus)
VT Ventral Tegmental Nucleus
ZI Zona Incerta
3 Oculomotor Nucleus
10 Dorsal Motor Nucleus of Vagus
12 Hypoglossal Nucleus

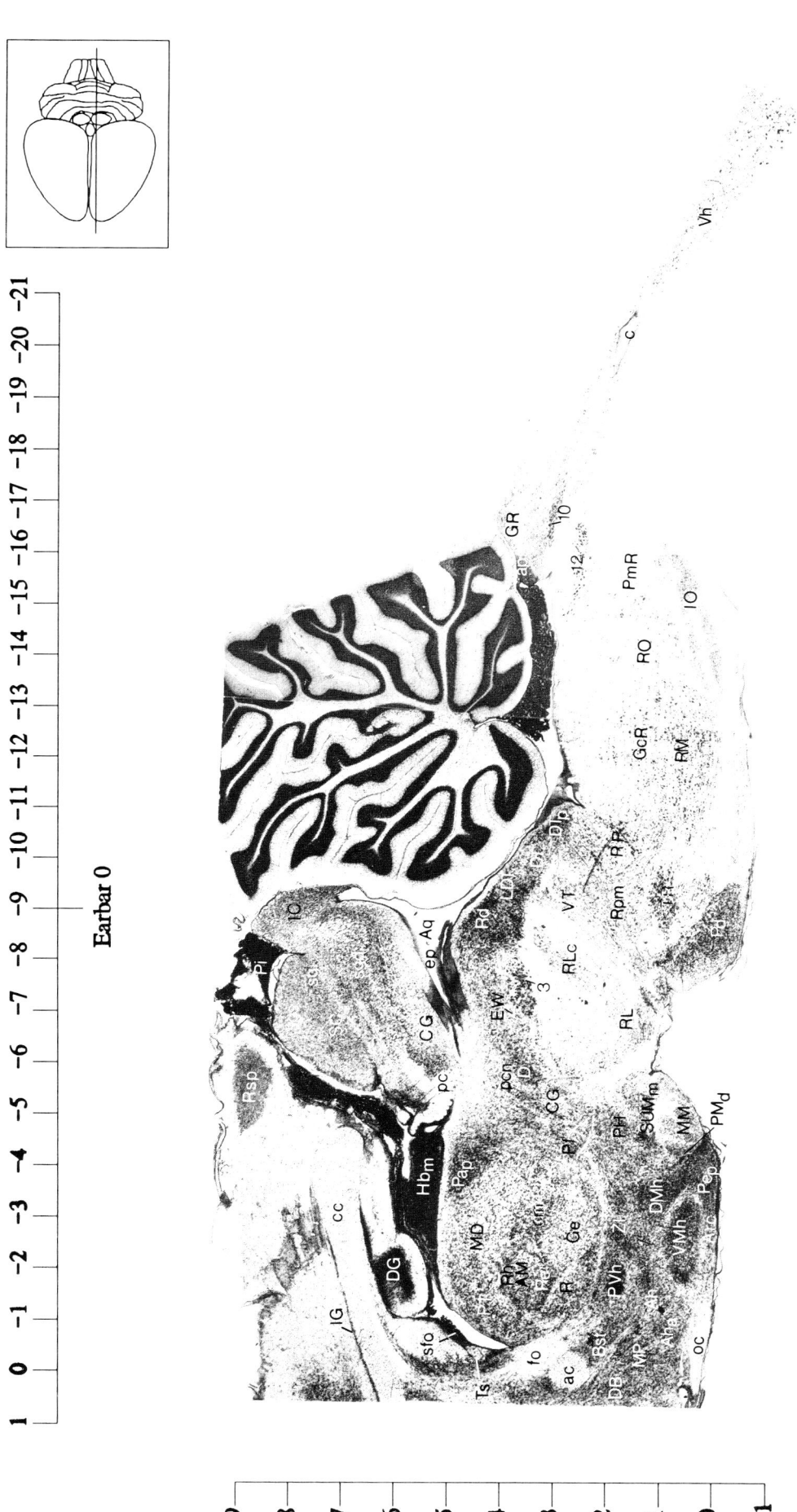

Earbar 0

1 0 -1 -2 -3 -4 -5 -6 -7 -8 -9 -10 -11 -12 -13 -14 -15 -16 -17 -18 -19 -20 -21

9 8 7 6 5 4 3 2 1 0 -1

Sagittal 0.6

P&W

-5 -6 -7 -8 -9 -10 -11 -12 -13 -14 -15

Plate 103

ac Anterior Commissure

AMi Interanteromedial Nucleus (Thalamus)

cc Corpus Callosum

chp Choroid Plexus

cm Central Medial Nucleus (Thalamus)

csc Commissure of the Superior Colliculus

dc Dorsal Column

DG Dentate Gyrus

fo Fornix

fo$_p$ Fornix (postcommissural)

fr Fasciculus Retroflexus

Ge Nucleus Gelatinosus (Nucleus Submedius) (Thalamus)

hc Habenula Commissure

IO Inferior Olive

mlf Medial Longitudinal Fascicle

oc Optic Chiasm

Pa Periventricular Nucleus (Thalamus)

pc Posterior Commissure

Pi Pineal Gland (Epiphysis)

PT Paratenial Nucleus (Thalamus)

pv Periventricular Fiber System

py Pyramidal Tract

pyx Decussation of Pyramidal Tract

Re Nucleus Reuniens (Thalamus)

Rh Rhomboid Nucleus (Thalamus)

scpx Decussation of the Superior Cerebellar Peduncle

sm Stria Medullaris

so Optic Layer of Superior Colliculus

swd Deep White Layer of Superior Colliculus

swi Intermediate White Layer of Superior Colliculus

tb Trapezoid Body

V4 Fourth Ventricle

vhc Ventral Hippocampal Commissure

zo Zonal Lamina of Superior Colliculus

Note. The continuity of the pineal gland with its epiphysial stalk can be traced back to the habenula in this fortuitous section, a relation that is difficult to visualize in reconstructing the point of attachment in other planes.

tc: Thalamic Commissures. The midline thalamic nuclei, sometimes called the *massa intermedia* as a group, are the site of numerous interthalamic commissures connecting the opposite sides of the thalamus. Most of these axons are intrinsic to the thalamus, but as can be seen in Plate 101, there are several myelinated bundles crossing the midline, including the long bundle to the anterior nuclei constituting the crossed component of the mammillothalamic tract (mt).

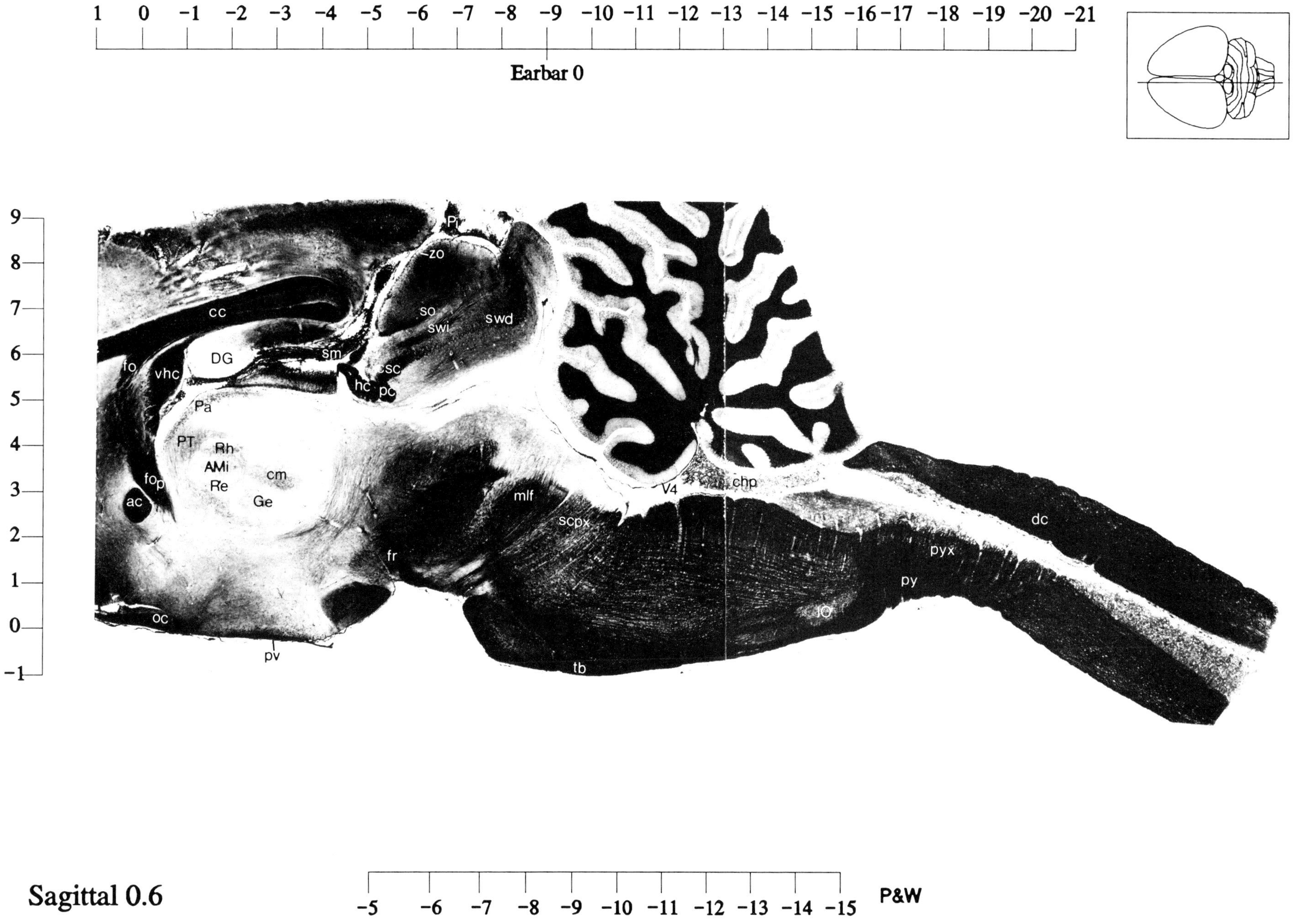

Earbar 0

1 0 -1 -2 -3 -4 -5 -6 -7 -8 -9 -10 -11 -12 -13 -14 -15 -16 -17 -18 -19 -20 -21

P&W

-5 -6 -7 -8 -9 -10 -11 -12 -13 -14 -15

Sagittal 0.6

9 8 7 6 5 4 3 2 1 0 -1

Plate 104

ac Anterior Commissure
AD Anterodorsal Nucleus (Thalamus)
AM Anteromedial Nucleus (Thalamus)
ap Area Postrema
Bst Bed Nucleus of Stria Terminalis
CG Central Gray
CU Cuneate Nucleus
DB Nucleus of the Diagonal Band
DG Dentate Gyrus
dMR Deep Mesencephalic Reticular Nucleus
DT Dorsal Tegmental Nucleus
Fa Fastigial (Medial) Cerebellar Nucleus
fo Fornix
fr Fasciculus Retroflexus
GcR Gigantocellular Reticular Nucleus
GR Gracile Nucleus
Hb_l Habenula Nucleus (lateral) (Thalamus)
IC Inferior Colliculus
Ic Intercalated Nucleus (Medulla)
Inc Interstitial Nucleus of Cajal
IO Inferior Olive
LDT Laterodorsal Tegmental Nucleus
LHA Lateral Hypothalamic Area
McR Magnocellular Reticular Nucleus (Medulla)
MD Mediodorsal Nucleus (Thalamus)
MM Medial Mammillary Nucleus (Hypothalamus)
oc Optic Chiasm
Ot Nucleus of Optic Tract
Pc Paracentral Nucleus (Thalamus)
pc Posterior Commissure
pcn Nucleus of Posterior Commissure
Pd Pontine Gray (Deep)
Pf Parafascicular Nucleus (Thalamus)
Ph Perihypoglossal Nucleus

Pi Pineal Gland (Epiphysis)
PL Posterior Limbic Area (Cortex)
PM_v Premammillary Nucleus (ventral) (Hypothalamus)
PmR Paramedian Reticular Nucleus
PpT Pedunculopontine Tegmental Nucleus
PR_c Pontine Reticular Nucleus (caudal)
PR_o Pontine Reticular Nucleus (oral)
PR_v Pontine Reticular Nucleus (ventral)
Ptm Medial Pretectal Area (Thalamus)
Pto Olivary Pretectal Nucleus (Thalamus)
R Reticular Nucleus (Thalamus)
Rdl Dorsal Raphé Nucleus (lateral extension)
Rum Red Nucleus (magnocellular)
Rup Red Nucleus (parvicellular)
s5c Spinal Trigeminal Nucleus (caudal)
SC Superior Colliculus
sgd Deep Gray Layer of Superior Colliculus
Sge Supragenual Nucleus
sgi Intermediate Gray Layer of Superior Colliculus
SI Substantia Innominata
sm Stria Medullaris
So Supraoptic Nucleus (Hypothalamus)
SPl Septal Nucleus (lateral)
SUM_l Supramammillary Nucleus (lateral)
TB Nucleus of the Trapezoid Body
TR Tegmental Reticular Nucleus
Tu Olfactory Tubercle
vhc Ventral Hippocampal Commissure
VM Ventromedial Nucleus (Thalamus)
Vm Medial Vestibular Nucleus
Vs Superior Vestibular Nucleus
VTA Ventral Tegmental Area (of Tsai)
ZI Zona Incerta
6 Abducens Nucleus
7g Genu of Facial Nerve
10 Dorsal Motor Nucleus of Vagus
12 Hypoglossal Nucleus

Note. The interface between the dorsal diencephalon and the superior colliculus reveals distinctive cytoarchitectural patterns that have been interpreted differently by numerous authors relying on transverse sections. Our usage of a *medial pretectal area* (Ptm) is not employed in other rat atlases (see Ptm: Medial Pretectal Area at Plate 25), and there are some serious discrepancies in the interpretation of the location of the nucleus of the optic tract (Ot) and the olivary pretectal nucleus (Pto) (refer to discussions of pretectal nuclei), rendering our labels susceptible to revision. Because many authors include the olivary pretectal nucleus in the nucleus of the optic tract, we have indicated them together as a tentative designation.

dMR: Deep Mesencephalic Reticular Nucleus. It is somewhat misleading to imply that this transitional region of the reticular formation containing some scattered large neurons constitutes a distinct nucleus or even a homogeneous zone. Paxinos and Watson ('86) label a broad region as DpMe without drawing limits; our usage approximates theirs as well as that of Swanson ('92), but this region is best recognized as a zone in fiber architecture. Berman ('68) referred to this large part of the midbrain reticular "formation" as the *central tegmental field*, and distinguished a *ventral retrorubral field* or *nucleus*, where the neurons are somewhat larger and some are dopaminergic (see Swanson, '82). Our failure to recognize subdivisions described by others reflects a conservative reticence in interpreting Nissl-stained material and should not be misconstrued as a rejection of other accounts of this region. It is our belief that newly emerging plans of organization based largely on chemoarchitectural patterns will prevail in establishing a new nomenclature.

pcn: Nucleus of the Posterior Commissure. These paired nuclei laterally straddling the posterior commissure are most readily delimited by their fiber architecture.

$PR_{o,v,c}$: Pontine Reticular Nucleus (oral, ventral, caudal). The part of the reticular formation between the caudal end of the midbrain and the genu of the facial nerve is usually referred to as the PR, although no clear boundaries with neighboring parts of the reticular formation can be found. Giant neurons become less frequent at progressively more rostral levels of the PR.

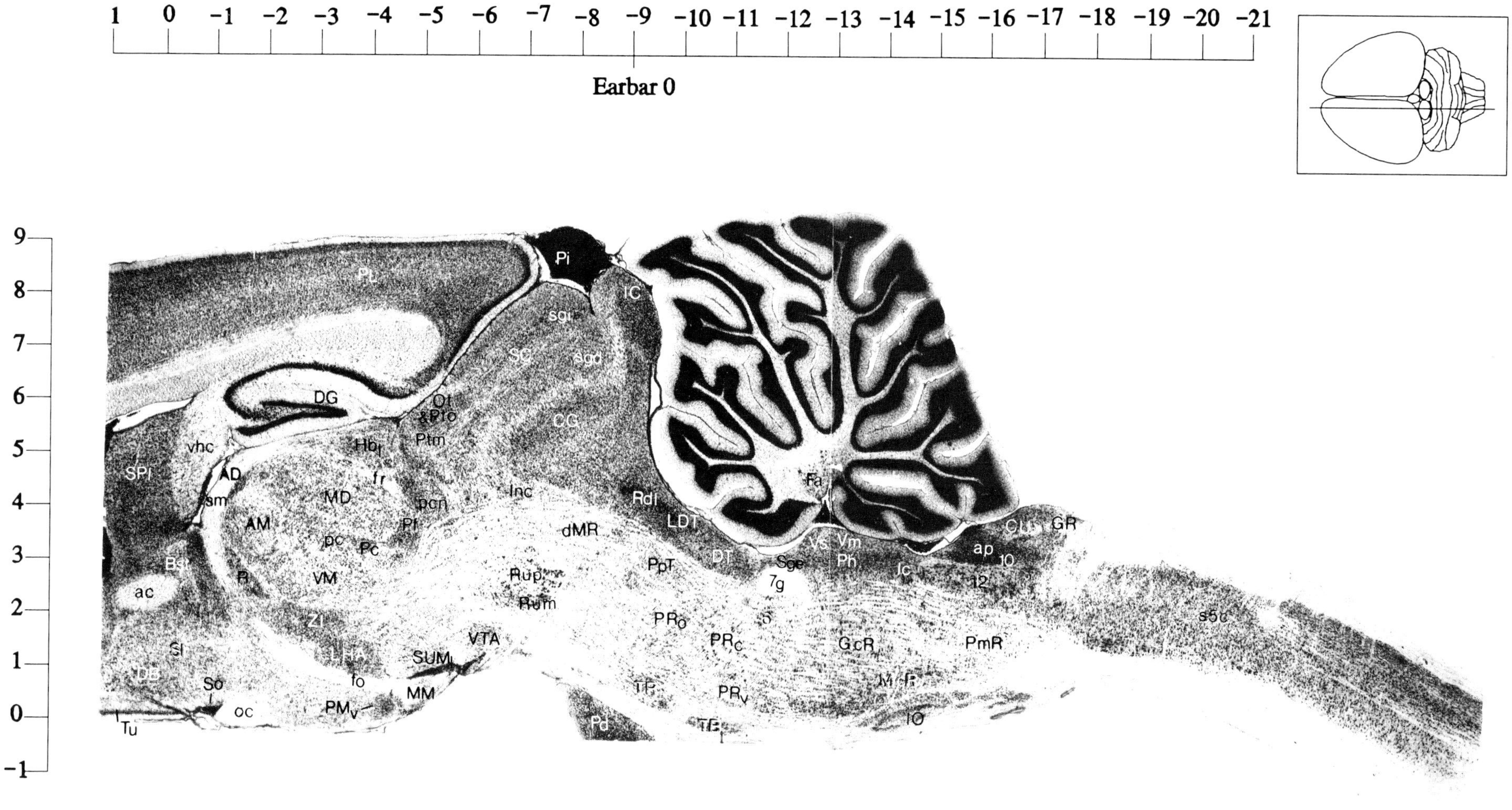

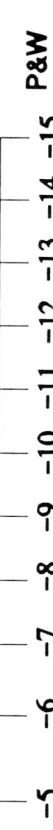

1 0 −1 −2 −3 −4 −5 −6 −7 −8 −9 −10 −11 −12 −13 −14 −15 −16 −17 −18 −19 −20 −21
Earbar 0
9 8 7 6 5 4 3 2 1 0 −1
Pi
PL
IC
sgi
SC
sgd
DG
Ot
&Pto
CG
Ptm
Hbl
vhc
SPl
AD
fr
sm
MD
pcn
Inc
Rdl
Fa
LDT
AM
Pf
dMR
pc
Pc
Vs
Vm
Cu
GR
ap
10
DT
Sge
Ph
lc
PpT
12
Bst
R
VM
Rup
7g
ac
Rum
Zl
PRo
PRc
GcR
PmR
s5c
Sl
LHA
SUMl
VTA
DB
So
fo
MM
TP
PRv
oc
PMv
Pd
TE
IO
Tu

Sagittal 0.9

−5 −6 −7 −8 −9 −10 −11 −12 −13 −14 −15
P&W

Plate 105

ac Anterior Commissure
CG Central Gray
cp Cerebral Peduncle
mfb Medial Forebrain Bundle
mlf Medial Longitudinal Fascicle
mt Mammillothalamic Tract
oc Optic Chiasm
Pta Anterior Pretectal Nucleus (Thalamus)
scpx Decussation of the Superior Cerebellar Peduncle
sm Stria Medullaris
so Optic Layer of Superior Colliculus
st Stria Terminalis
swd Deep White Layer of Superior Colliculus
swi Intermediate White Layer of Superior Colliculus
tb Trapezoid Body
Vm Medial Vestibular Nucleus
Vs Superior Vestibular Nucleus
zo Zonal Lamina of Superior Colliculus

ac: Anterior Commissure. This large myelinated commissure, rostral to the thalamus, contains a caudal component extending laterally toward the amygdala and a rostral component that can be traced forward to the olfactory bulb, serving to connect the bulb of both sides (Plate 86). This latter component is sometimes called the *medial olfactory tract*, based on a convention derived from gross dissection of the human brain in which an intermediate tract ending in the olfactory tubercle is also recognized. The posterior portion of the ac contributes to the posterior limb connecting the "temporal" region, largely the amygdala of both sides, and is distinct because it contains less myelin.

mlf: Medial Longitudinal Fascicle. Ventral parts of this fiber system blend indistinguishably with fibers in adjacent parts of the tegmentum or reticular formation. Rostrally, the mlf forms the ventromedial border of the midbrain central gray and contains primarily ascending projections from the vestibular nuclei to the abducens, trochlear, and oculomotor nuclei, as well as fibers interrelating these three motor nuclei controlling the eyes. Caudally, fibers from many parts of the brain stem descend to the spinal cord in and around the mlf. As is particularly evident in Plate 105, neurons are found scattered among fibers along the length of the mlf. The history of names applied to groups of these cells is very complex, with substantial discord. For example, in the late 19th century what we now call the interstitial nucleus of Cajal was often referred to as the *interstitial nucleus of the medial longitudinal fascicle.*

sm: Stria Medullaris. This fiber tract converges from the posterior division of the septal region and from the preoptic region to form a discrete tract that extends along the dorsomedial surface of the thalamus to end in the habenula. Some fibers continue with the fasciculus retroflexus to the interpeduncular nucleus, and others cross in the habenular commissure. It should not be confused with the *stria medullaris* of human anatomy, which comes obliquely along the floor of the medullary fourth ventricle.

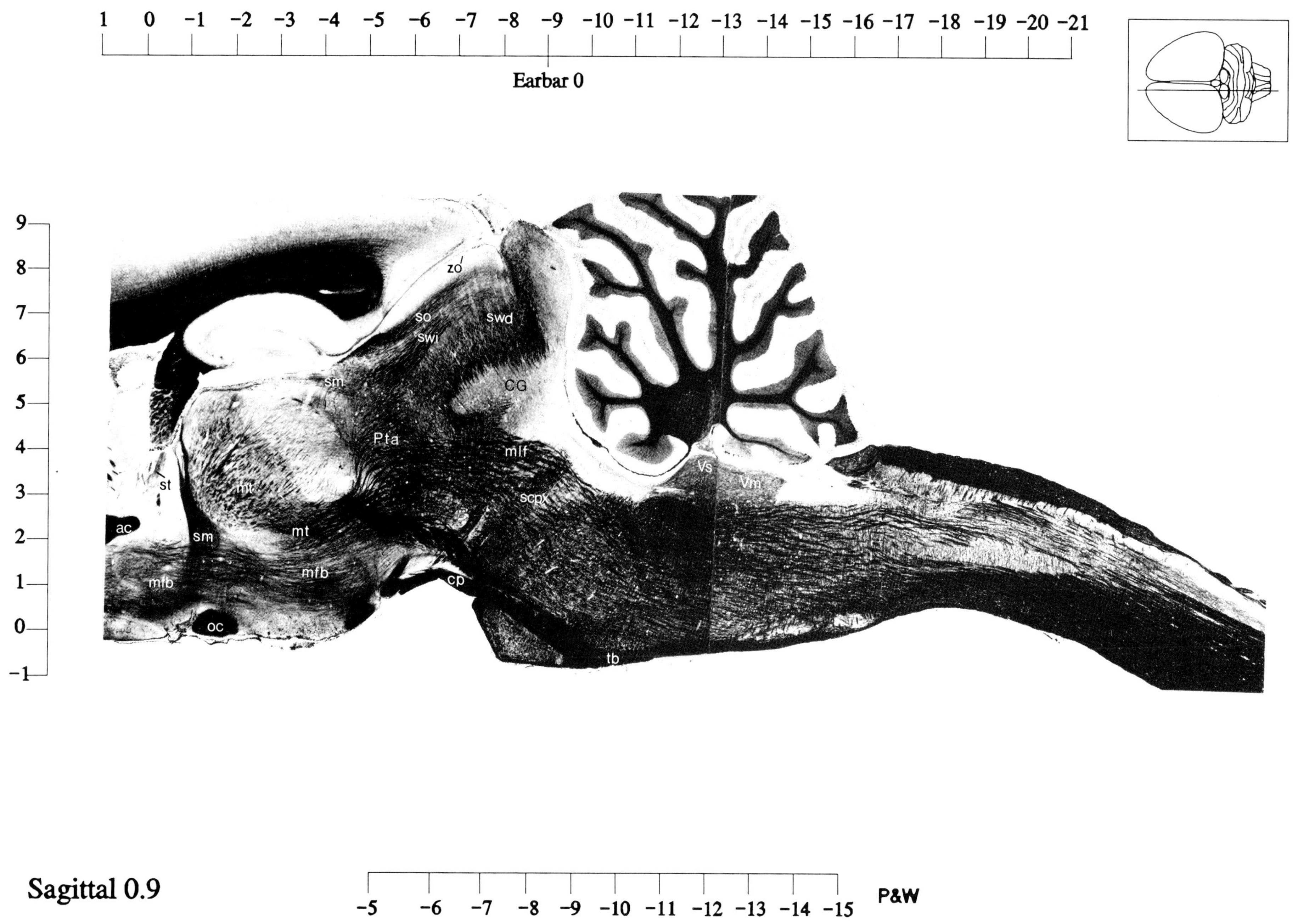

Earbar 0

1 0 -1 -2 -3 -4 -5 -6 -7 -8 -9 -10 -11 -12 -13 -14 -15 -16 -17 -18 -19 -20 -21

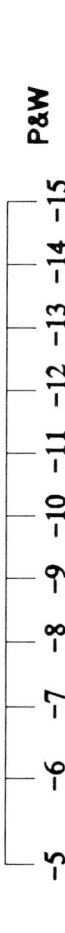

zo
SWd
CG
SO
SWi
Vm
sm
Pta
mlf
Vis
scp
tb
mt
mfb
cp
mt
sm
oc
st
mfb
ac

9 8 7 6 5 4 3 2 1 0 -1

Sagittal 0.9

-5 -6 -7 -8 -9 -10 -11 -12 -13 -14 -15

P&W

Plate 106

AD Anterodorsal Nucleus (Thalamus)
AM Anteromedial Nucleus (Thalamus)
ap Area Postrema
AV Anteroventral Nucleus (Thalamus)
Bst Bed Nucleus of Stria Terminalis
CdP Caudoputamen
CG Central Gray
CM Centromedial Nucleus (Thalamus)
cp Cerebral Peduncle
CU Cuneate Nucleus
DB Nucleus of the Diagonal Band
DG Dentate Gyrus
dMR Deep Mesencephalic Reticular Nucleus
Fa Fastigial (Medial) Cerebellar Nucleus
fr Fasciculus Retroflexus
GcR Gigantocellular Reticular Nucleus
Hb₁ Habenula Nucleus (lateral) (Thalamus)
IC Inferior Colliculus
Inc Interstitial Nucleus of Cajal
IO Inferior Olive
LDT Laterodorsal Tegmental Nucleus
LHA Lateral Hypothalamic Area
LPO Lateral Preoptic Area (Hypothalamus)
McR Magnocellular Reticular Nucleus (Medulla)
MD Mediodorsal Nucleus (Thalamus)
ml Medial Lemniscus
ms5 Mesencephalic Nucleus of the Trigeminal
ot Optic Tract
pc Posterior Commissure
Pd Pontine Gray (Deep)
Pf Parafascicular Nucleus (Thalamus)
Ph Perihypoglossal Nucleus
Pi Pineal Gland (Epiphysis)
PMᵥ Premammillary Nucleus (ventral) (Hypothalamus)

PmR Paramedian Reticular Nucleus
PpT Pedunculopontine Tegmental Nucleus
PRc Pontine Reticular Nucleus (caudal)
PRo Pontine Reticular Nucleus (oral)
PRᵥ Pontine Reticular Nucleus (ventral)
R Reticular Nucleus (Thalamus)
Rdl Dorsal Raphé Nucleus (lateral extension)
Rum Red Nucleus (magnocellular)
Rup Red Nucleus (parvicellular)
s5c Spinal Trigeminal Nucleus (caudal)
SC Superior Colliculus
sgd Deep Gray Layer of Superior Colliculus
Sge Supragenual Nucleus
sgi Intermediate Gray Layer of Superior Colliculus
sgs Superficial Gray Layer of Superior Colliculus
SI Substantia Innominata
sm Stria Medullaris
SNc Substantia Nigra (compact)
SNr Substantia Nigra (reticular)
So Supraoptic Nucleus (Hypothalamus)
soc Supraoptic Commissure
SPl Septal Nucleus (lateral)
Sub Subiculum
TB Nucleus of the Trapezoid Body
TM Tuberomammillary Nucleus (Hypothalamus)
TSₘ Nucleus of the Solitary Tract (medial)
TU Tuberal Nucleus (Hypothalamus)
Tu Olfactory Tubercle
vhc Ventral Hippocampal Commissure
VM Ventromedial Nucleus (Thalamus)
Vm Medial Vestibular Nucleus
VR Ventral Reticular Nucleus
Vs Superior Vestibular Nucleus
ZI Zona Incerta
7g Genu of Facial Nerve
10 Dorsal Motor Nucleus of Vagus

Ru: Red Nucleus. The red nucleus (nucleus ruber) derives its name from the natural pigment with iron in its heme prosthetic group in the human brain, where it is conventionally divided into magnocellular and parvicellular divisions. The pigment is not evident in the rat brain, where subdivision based on cell size is less evident than in larger mammals, but the pattern is detectable with the parvicellular portion lying rostral, dorsal, and slightly more lateral in relation to the magnocellular division. The cytoarchitecture was described by Brown ('74) and is depicted in detail in all three axes in excellent accounts by Brown ('74) and Reid et al. ('75). A retrorubral nucleus behind the red nucleus is recognized in several descriptions and is easily seen in sagittal sections (see Plate 106), but we have not employed the term so as to avoid the implication that it is functionally related to the red nucleus.

TM: Tuberomammillary Nucleus (Hypothalamus). This term derives from the retention of the gross anatomical term *tuber cinereum* in the Nomina Anatomica referred to as the *gray (ashen) swelling*, located rostral to the mammillary bodies and behind the optic chiasm forming the floor of the hypothalamic third ventricle. In the human brain the lateral tuberal and tuberomammillary nuclei are embedded in a more distinct protuberance, and the homology in rat is arguable (see Nauta and Haymaker, '69). Malone ('10, '14) identified this heterogeneous, relatively dispersed group of large neurons in the human, monkey, lemur and cat hypothalamus, and it was subsequently found in an even wider range of mammals (see Diepen, '62). With the advent of immunohistochemical methods for localizing histidine decarboxylase, current usage in the rat typically refers to histaminergic neurons in the caudal half of the traditional TM (forming a cradle along the base of the brain for the ventral half of the mammillary body) as the *tuberomammillary nucleus* (see Köhler et al., '85), while the rostral half of the traditional TM (in the far lateral hypothalamic area adjacent to the cerebral peduncle) has recently been called the *lateral tuberal nucleus* (Bleier et al., '79) or *magnocellular nucleus of the lateral hypothalamus* (Paxinos and Watson, '86) (see TU: Tuberal Nucleus at Plate 107). The large neurons along the base of the brain at one time were mistakenly thought to form a caudal part of the magnocellular neurosecretory system (a caudal extension of the retrochiasmatic part of the supraoptic nucleus).

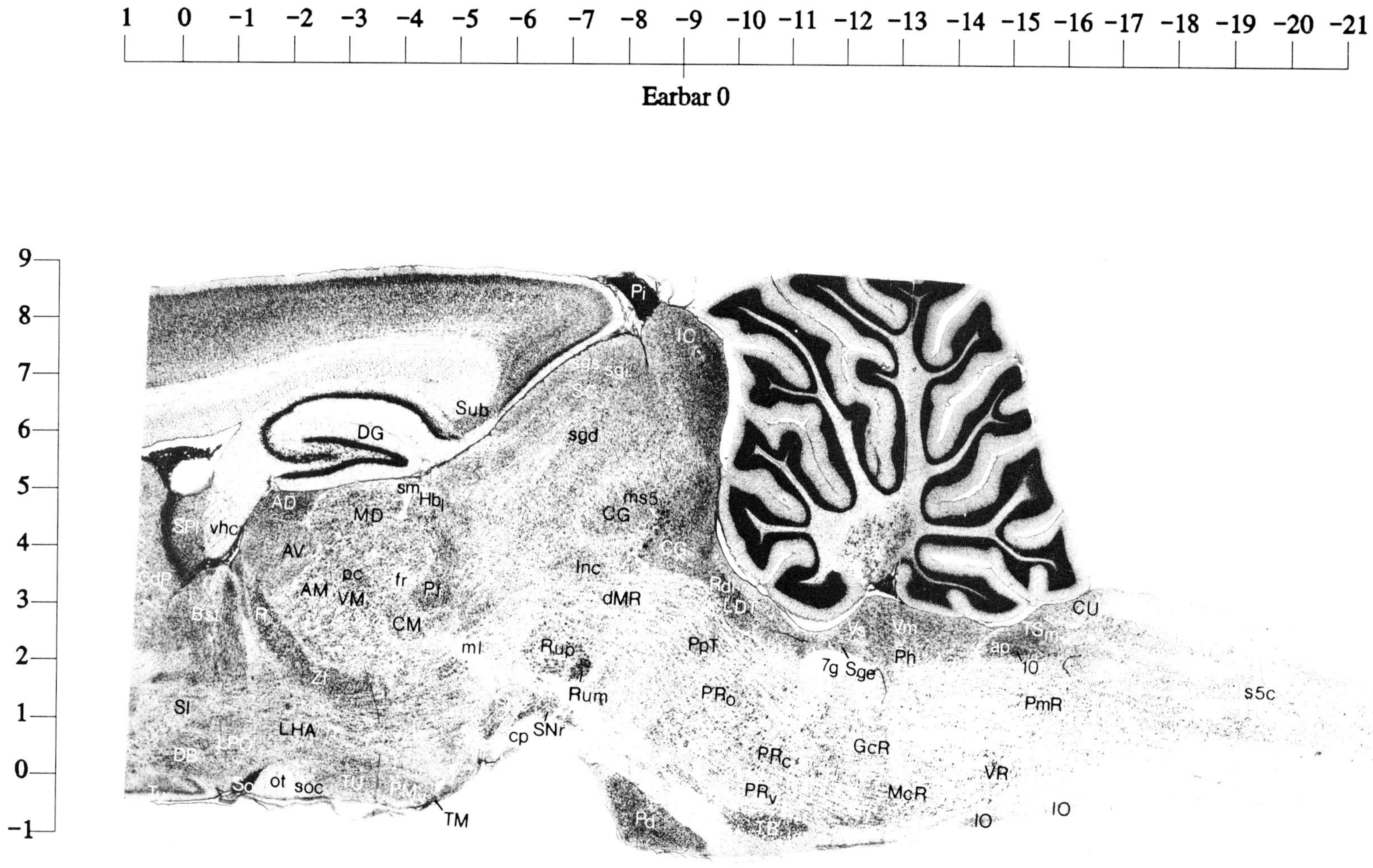

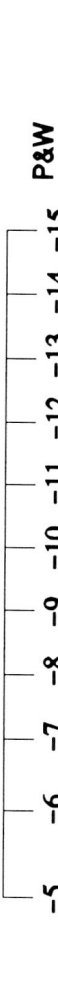

1 0 −1 −2 −3 −4 −5 −6 −7 −8 −9 −10 −11 −12 −13 −14 −15 −16 −17 −18 −19 −20 −21
Earbar 0
9 8 7 6 5 4 3 2 1 0 −1
Pi
IC
Sub
sgd
DG
sm
Hbl
AD
MD
ms5
CG
SPi
vhc
AV
Inc
pc
AM
VM
fr
Pf
dMR
CM
ml
Rub
PpT
7g
Sge
Ph
CU
ap
10
Rum
PRo
PmR
s5c
SI
LHA
cp
SNr
PRc
GcR
VR
ot
soc
TM
PRv
McR
IO
IO
−5 −6 −7 −8 −9 −10 −11 −12 −13 −14 −15
P&W

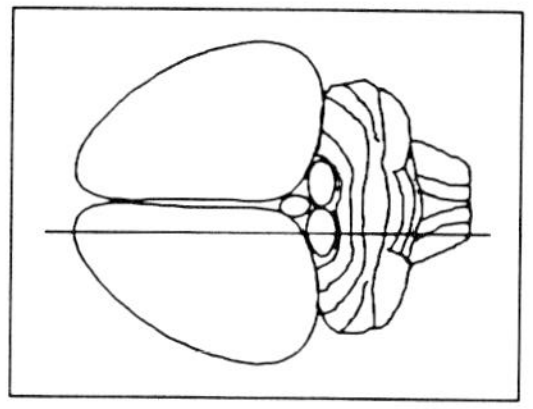

Sagittal 1.2

Plate 107

AD Anterodorsal Nucleus (Thalamus)
AM Anteromedial Nucleus (Thalamus)
AV Anteroventral Nucleus (Thalamus)
Bst Bed Nucleus of Stria Terminalis
CG Central Gray
cl Central Lateral Nucleus (Thalamus)
CU Cuneate Nucleus
DB Nucleus of the Diagonal Band
DG Dentate Gyrus
dMR Deep Mesencephalic Reticular Nucleus
Fa Fastigial (Medial) Cerebellar Nucleus
Ge Nucleus Gelatinosus (Nucleus Submedius) (Thalamus)
GR Gracile Nucleus
Hb₁ Habenula Nucleus (lateral) (Thalamus)
IC Inferior Colliculus
IO Inferior Olive
LC Locus Cerulus
LD Lateral Dorsal Nucleus (Thalamus)
LHA Lateral Hypothalamic Area
LPO Lateral Preoptic Area (Hypothalamus)
MD Mediodorsal Nucleus (Thalamus)
ml Medial Lemniscus
mp Mammillary Peduncle
ms5 Mesencephalic Nucleus of the Trigeminal
Olep External Periolivary Nucleus
ot Optic Tract
PcR Parvicellular Reticular Nucleus
Pd Pontine Gray (Deep)

Pf Parafascicular Nucleus (Thalamus)
PmR Paramedian Reticular Nucleus
PpT Pedunculopontine Tegmental Nucleus
PR_c Pontine Reticular Nucleus (caudal)
PR_o Pontine Reticular Nucleus (oral)
Pta Anterior Pretectal Nucleus (Thalamus)
Ptp Posterior Pretectal Nucleus (Thalamus)
R Reticular Nucleus (Thalamus)
Rup Red Nucleus (parvicellular)
s5c Spinal Trigeminal Nucleus (caudal)
SC Superior Colliculus
sgd Deep Gray Layer of Superior Colliculus
sgi Intermediate Gray Layer of Superior Colliculus
sgs Superficial Gray Layer of Superior Colliculus
SI Substantia Innominata
SNc Substantia Nigra (compact)
SNr Substantia Nigra (reticular)
So Supraoptic Nucleus (Hypothalamus)
sPf Subparafascicular Nucleus (Thalamus)
ST Subthalamic Nucleus
Sub Subiculum
TB Nucleus of the Trapezoid Body
TS_m Nucleus of Solitary Tract (medial)
TU Tuberal Nucleus (Hypothalamus)
Vl Lateral Vestibular Nucleus
VM Ventromedial Nucleus (Thalamus)
Vm Medial Vestibular Nucleus
Vs Superior Vestibular Nucleus
ZI Zona Incerta
7 Facial Nucleus
7g Genu of Facial Nerve

PpT: Pedunculopontine Tegmental Nucleus. This relatively poorly defined region in Nissl preparations was identified in the human brain by Jacobsohn ('09) and Olszewski and Baxter ('54), and appears to be outlined rather clearly by a subpopulation of cholinergic neurons in the rat (Rye et al., '87). It is associated topographically with lateral (nondecussating) parts of the superior cerebellar peduncle.

TU: Tuberal Nucleus (Hypothalamus) (also called Lateral Tuberal Nucleus). This relatively homogeneous group of small to medium-sized neurons in the rat was identified by Malone ('10) in the human brain and then found in a wide range of mammals. In many species it forms clear medial and lateral condensations (see Clark, '38; Diepen, '62; Nauta and Haymaker, '69), although this feature is less obvious in the rat, where it is characterized precisely by estrogen receptor-expressing neurons (Simerly et al., '90). Recent events in the naming of this cell group present a nice example of why it is occasionally so difficult to compare results in the neuroanatomical literature. The original name reflects a location along the base of the tuber cinereum, and later Malone ('14) also referred to it as the *lateral tuberal nucleus* because it in fact lies along the base of the lateral tuberal eminence on either side of the median (tuberal) eminence. Unfortunately, Bleier et al. ('79) renamed this cell group the *medial tuberal nucleus* and called the rostral end of the traditional tuberomammillary nucleus, the *lateral tuberal nucleus* – designations adopted by Geeraedts et al. ('90). Paxinos and Watson ('86) also referred to the TU as the *medial tuberal nucleus*, although they separated out the lateral fifth or so as a *terete hypothalamic nucleus* and did not recognize the rostral half of the TU; they also referred to the rostral tuberomammillary nucleus as the *magnocellular nucleus of the lateral hypothalamus* (see TM: Tuberomammillary Nucleus at Plate 106).

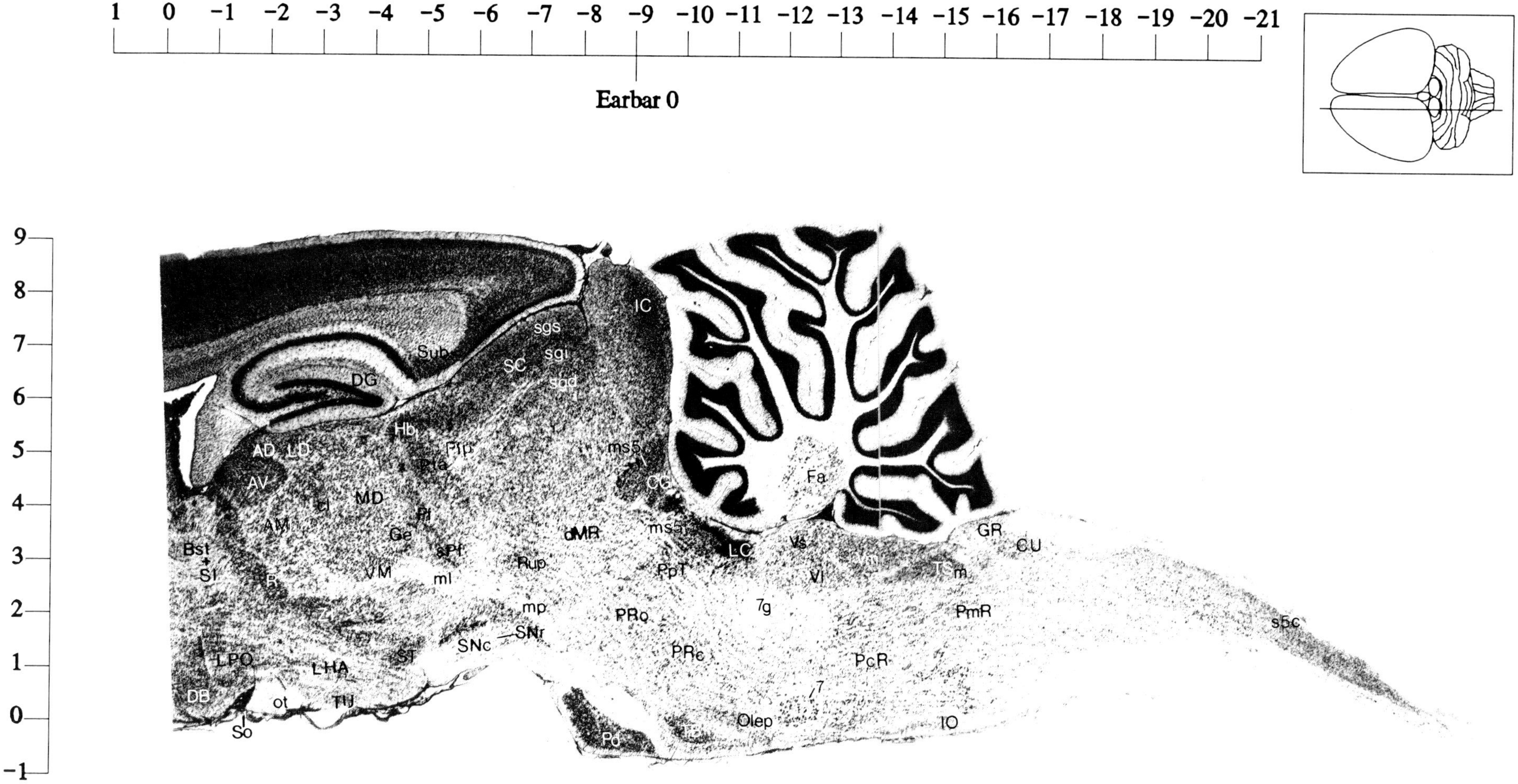

Earbar 0

1 0 -1 -2 -3 -4 -5 -6 -7 -8 -9 -10 -11 -12 -13 -14 -15 -16 -17 -18 -19 -20 -21

9 8 7 6 5 4 3 2 1 0 -1

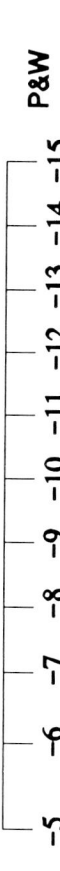

GR CU
Vm
PmR
sp5c

Fa
Vs
VI
IC
7g
mss
ms5
Pp5
PCR
7
IO
Olep
PRo
PRe
P5

IC
sgs
sgi
sgd
dMR
Rllp
mp
SNr
PRo
Po

Sub
PtP
Pta
Rl
SPt
SNc

HD
Ge
ml
ST

DG
MD
VM
LHA
FII

AD LD
AM
Rt
LPO
ot
So

Bst
SI
DB

Sagittal 1.5

P&W

-5 -6 -7 -8 -9 -10 -11 -12 -13 -14 -15

Plate 108

ac Anterior Commissure
CdP Caudoputamen
cp Cerebral Peduncle
dc Dorsal Column
fr Fasciculus Retroflexus
GP Globus Pallidus (external or lateral)
Hb₁ Habenula Nucleus (lateral) (Thalamus)
mcp Middle Cerebellar Peduncle
mfb Medial Forebrain Bundle
ml Medial Lemniscus
mp Mammillary Peduncle
mt Mammillothalamic Tract
ot Optic Tract
s5c Spinal Trigeminal Nucleus (caudal)
s5i Spinal Trigeminal Nucleus (interpolar)
scp Superior Cerebellar Peduncle
sm Stria Medullaris
so Optic Layer of Superior Colliculus
soc Supraoptic Commissure
swd Deep White Layer of Superior Colliculus
swi Intermediate White Layer of Superior Colliculus
tb Trapezoid Body
ts Solitary Tract
VB Ventrobasal Nuclear Complex (Thalamus)
zo Zonal Lamina of Superior Colliculus
3n Oculomotor Nerve

mfb: Medial Forebrain Bundle. This term is derived from comparative neuroanatomy and was contrasted to the lateral forebrain bundle or internal capsule–cerebral peduncle fiber system. In silver-stained frontal sections through the diencephalon it is clear that the medial forebrain bundle is a medial extension of what Cajal thought of as the cerebral peduncle, although the former is much more condensed than the latter, which has an extensive interstitial or bed nucleus (the lateral hypothalamic and lateral preoptic areas). The medial forebrain bundle is by far the most complex fiber system in the brain with numerous ascending and descending components (see Nieuwenhuys et al., '82; Swanson, '87), and is not a single bundle in the strict sense. Most of these components converge in the lateral zone of the hypothalamus where the medial forebrain bundle can be recognized in fiber stains. However, its fibers diverge rostrally to enter all parts of the telencephalon, and extend caudally as far as the tip of the filum terminale. A particularly massive component of the medial forebrain bundle can be traced rostrally into parts of the substantia innominata just below the olfactory tubercle as individual fascicles.

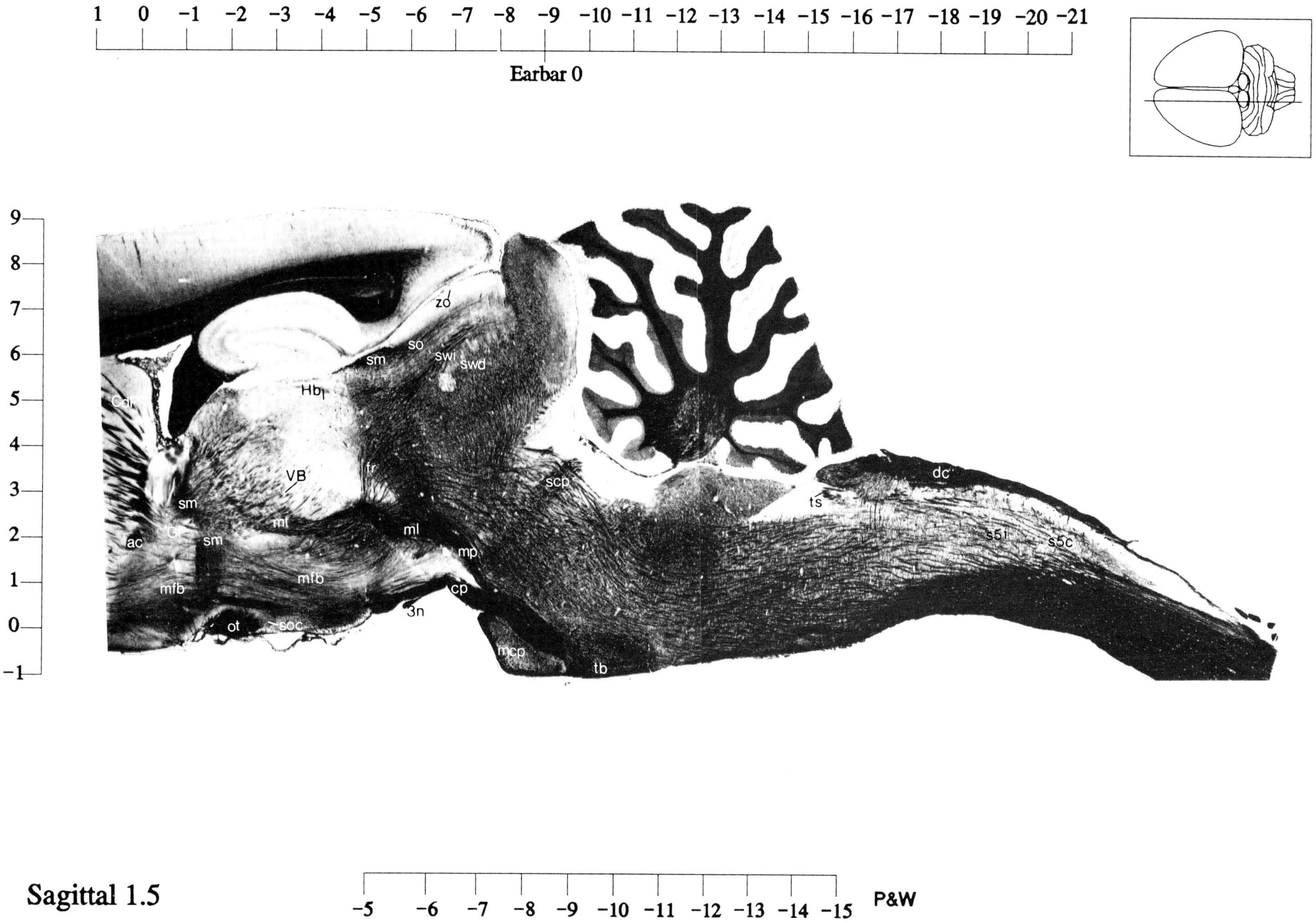

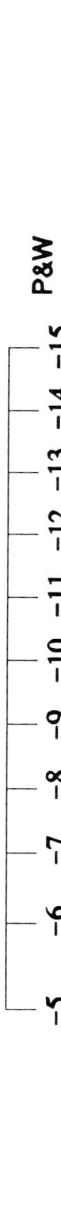

Earbar 0

1 0 −1 −2 −3 −4 −5 −6 −7 −8 −9 −10 −11 −12 −13 −14 −15 −16 −17 −18 −19 −20 −21

9 8 7 6 5 4 3 2 1 0 −1

Sagittal 1.5

P&W

−5 −6 −7 −8 −9 −10 −11 −12 −13 −14 −15

Plate 109

Ast Area Striata (Cortex)
AV Anteroventral Nucleus (Thalamus)
CA$_1$ Hippocampal Area CA$_1$ (Ammon's Horn)
CA$_3$ Hippocampal Area CA$_3$ (Ammon's Horn)
CdP Caudoputamen
CU Cuneate Nucleus
Cun Cuneiform Nucleus
DB$_h$ Nucleus of the Diagonal Band (horizontal)
DG Dentate Gyrus
dMR Deep Mesencephalic Reticular Nucleus
Fa Fastigial (Medial) Cerebellar Nucleus
GP Globus Pallidus (external part)
IC Inferior Colliculus
Int Interpositus (Intermediate or Interposed) Cerebellar Nucleus
LD Lateral Dorsal Nucleus (Thalamus)
LHA Lateral Hypothalamic Area
LP Lateral Posterior Nucleus (Thalamus)
LR Lateral Reticular Nucleus
ms5 Mesencephalic Nucleus of the Trigeminal
Olep External Periolivary Nucleus
Olsp Superior Paraolivary Nucleus
ot Optic Tract
PcR Parvicellular Reticular Nucleus
Pd Pontine Gray (Deep)
Po Posterior Group (Thalamus)
PpT Pedunculopontine Tegmental Nucleus
PR$_c$ Pontine Reticular Nucleus (caudal)

PR$_o$ Pontine Reticular Nucleus (oral)
Pta Anterior Pretectal Nucleus (Thalamus)
Ptp Posterior Pretectal Nucleus (Thalamus)
R Reticular Nucleus (Thalamus)
s5c Spinal Trigeminal Nucleus (caudal)
s5i Spinal Trigeminal Nucleus (interpolar)
SC Superior Colliculus
sgd Deep Gray Layer of Superior Colliculus
sgi Intermediate Gray Layer of Superior Colliculus
sgs Superficial Gray Layer of Superior Colliculus
SI Substantia Innominata
SNc Substantia Nigra (compact)
SNr Substantia Nigra (reticular)
So Supraoptic Nucleus (Hypothalamus)
SomS Somatosensory Cortex
sPf Subparafascicular Nucleus (Thalamus)
ST Subthalamic Nucleus
STa Subthalamic Nucleus (accessory part)
Sub Subiculum
TS Nucleus of Solitary Tract
ts Solitary Tract
Tu Olfactory Tubercle
VB Ventrobasal Nuclear Complex (Thalamus)
VL Ventrolateral Nucleus (Thalamus)
Vl Lateral Vestibular Nucleus
VM Ventromedial Nucleus (Thalamus)
Vm Medial Vestibular Nucleus
Vs Superior Vestibular Nucleus
X Nucleus X
Y Nucleus Y
ZI Zona Incerta
7 Facial Nucleus

Pta: Anterior Pretectal Nucleus (Thalamus) (also called Nucleus Pretectalis Anterior). The principal and most distinct component of the pretectal group, extending into the rostral midbrain. Its limits are clearly evident in its distinctive fiber architecture in all three planes consequent to its encapsulation, except perhaps at its medial border. Although easily recognized, it has been called the *nucleus praebigeminalis* (Cajal, '11), the *nucleus praetectalis* (Le Gros Clark, '32), the *nucleus praebigeminalis lateralis* (M. Rose, '35), the *nucleus praetectalis and nucleus posterior* (Kuhlenbeck and Miller, '42) and the *nucleus praetectalis anterior* (Rose and Woolsey, '43). It is illustrated in three planes by Siminoff et al. ('67), who recognize a dorsal, compact division and a ventral, loose portion that have been called the *nuclei principalis* and *profundus*, respectively (Bucher and Nauta, '54; Hayhow et al., '60). The descriptions by Scalia ('72) and Giolli et al. ('89) are useful in understanding the complexities of this region.

Ptp: Posterior Pretectal Nucleus (Thalamus). This is often considered part of the mesencephalon because it fuses with the deepest layer of the superior colliculus, but it is considered thalamic by others on ontogenetic grounds due to its apparent derivation from the epithalamic plate (J. E. Rose, '42a). It probably constitutes much of the *nucleus praebigeminalis medialis*, identified in rabbit by M. Rose ('35), and the *nucleus lentiformis mesencephali parvocellularis and sublentiformis* of Kuhlenbeck and Miller ('42). Much of the difficulty derives from attempting delimitation in transverse sections but it is more easily recognized in horizontal sections (see Rose and Woolsey, '43, in rabbit and Siminoff et al., '67, in rat). There may be difficulty in reconciling our delimitation with other accounts (e.g., Scalia, '72), which suggests that our usage should be applied with caution.

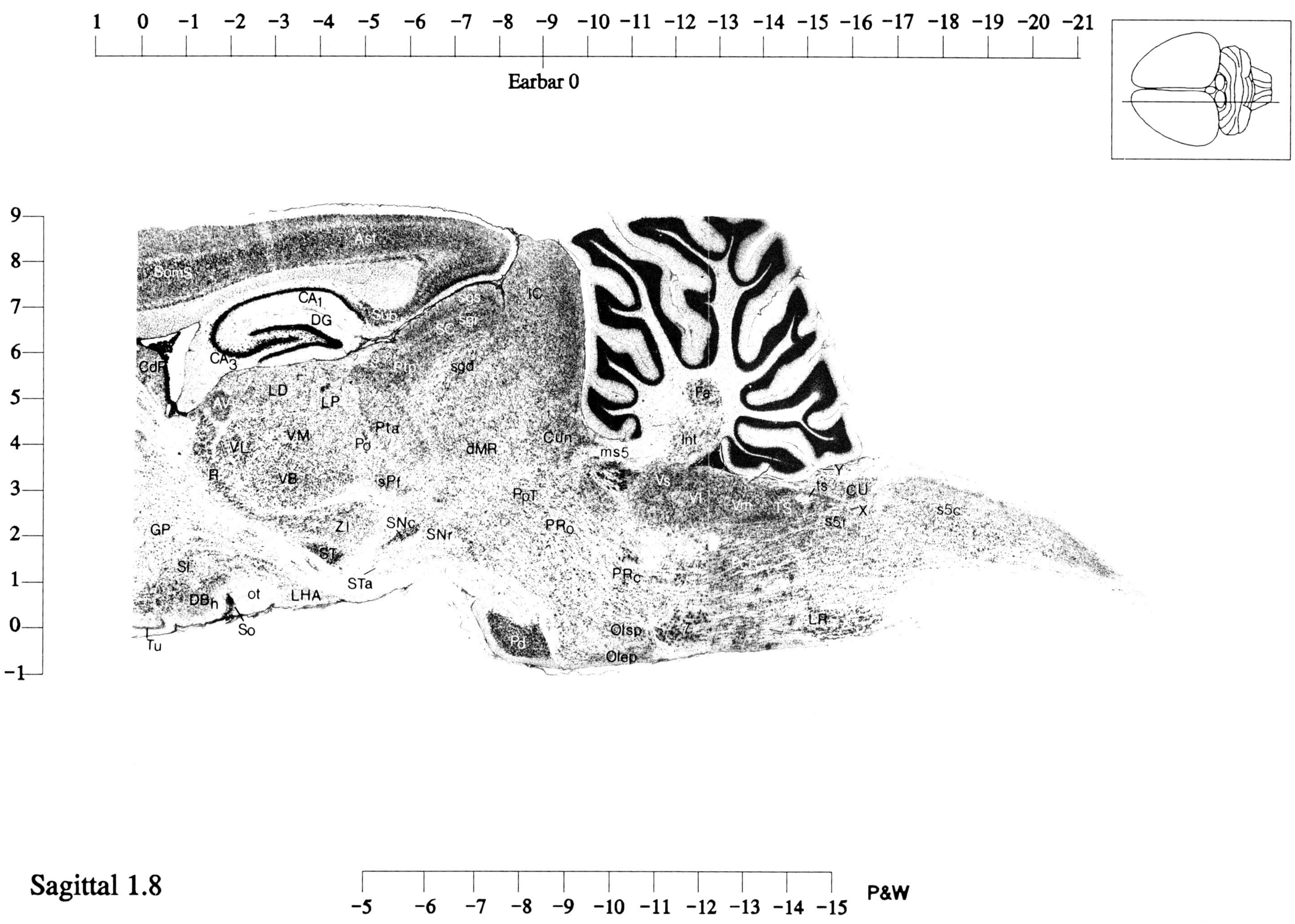

Earbar 0

9 8 7 6 5 4 3 2 1 0 -1

1 0 -1 -2 -3 -4 -5 -6 -7 -8 -9 -10 -11 -12 -13 -14 -15 -16 -17 -18 -19 -20 -21

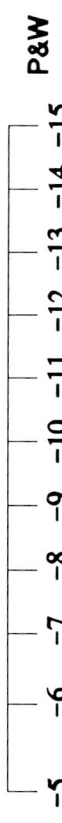

Sagittal 1.8

P&W

-5 -6 -7 -8 -9 -10 -11 -12 -13 -14 -15

Plate 110

ac Anterior Commissure
alv Alveus
CdP Caudoputamen
cp Cerebral Peduncle
db Diagonal Band of Broca
GP Globus Pallidus (external or lateral)
IC Inferior Colliculus
ic Internal Capsule
Int Interpositus (Intermediate or Interposed) Cerebellar Nucleus
LD Lateral Dorsal Nucleus (Thalamus)
LP Lateral Posterior Nucleus (Thalamus)
LV Lateral Ventricle
mfb Medial Forebrain Bundle
ml Medial Lemniscus
ot Optic Tract
PB$_l$ Parabrachial Nucleus (lateral)
PB$_m$ Parabrachial Nucleus (medial)
Po(m) Posterior Group (medial) (Thalamus)
s5c Spinal Trigeminal Nucleus (caudal)
s5i Spinal Trigeminal Nucleus (interpolar)
SC Superior Colliculus
scp Superior Cerebellar Peduncle
so Optic Layer of Superior Colliculus
st Stria Terminalis
swd Deep White Layer of Superior Colliculus
swi Intermediate White Layer of Superior Colliculus
tb Trapezoid Body
ts Solitary Tract
VB Ventrobasal Nuclear Complex (Thalamus)
vhc Ventral Hippocampal Commissure
zo Zonal Lamina of Superior Colliculus
3n Oculomotor Nerve
5t Trigeminal Tract
7n Facial Nerve

alv: Alveus. This fiber tract is the deep white matter of Ammon's horn and the subiculum, and thus is analogous to the external capsule of the isocortex. It condenses into the fimbria of the fornix.

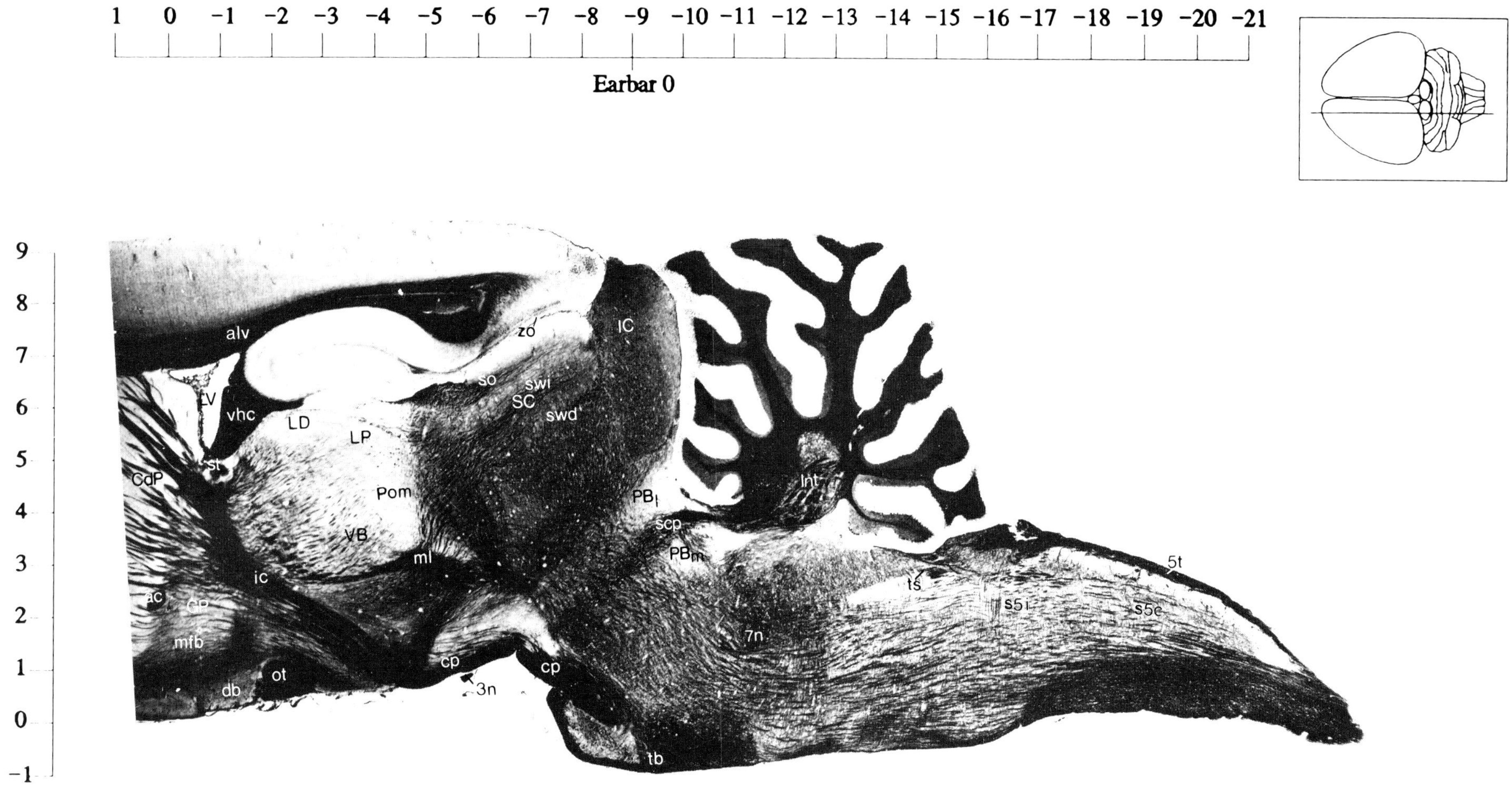

Earbar 0

1 0 -1 -2 -3 -4 -5 -6 -7 -8 -9 -10 -11 -12 -13 -14 -15 -16 -17 -18 -19 -20 -21

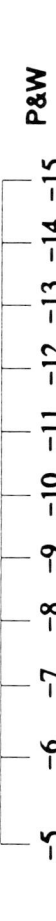

9
8
7
6
5
4
3
2
1
0
-1

alv
fV
vhc
LD
st
zo'
so
SC
swl
swd'
Pom
LP
VB
IC
iC
cp
ml
ic
GP
iC
mfb
cp
3n
ot
db
PBl
scp
PBm
lht
7n
tb
ts
s5i
5t
cs5

Sagittal 1.8   P&W

-5 -6 -7 -8 -9 -10 -11 -12 -13 -14 -15

Plate 111

AAA Anterior Amygdaloid Area
Ast Area Striata (Cortex)
Aud Auditory Area (Cortex)
CA₁ Hippocampal Area CA₁ (Ammon's Horn)
CdP Caudoputamen
cl Central Lateral Nucleus (Thalamus)
CO Cochlear Nucleus
CUe External Cuneate Nucleus
Cun Cuneiform Nucleus
DG Dentate Gyrus
dMR Deep Mesencephalic Reticular Nucleus
ET Entopeduncular Nucleus
GP Globus Pallidus (external or lateral)
IC Inferior Colliculus
Int Interpositus (Intermediate or Interposed) Cerebellar Nucleus
LD Lateral Dorsal Nucleus (Thalamus)
Lot Nucleus of Lateral Olfactory Tract
LP Lateral Posterior Nucleus (Thalamus)
LR Lateral Reticular Nucleus
m5 Motor Trigeminal Nucleus
MeA Medial Nucleus of the Amygdala
Olep External Periolivary Nucleus
Olsp Superior Paraolivary Nucleus
Ot Nucleus of Optic Tract
PB$_l$ Parabrachial Nucleus (lateral)
PB$_m$ Parabrachial Nucleus (medial)
PcR Parvicellular Reticular Nucleus

Pd Pontine Gray (Deep)
Po Posterior Group (Thalamus)
PpT Pedunculopontine Tegmental Nucleus
Pta Anterior Pretectal Nucleus (Thalamus)
Ptp Posterior Pretectal Nucleus (Thalamus)
R Reticular Nucleus (Thalamus)
S5 Supratrigeminal Nucleus
s5c Spinal Trigeminal Nucleus (caudal)
s5i Spinal Trigeminal Nucleus (interpolar)
SC Superior Colliculus
sgi Intermediate Gray Layer of Superior Colliculus
sgs Superficial Gray Layer of Superior Colliculus
SI Substantia Innominata
SNc Substantia Nigra (compact)
SNr Substantia Nigra (reticular)
SO$_l$ Superior Olivary Complex (lateral)
sPf Subparafascicular Nucleus (Thalamus)
ST Subthalamic Nucleus
Sub Subiculum
Tu Olfactory Tubercle
VB$_l$ Ventrobasal Nuclear Complex (lateral) (Thalamus)
VB$_m$ Ventrobasal Nuclear Complex (medial) (Thalamus)
vhc Ventral Hippocampal Commissure
Vl Lateral Vestibular Nucleus
Vs Superior Vestibular Nucleus
Vsp Spinal Vestibular Nucleus
ZI Zona Incerta
7 Facial Nucleus
7n Facial Nerve

AAA: Anterior Amygdaloid Area. The rostral pole of the amygdala lateral to the magnocellular preoptic nucleus is generally given this designation, following the descriptions by Gurdjian ('27 and '28, and Brodal, '47). There are substantial differences in the way it is outlined in various atlases and descriptions, but the surrounding nuclei are, for the most part, easily defined. Paxinos and Watson ('86) avoid outlines but divide this area into dorsal, ventral, and shell divisions, apparently on topographic grounds.

GP and ET: Globus Pallidus and Entopeduncular Nucleus. In primates the globus pallidus is divided into external or lateral and internal or medial segments or parts, whereas in rodents the presumably equivalent parts are commonly referred to as the globus pallidus (lateral) and entopeduncular nucleus (medial), although some authors have applied the more general terminology to rodents (e.g., Swanson, '92). In rodents there is no clear medial (internal) medullary lamina separating the two parts, and in the absence of detailed cytoarchitectonic and connectional work, the entopeduncular nucleus is typically regarded as lying within the cerebral peduncle and internal capsule. In the rat, certain parts of the globus pallidus and entopeduncular nucleus are invaded (and/or bordered) by ascending cholinergic neurons of the substantia innominata (see Gritti et al., '93).

PB: Parabrachial Nuclei. This nuclear complex, pierced longitudinally by the superior cerebellar peduncle (brachium conjunctivum), is strikingly heterogeneous in architecture and functional connections. Description in relation to its location surrounding the brachium has been traditional, with main dorsolateral and ventromedial wings usually called lateral and medial or dorsal and ventral, respectively, and a separate large-celled ventrolateral nucleus generally known only by its eponym, the *Kölliker-Fuse nucleus*. These nuclei receive principally visceral, gustatory, and somatic inputs in largely segregated sectors (see Saper and Loewy, '80). A more complex subdivision in the rat is provided in a comprehensive review by Fulwiler and Saper ('84), but this remains somewhat controversial (see Kruger and Mantyh, '89, for review).

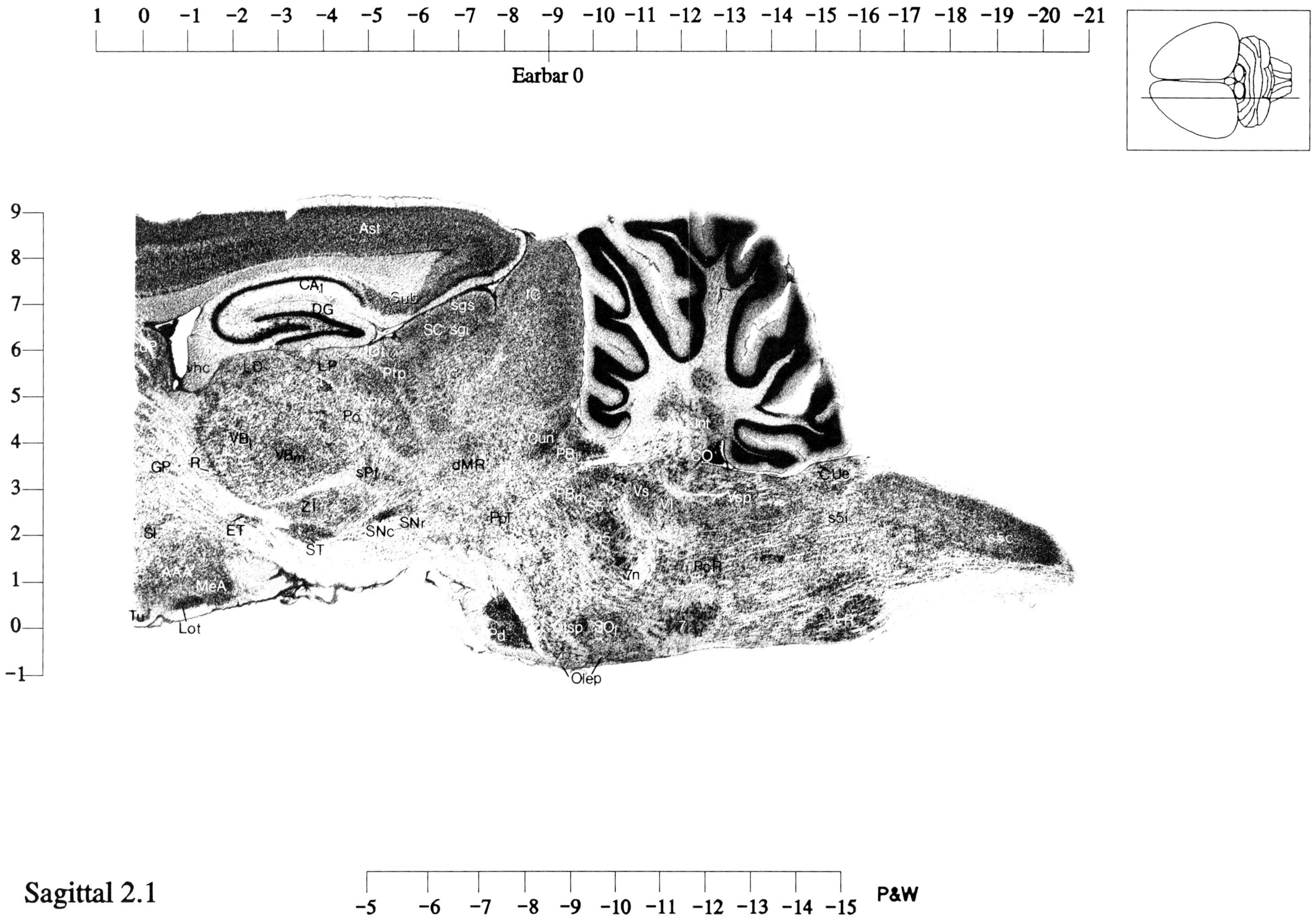

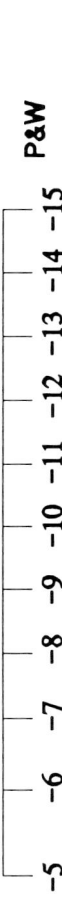

Earbar 0

1 0 -1 -2 -3 -4 -5 -6 -7 -8 -9 -10 -11 -12 -13 -14 -15 -16 -17 -18 -19 -20 -21

9 8 7 6 5 4 3 2 1 0 -1

CA1 DG Asl Sulg sgs SC sgi Pro Pg CG CM VBm sPM dMR Vs PBL Cun PnT SNr SNc ST ZI ET VB R GP St MeA Lot Tu AvX PnR Vn 7 sp SO Olep 5d s5 Cue O

Sagittal 2.1

-5 -6 -7 -8 -9 -10 -11 -12 -13 -14 -15

P&W

Plate 112

ac Anterior Commissure
CA₁ Hippocampal Area CA₁ (Ammon's Horn)
CA₂ Hippocampal Area CA₂ (Ammon's Horn)
CA₃ Hippocampal Area CA₃ (Ammon's Horn)
cc Corpus Callosum
CdP Caudoputamen
cp Cerebral Peduncle
DG Dentate Gyrus
dhc Dorsal Hippocampal Commissure
eml External Medullary Lamina
fi Fimbria
GP Globus Pallidus (external or lateral)
IC Inferior Colliculus
ic Internal Capsule
icp Inferior Cerebellar Peduncle
lot Lateral Olfactory Tract
mfb Medial Forebrain Bundle
ml Medial Lemniscus
ot Optic Tract

PB₁ Parabrachial Nucleus (lateral)
PBₘ Parabrachial Nucleus (medial)
Pir Piriform Area (Cortex)
Pta Anterior Pretectal Nucleus (Thalamus)
Ptp Posterior Pretectal Nucleus (Thalamus)
s5c Spinal Trigeminal Nucleus (caudal)
SC Superior Colliculus
scp Superior Cerebellar Peduncle
sm Stria Medullaris
SO₁ Superior Olivary Complex (lateral)
st Stria Terminalis
ts Solitary Tract
VB₁ Ventrobasal Nuclear Complex (lateral) (Thalamus)
VBₘ Ventrobasal Nuclear Complex (medial) (Thalamus)
vhc Ventral Hippocampal Commissure
Vs Superior Vestibular Nucleus
Vsp Spinal Vestibular Nucleus
ZI Zona Incerta
5t Trigeminal Tract
7 Facial Nucleus
7n Facial Nerve
8n Vestibular-Cochlear (Acoustic) Nerve

dhc: Dorsal Hippocampal Commissure. This fiber tract connects retrohippocampal cortical areas in the two hemispheres. Thus, it may be thought of as a part of the callosal system that comes to lie ventral to the splenium of the corpus callosum in the rat because of the complex infolding of the medial temporal cortex (see Blackstad, '56). The angular bundle contains retrohippocampal commissural fibers before they approach the midline, and thus constitutes the white matter or external capsule for the retrohippocampal areas.

VB: Ventrobasal Nuclear Complex (Thalamus) (also called Nucleus Ventralis Posterior or Ventroposterior Nucleus, usually subdivided into medial and lateral divisions). The nomenclature for this nucleus is discussed in some detail in the Introduction. Subdivision can be achieved employing fiber architecture as a guide, and in many species a fibrous lamina separates the trigeminal, cervical, and lumbar tactile projections, accounting for general acceptance of medial and lateral components as separate nuclei. In the rat, the tactile region is recognizable as a separate entity on a cytoarchitectonic basis, most of it devoted to trigeminal representation; this is demonstrated by axonal tracing studies and is illustrated in cell and fiber architecture in two planes by Feldman and Kruger ('80).

vhc: Ventral Hippocampal Commissure. It lies between the rostral end of the hippocampus and the caudal end of the septal region, and interconnects the dentate gyrus and Ammon's horn in the two hemispheres. Thus, it corresponds to the corpus callosum for the hippocampus.

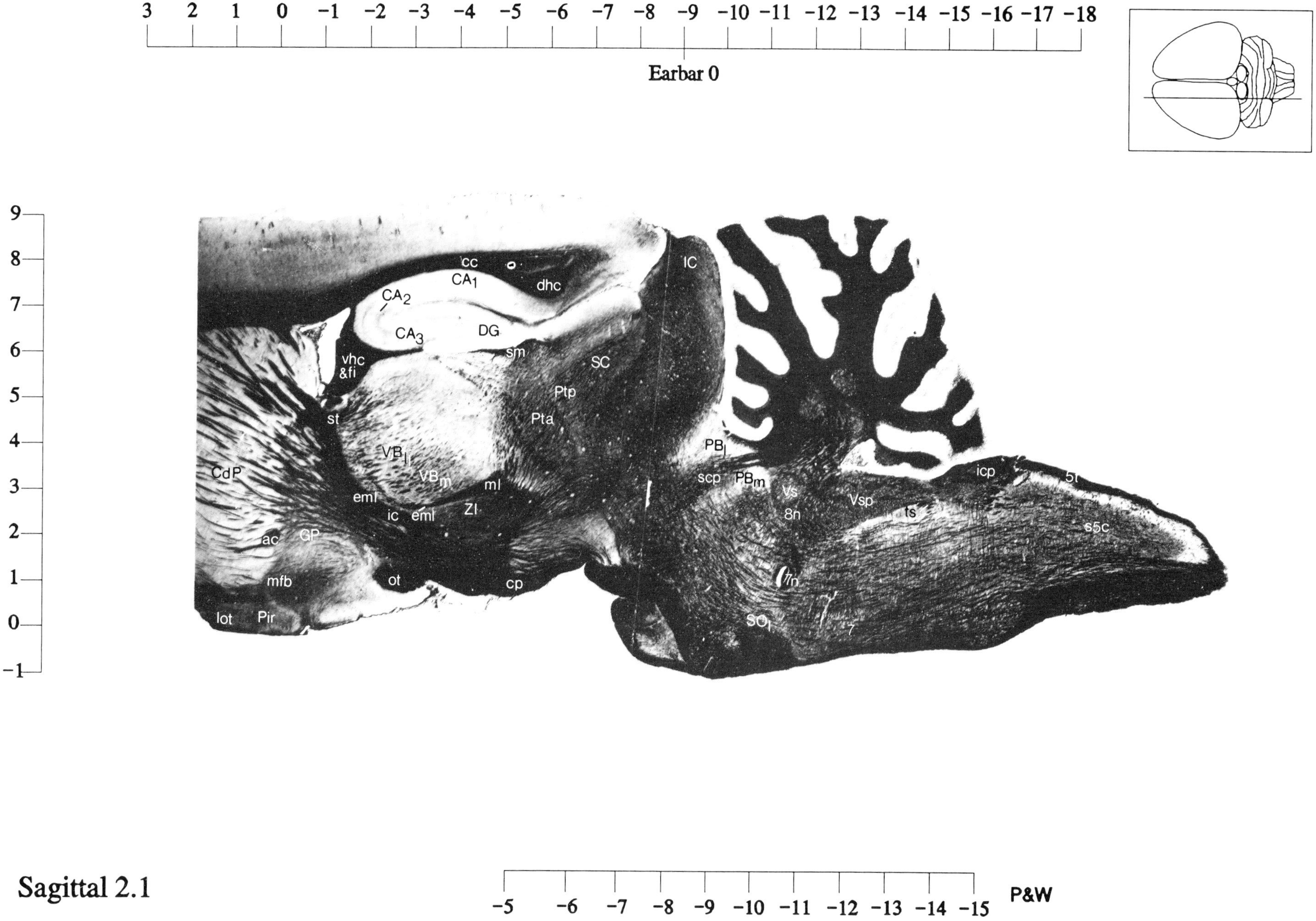

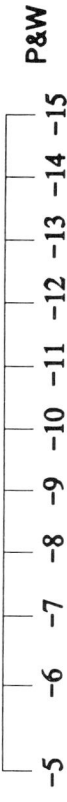

Earbar 0

3 2 1 0 -1 -2 -3 -4 -5 -6 -7 -8 -9 -10 -11 -12 -13 -14 -15 -16 -17 -18

9 8 7 6 5 4 3 2 1 0 -1

cc
CA1
dhc
SC
IC
DG
sm
Ptp
Pta
CA2
CA3
VB l
VBm
ml
Zl
cp
vhc
&fi
st
ic
eml
ot
CdP
ac
GP
mfb
Pir
lot
5t
icp
ts
Vsp
Vs
8n
7n
SO
PBm
PBl
scp
65c

P&W

Sagittal 2.1

-5 -6 -7 -8 -9 -10 -11 -12 -13 -14 -15

Plate 113

AHZ Amygdalo-Hippocampal Area
AMB Nucleus Ambiguus
Ast Area Striata (Cortex)
CA₁ Hippocampal Area CA_1 (Ammon's Horn)
Cun Cuneiform Nucleus
DG Dentate Gyrus
dMR Deep Mesencephalic Reticular Nucleus
IC Inferior Colliculus
Int Interpositus (Intermediate or Interposed) Cerebellar Nucleus
LD Lateral Dorsal Nucleus (Thalamus)
Lot Nucleus of Lateral Olfactory Tract
LP Lateral Posterior Nucleus (Thalamus)
LR_m Lateral Reticular Nucleus (medial)
LR_p Lateral Reticular Nucleus (posterior)
LV Lateral Ventricle
m5 Motor Trigeminal Nucleus
MaPO Magnocellular Preoptic Nucleus (Hypothalamus)
MeA Medial Nucleus of the Amygdala
Oc Occipital Area (Cortex)
Olep External Periolivary Nucleus

Olsp Superior Paraolivary Nucleus
PB_l Parabrachial Nucleus (lateral)
PB_m Parabrachial Nucleus (medial)
PcR Parvicellular Reticular Nucleus
Pd Pontine Gray (Deep)
Po Posterior Group (Thalamus)
PpT Pedunculopontine Tegmental Nucleus
Pta Anterior Pretectal Nucleus (Thalamus)
Ptp Posterior Pretectal Nucleus (Thalamus)
R Reticular Nucleus (Thalamus)
s5c Spinal Trigeminal Nucleus (caudal)
SC Superior Colliculus
SI Substantia Innominata
SO_l Superior Olivary Complex (lateral)
SomS Somatosensory Cortex
ST Subthalamic Nucleus
Sub Subiculum
VB_l Ventrobasal Nuclear Complex (lateral) (Thalamus)
VB_m Ventrobasal Nuclear Complex (medial) (Thalamus)
VI Lateral Vestibular Nucleus
Vs Superior Vestibular Nucleus
Vsp Spinal Vestibular Nucleus
ZI Zona Incerta
7 Facial Nucleus
7n Facial Nerve

AMB: Nucleus Ambiguus. This nucleus consists of a thin rostro-caudal string of large neurons extending the length of the medulla to merge with the ventral horn motoneurons at the spinal C_1 level. In transverse section only a few cells are usually encountered and, as the name implies, identification can be ambiguous in some sections. The larger cells, which innervate muscles of branchial origin, resemble other somatic motoneurons, but the more rostral components contain slightly smaller scattered cells belonging to the brain-stem parasympathetic preganglionic system. The spinal accessory components and each muscle group have been traced by retrograde labeling techniques, but there are some disagreements concerning the exact location of the parasympathetics – for example, the inferior and superior salivatory and lacrimal nuclei. We have labeled the group AMB because the separate components cannot be distinguished easily in Nissl-stained sections. The original literature is extensive and should be consulted for details.

Int: Interpositus (Intermediate) Cerebellar Nucleus. The middle deep cerebellar nucleus has been divided into main (dorsally) and parvicellular (ventrolaterally) regions in the rat (Korneliussen, '68). However, the Int blends imperceptibly with the lateral or dentate nucleus at many levels. In human anatomy the Int is often divided into globose and emboliform nuclei, and presumably corresponding posterior and anterior regions of the Int have been mentioned in the rat by Voogd et al. ('85). They divide this nuclear group into a small, densely packed posterior division and an anterior portion that possesses a "dorsomedial crest" and a "dorsolateral hump."

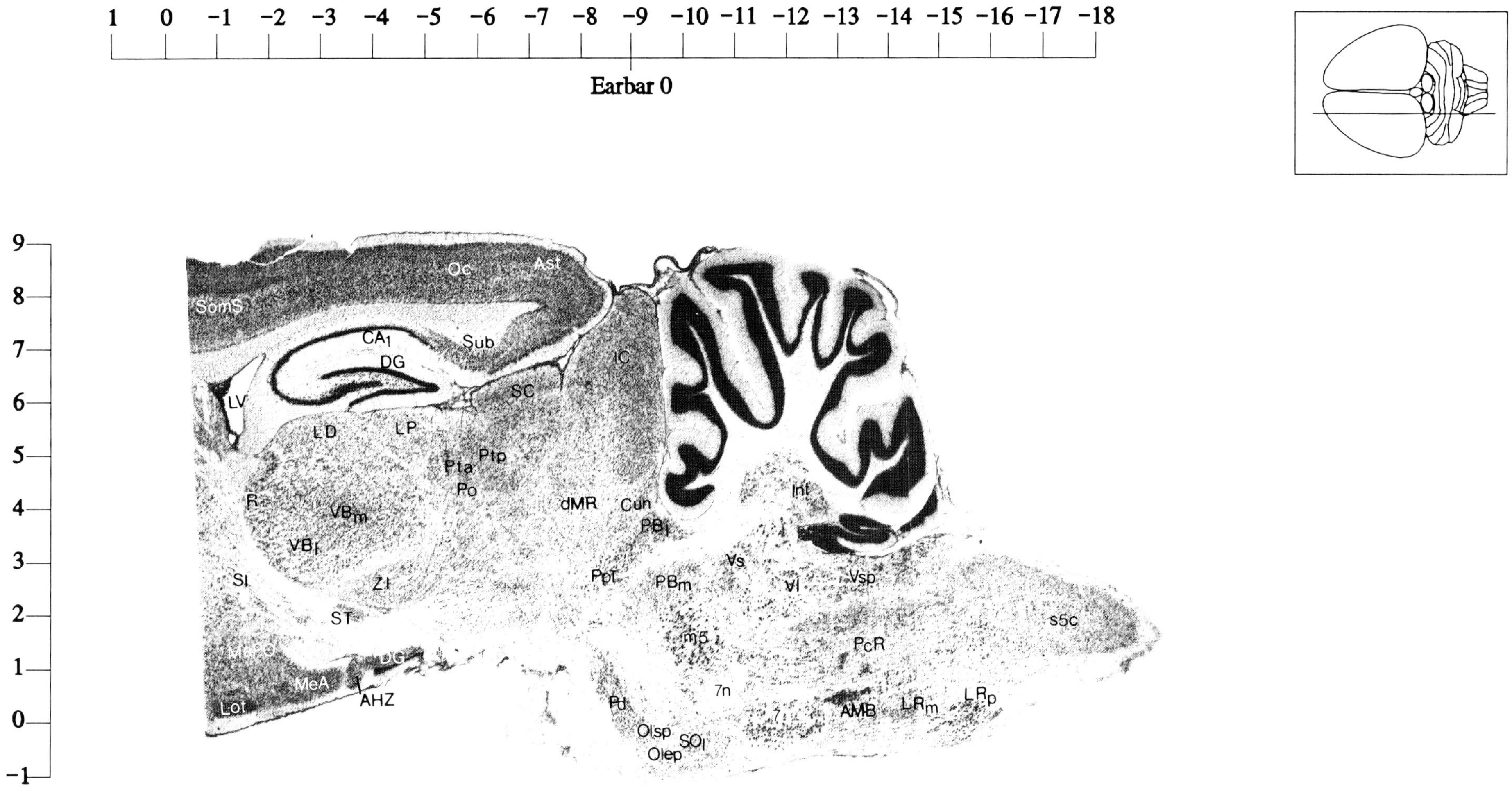

Earbar 0

1 0 -1 -2 -3 -4 -5 -6 -7 -8 -9 -10 -11 -12 -13 -14 -15 -16 -17 -18

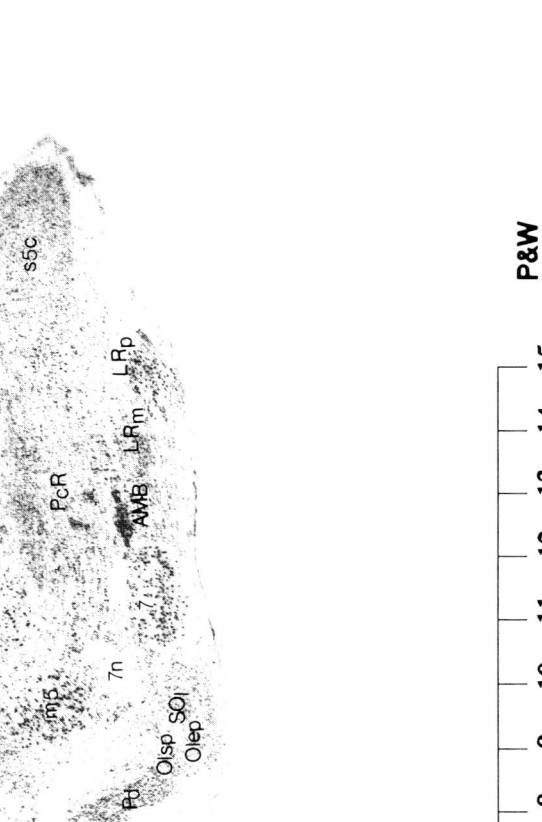

9 8 7 6 5 4 3 2 1 0 -1

Sagittal 2.4

P&W

-5 -6 -7 -8 -9 -10 -11 -12 -13 -14 -15

Plate 114

fi: Fimbria. This fiber tract derives from the alveus of Ammon's horn to form the fornix and can be viewed as analogous to the internal capsule of the isocortex.

S5: Supratrigeminal Nucleus. This term is attributed to Lorente de Nó's ('22) description in the mouse of a sector adjoining the dorsomedial sector of the principal trigeminal nucleus (Pr5) (see Åstrom, '53; Torvik, '56). The S5 is sometimes called the *dorsomedial principal trigeminal nucleus* or is interpreted as a separate medial sector. It contains larger and more dispersed cells than Pr5 in the rat and it is considered a "reticular zone" by some (e.g., Mizuno, '70), but there is some evidence in the cat for a nucleus in this region that projects bilaterally to the thalamus and receives axons from the oral cavity (Kruger et al., '77). The functional status of this region in the rat brain, in which an ipsilateral trigeminal projection to thalamus has not been demonstrated (Mantle-St. John and Tracey, '87), remains enigmatic, although a distinctive nucleus capping the motor trigeminal nucleus appears to be a consistent mammalian feature. We identify it as part of the sensory trigeminal nuclear complex because it constitutes a distinct, heavily labeled nucleus when labeled by anterograde axonal transport from the trigeminal ganglion (Kruger et al., '77).

Zl: Zona Incerta. See R: Reticular Nucleus (Plate 117).

ac Anterior Commissure
AHZ Amygdalo-Hippocampal Area
CA₁ Hippocampal Area CA₁ (Ammon's Horn)
CA₂ Hippocampal Area CA₂ (Ammon's Horn)
CA₃ Hippocampal Area CA₃ (Ammon's Horn)
CdP Caudoputamen
cp Cerebral Peduncle
fi Fimbria
fr Fasciculus Retroflexus
GP Globus Pallidus (external or lateral)
IC Inferior Colliculus
ic Internal Capsule
ll Lateral Lemniscus
lot Lateral Olfactory Tract
mcp Middle Cerebellar Peduncle
mfb Medial Forebrain Bundle
ml Medial Lemniscus
ot Optic Tract
pa5 Paratrigeminal Nucleus
Pir Piriform Area (Cortex)
Pr5 Principal Sensory Trigeminal Nucleus
Rsp Retrosplenial Area (Cortex)
S5 Supratrigeminal Nucleus
s5c Spinal Trigeminal Nucleus (caudal)
s5i Spinal Trigeminal Nucleus (interpolar)
s5o Spinal Trigeminal Nucleus (oral)
SC Superior Colliculus
scp Superior Cerebellar Peduncle
sg Substantia Gelatinosa
sm Stria Medullaris
st Stria Terminalis
VB₁ Ventrobasal Nuclear Complex (lateral) (Thalamus)
VBₘ Ventrobasal Nuclear Complex (medial) (Thalamus)
vhc Ventral Hippocampal Commissure
Zl Zona Incerta
5r Trigeminal Nerve Root
7 Facial Nucleus
7n Facial Nerve

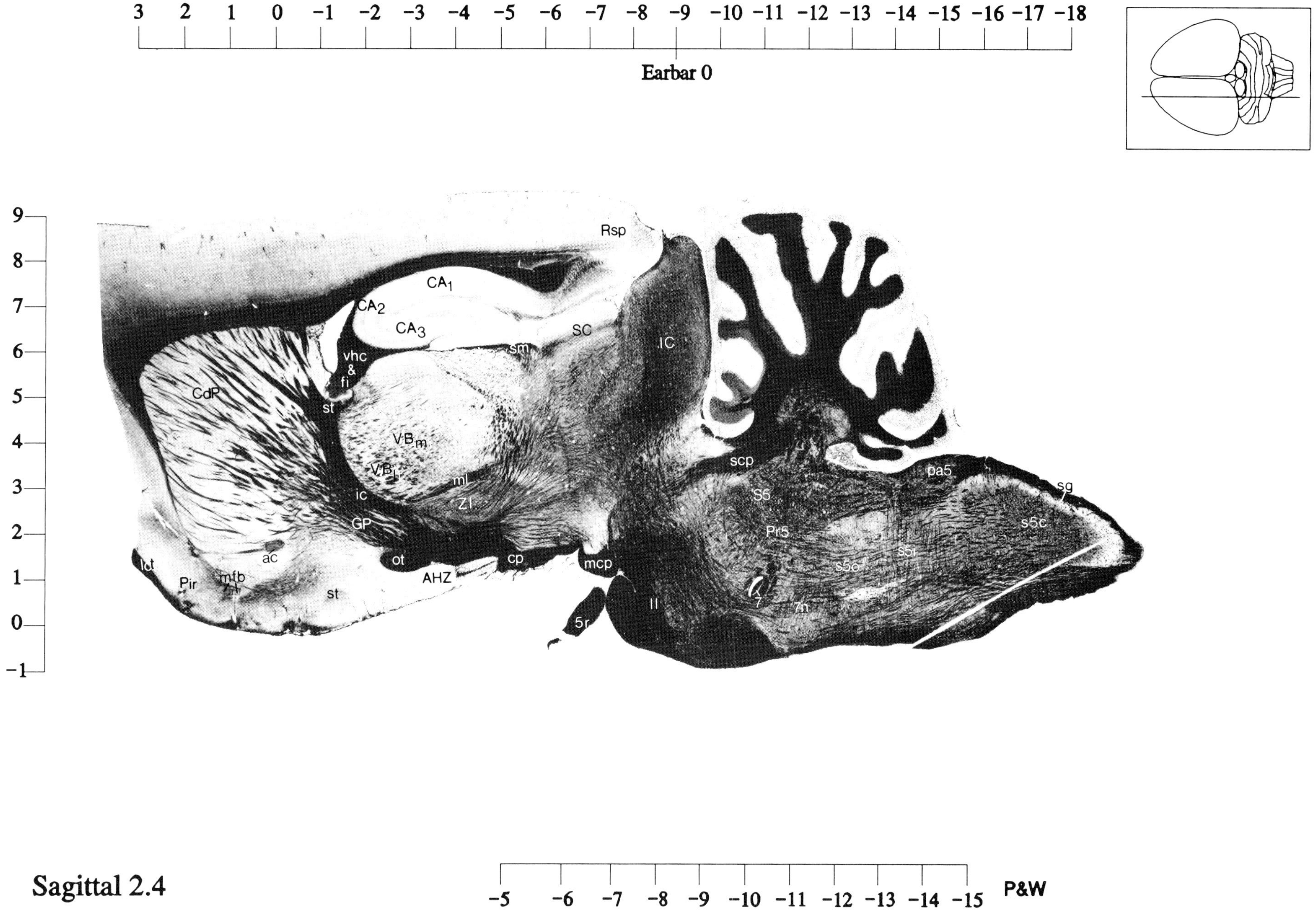

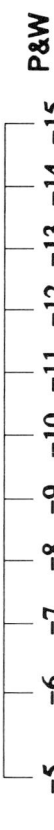

Earbar 0

3 2 1 0 −1 −2 −3 −4 −5 −6 −7 −8 −9 −10 −11 −12 −13 −14 −15 −16 −17 −18

9 8 7 6 5 4 3 2 1 0 −1

Rsp

SC

sm

CA₁

CA₂

CA₃

vhc & fi

st

CdP

ac

mfb

Pir

lot

V₁B₋ₘ

V₁B₋ᵢ

ic

GP

st

ml

ZI

ot

AHZ

cp

IC

scp

S5

Pr5

7n

7

ll

mcp

5r

sg

s5c

pa5

S5r

S5o

−5 −6 −7 −8 −9 −10 −11 −12 −13 −14 −15

P&W

Sagittal 2.4

Plate 115

AHZ Amygdalo-Hippocampal Area
Ap Anterior Pituitary (adenohypophysis)
Ast Area Striata (Cortex)
BI Nucleus of the Brachium Inferior Colliculus
BIs Nucleus of the Brachium Inferior Colliculus (subbrachial sector)
CA₁ Hippocampal Area CA₁ (Ammon's Horn)
CA₃ Hippocampal Area CA₃ (Ammon's Horn)
CoA$_a$ Cortical Nucleus of the Amygdala (anterior)
CoA$_p$ Cortical Nucleus of the Amygdala (posterior)
Cun Cuneiform Nucleus
DG Dentate Gyrus
IC Inferior Colliculus
Int Interpositus (Intermediate or Interposed) Cerebellar Nucleus
LD Lateral Dorsal Nucleus (Thalamus)
LGd Dorsal Lateral Geniculate Nucleus (Thalamus)
LL$_d$ Nucleus of Lateral Lemniscus (dorsal)
LL$_v$ Nucleus of Lateral Lemniscus (ventral)
LP Lateral Posterior Nucleus (Thalamus)
m5 Motor Trigeminal Nucleus

MeA Medial Nucleus of the Amygdala
PB$_l$ Parabrachial Nucleus (lateral)
Pbg Parabigeminal Nucleus
Pd Pontine Gray (Deep)
Pir Piriform Area (Cortex)
Po Posterior Group (Thalamus)
PpT Peduncolopontine Tegmental Nucleus
R Reticular Nucleus (Thalamus)
S5 Supratrigeminal Nucleus
s5c Spinal Trigeminal Nucleus (caudal)
s5i Spinal Trigeminal Nucleus (interpolar)
s5o Spinal Trigeminal Nucleus (oral)
Sag Nucleus Sagulum
SC Superior Colliculus
SO$_l$ Superior Olivary Complex (lateral)
ST Subthalamic Nucleus
Sub Subiculum
VB$_l$ Ventrobasal Nuclear Complex (lateral) (Thalamus)
VB$_m$ Ventrobasal Nuclear Complex (medial) (Thalamus)
Vl Lateral Vestibular Nucleus
Vs Superior Vestibular Nucleus
Vsp Spinal Vestibular Nucleus
ZI Zona Incerta
5r Trigeminal Root
5s Trigeminal Nerve (sensory or major division)
7 Facial Nucleus

LP: Lateral Posterior Nucleus (Thalamus). The lateral and medial borders of this nucleus are distinct, but there is substantial confusion about the location and naming of other boundary zones. This caudal portion of the lateral nuclear group is especially large in the human brain and is generally subdivided in all large brains to include a caudal protuberance or "pillow" called the *pulvinar nucleus*, which is further divided into separate nuclei in most descriptions of a primate thalamus. Such subdivision is rarely employed in the rat, where a pulvinar nucleus is questionable at best, but it should be possible to separate various components based on the fiber architecture, which enables separation in the rostrocaudal plane from the laterodorsal (LD) nucleus. Discrepancies in interpreting the ventral limits abound in the relevant literature.

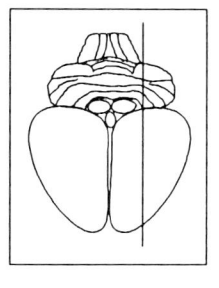

Earbar 0

3 2 1 0 -1 -2 -3 -4 -5 -6 -7 -8 -9 -10 -11 -12 -13 -14 -15 -16 -17 -18

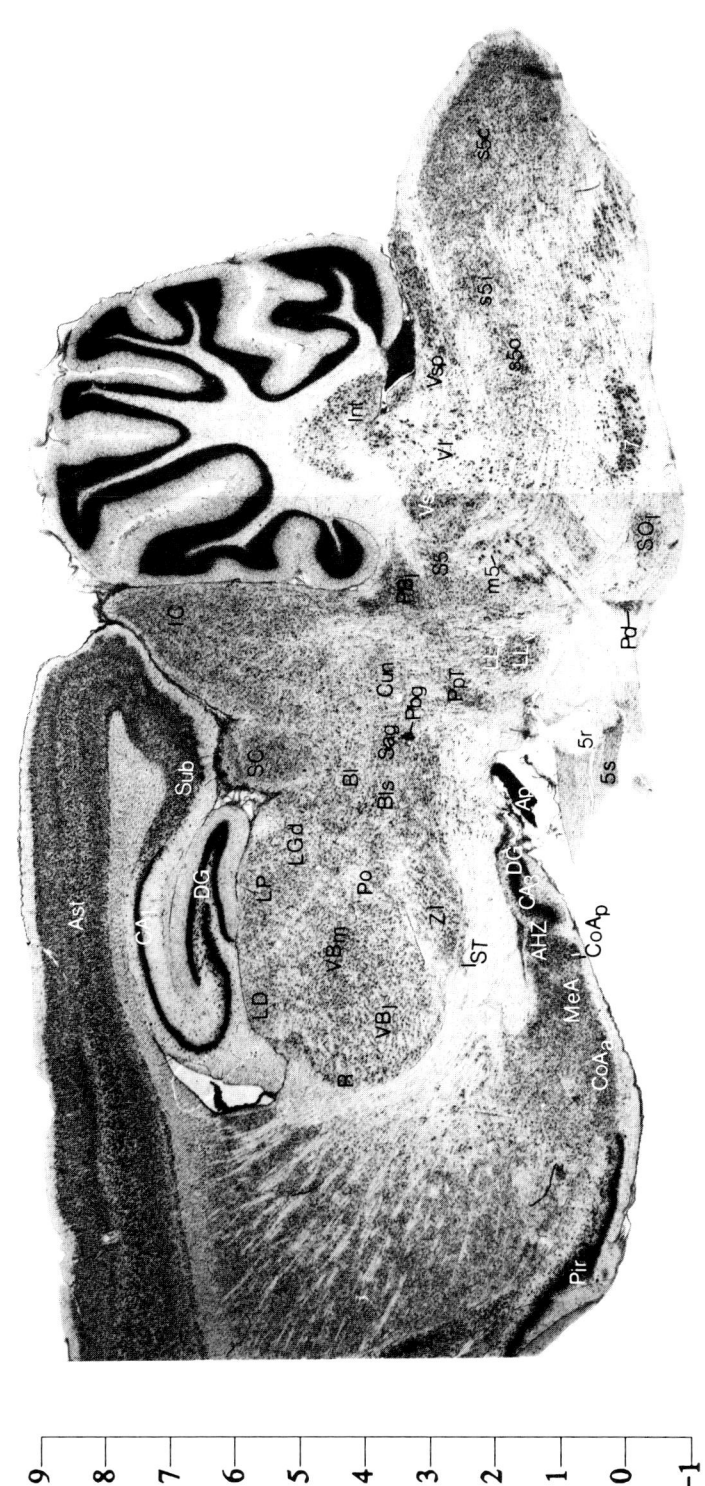

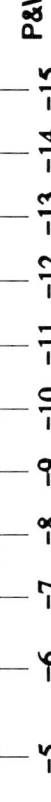

9 8 7 6 5 4 3 2 1 0 -1

Sagittal 2.7

P&W

-5 -6 -7 -8 -9 -10 -11 -12 -13 -14 -15

Plate 116

ac Anterior Commissure

AHZ Amygdalo-Hippocampal Area

CA₁ Hippocampal Area CA₁ (Ammon's Horn)

CA₂ Hippocampal Area CA₂ (Ammon's Horn)

CA₃ Hippocampal Area CA₃ (Ammon's Horn)

CdP Caudoputamen

cp Cerebral Peduncle

eml External Medullary Lamina

fi Fimbria

fr Fasciculus Retroflexus

GP Globus Pallidus (external or lateral)

IC Inferior Colliculus

ic Internal Capsule

icp Inferior Cerebellar Peduncle

Int Interpositus (Intermediate or Interposed) Cerebellar Nucleus

lot Lateral Olfactory Tract

mcp Middle Cerebellar Peduncle

mfb Medial Forebrain Bundle

ml Medial Lemniscus

ot Optic Tract

pa5 Paratrigeminal Nucleus

Pir Piriform Area (Cortex)

Po Posterior Group (Thalamus)

s5c Spinal Trigeminal Nucleus (caudal)

SC Superior Colliculus

scp Superior Cerebellar Peduncle

sg Substantia Gelatinosa

sm Stria Medullaris

st Stria Terminalis

VB₁ Ventrobasal Nuclear Complex (lateral) (Thalamus)

VBₘ Ventrobasal Nuclear Complex (medial) (Thalamus)

vhc Ventral Hippocampal Commissure

Vl Lateral Vestibular Nucleus

Vsp Spinal Vestibular Nucleus

5m Trigeminal Nerve Root (motor or minor division)

5n Trigeminal Nerve

5r Trigeminal Nerve Root

5t Trigeminal Tract

7n Facial Nerve

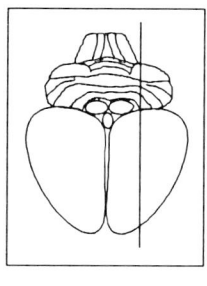
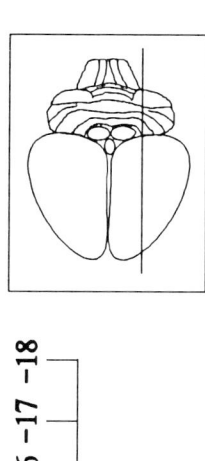

4 3 2 1 0 −1 −2 −3 −4 −5 −6 −7 −8 −9 −10 −11 −12 −13 −14 −15 −16 −17 −18

Earbar 0

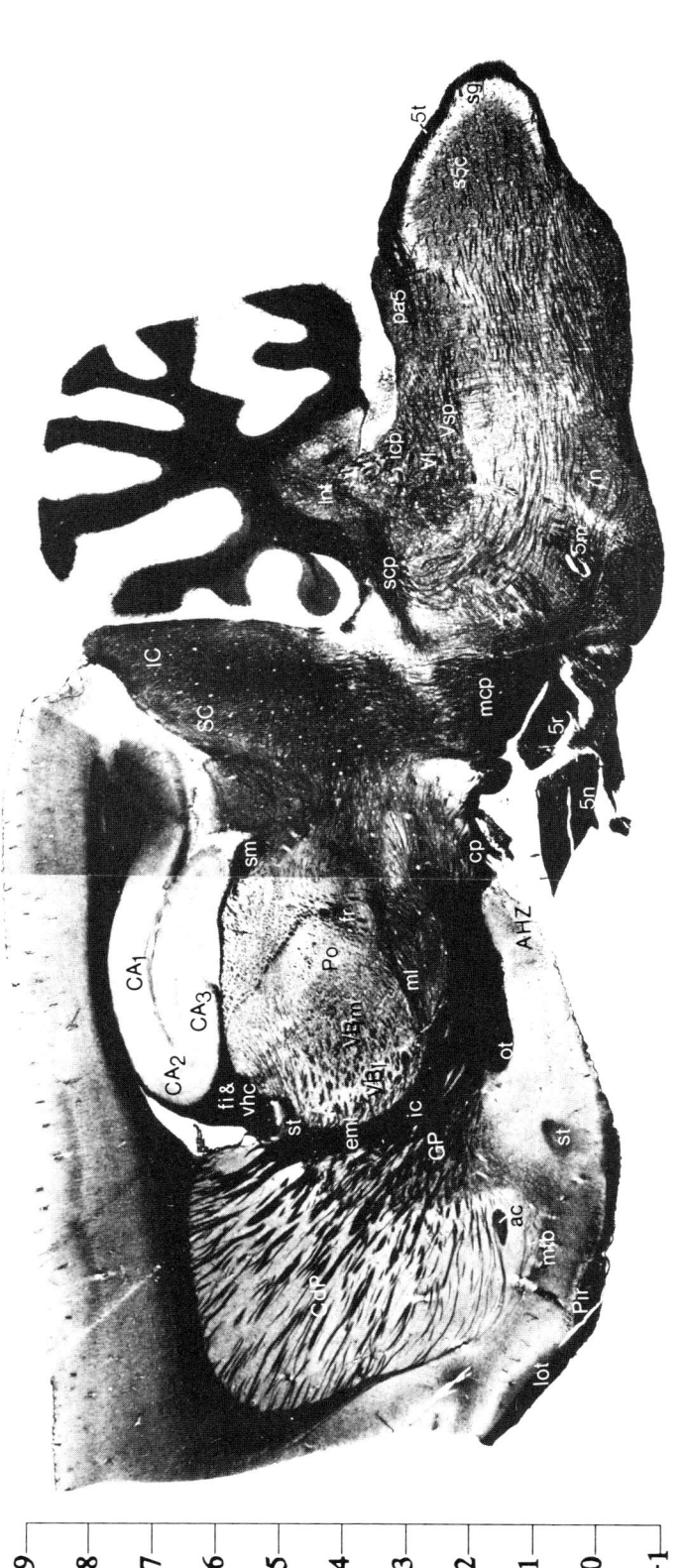

9 8 7 6 5 4 3 2 1 0 −1

Sagittal 2.7

P&W

−5 −6 −7 −8 −9 −10 −11 −12 −13 −14 −15

Plate 117

AHZ Amygdalo-Hippocampal Area
Ap Anterior Pituitary (adenohypophysis)
Ast Area Striata (Cortex)
BI Nucleus of the Brachium Inferior Colliculus
BIs Nucleus of the Brachium Inferior Colliculus (subbrachial sector)
BMA Basomedial Nucleus of the Amygdala
CA₁ Hippocampal Area CA_1 (Ammon's Horn)
CA₃ Hippocampal Area CA_3 (Ammon's Horn)
CdP Caudoputamen
CeA Central Nucleus of the Amygdala
CO_d Cochlear Nucleus (dorsal)
CoA Cortical Nucleus of the Amygdala
CoA_p Cortical Nucleus of the Amygdala (posterior)
CUe External Cuneate Nucleus
De Dentate (Lateral) Cerebellar Nucleus
DG Dentate Gyrus
ENT Entorhinal Area
EP Endopiriform Nucleus
GP Globus Pallidus (external or lateral)
IA Intercalated Mass of the Amygdala
IC Inferior Colliculus
LD Lateral Dorsal Nucleus (Thalamus)
LGd Dorsal Lateral Geniculate Nucleus (Thalamus)
LL_d Nucleus of Lateral Lemniscus (dorsal)
LL_v Nucleus of Lateral Lemniscus (ventral)

lot Lateral Olfactory Tract
LP Lateral Posterior Nucleus (Thalamus)
LR Lateral Reticular Nucleus
MeA Medial Nucleus of the Amygdala
MG Medial Geniculate Nucleus
Oc Occipital Area (Cortex)
Of Orbitofrontal Area (Cortex)
Pam Periamygdaloid Cortex
PB_l Parabrachial Nucleus (lateral)
Pbg Parabigeminal Nucleus
Pir Piriform Area (Cortex)
Po Posterior Group (Thalamus)
Pr5 Principal Sensory Trigeminal Nucleus
Ptp Posterior Pretectal Nucleus (Thalamus)
R Reticular Nucleus (Thalamus)
S5 Supratrigeminal Nucleus
s5c Spinal Trigeminal Nucleus (caudal)
s5i Spinal Trigeminal Nucleus (interpolar)
s5o Spinal Trigeminal Nucleus (oral)
Sag Nucleus Sagulum
SC Superior Colliculus
SomS Somatosensory Cortex
STF Striatal Fundus
Sub Subiculum
VB Ventrobasal Nuclear Complex (Thalamus)
VI Lateral Vestibular Nucleus
Vsp Spinal Vestibular Nucleus
X Nucleus X
ZI Zona Incerta
5r Trigeminal Root
5s Trigeminal Nerve (sensory or major division)
7 Facial Nucleus

R: Reticular Nucleus (Thalamus) (also called Nucleus Reticularis). This nuclear complex is embedded in the distinctive fiber bundle constituting the external medullary lamina. It differentiates early in development from the embryonic ventral thalamic plate before the large middle plate, called *dorsal thalamus*, becomes subdivided. There have been several proposals for subdivision of this heterogeneous entity, but this is difficult in the rat and a general principle for nomenclature is lacking. The rostral pole is distinctive in sagittal sections but is generally not allotted separate status, despite evidence of connections that would enable subdivision. The ventromedial and caudal portions merge gradually with the contiguous zona incerta. We have not subdivided the zona incerta, although some authors indicate a separate "subincertal" nucleus (e.g., Paxinos and Watson, '86).

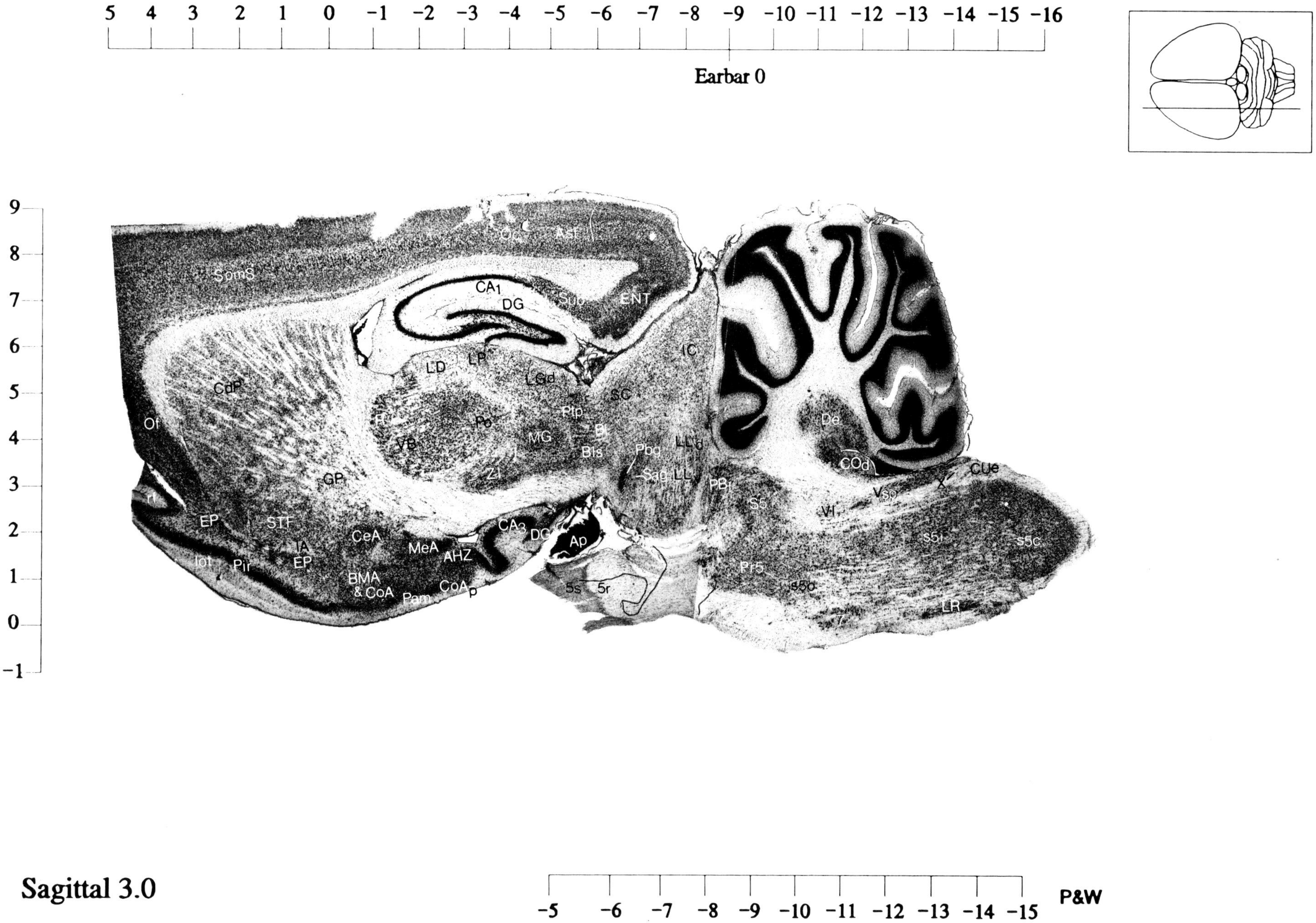

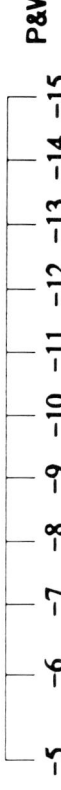

Earbar 0

5 4 3 2 1 0 -1 -2 -3 -4 -5 -6 -7 -8 -9 -10 -11 -12 -13 -14 -15 -16

9
8
7
6
5
4
3
2
1
0
-1

P&W

-5 -6 -7 -8 -9 -10 -11 -12 -13 -14 -15

Sagittal 3.0

Plate 118

ac Anterior Commissure
alv Alveus
Ap Anterior Pituitary (adenohypophysis)
bic Brachium of Inferior Colliculus
CA₁ Hippocampal Area CA₁ (Ammon's Horn)
CA₂ Hippocampal Area CA₂ (Ammon's Horn)
CA₃ Hippocampal Area CA₃ (Ammon's Horn)
CdP Caudoputamen
cp Cerebral Peduncle
eml External Medullary Lamina
fi Fimbria
GP Globus Pallidus (external or lateral)
IC Inferior Colliculus
ic Internal Capsule
icp Inferior Cerebellar Peduncle
iml Internal Medullary Lamina
ll Lateral Lemniscus
lot Lateral Olfactory Tract
ml Medial Lemniscus
ot Optic Tract
rf Rhinal Fissure
SC Superior Colliculus
sg Substantia Gelatinosa
sm Stria Medullaris
SomS Somatosensory Cortex
st Stria Terminalis
VB Ventrobasal Nuclear Complex (Thalamus)
5n Trigeminal Nerve
5r Trigeminal Nerve Root
5t Trigeminal Tract

bic: Brachium of the Inferior Colliculus. A myelinated tract extending from the central inferior colliculus to the thalamic medial geniculate body. It is relatively compact and identifiable throughout most of its course.

5m,n,r,s: Trigeminal Nerve and Root. Before entering the brain stem, the trigeminal nerve exhibits a marked alteration in staining properties, and in sagittal sections (Plates 117–120) a "glial dome" is evident denoting the transition from nerve to root. The peripheral nerve portion contains Schwann cells, and the root (central portion) contains glial supportive cells with thicker myelin and heavier staining in the nerve than in the root (Maxwell et al., '69). In the horizontal plane, it is often feasible to identify the separate small motor division (5m) from the larger sensory division (5s) corresponding to the *portio minor* and *major* of the human nerve.

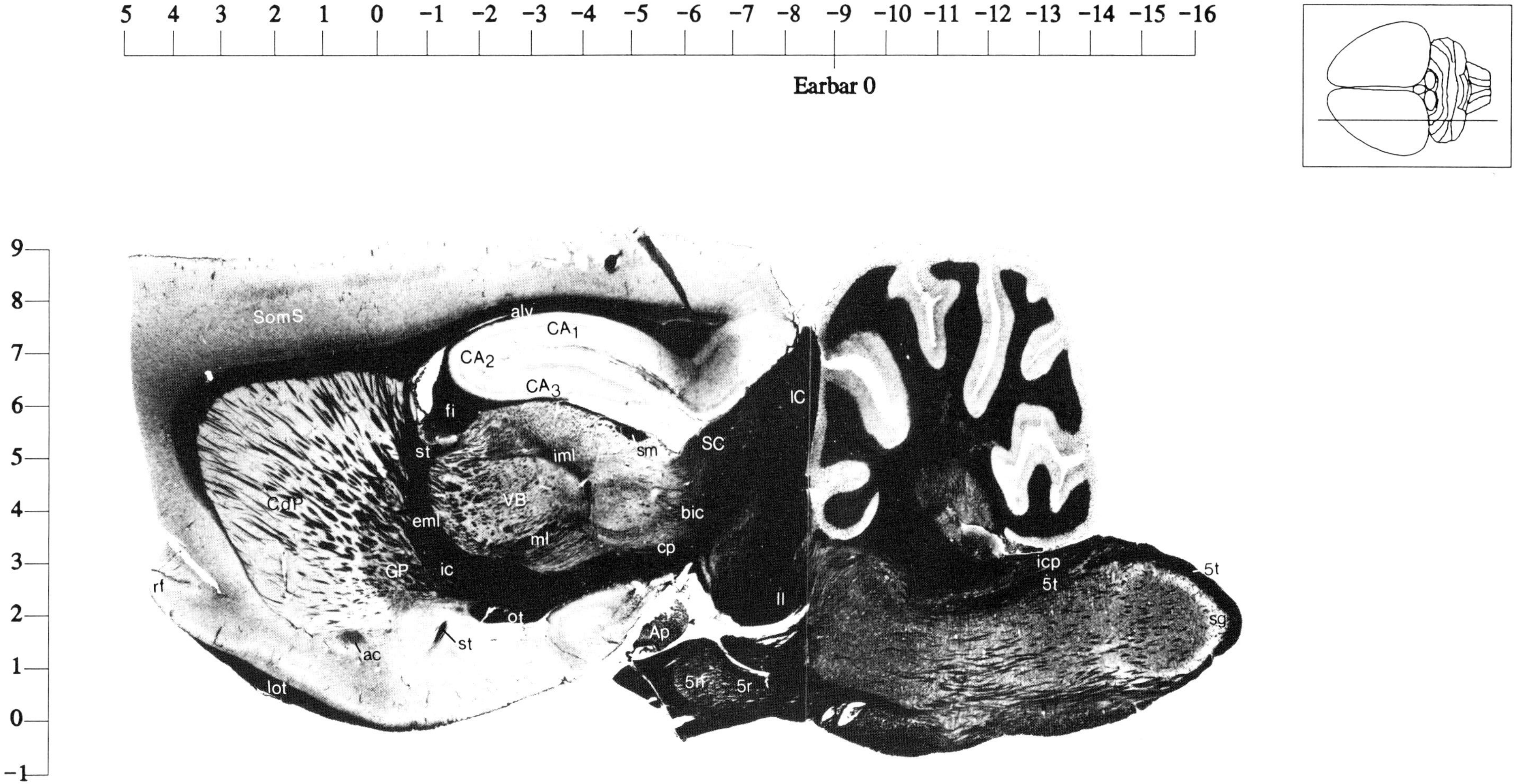

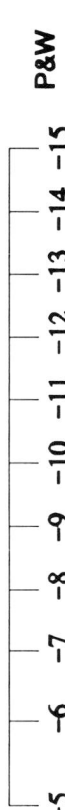

Earbar 0

5 4 3 2 1 0 −1 −2 −3 −4 −5 −6 −7 −8 −9 −10 −11 −12 −13 −14 −15 −16

SomS

CA₂
CA₁
alv
CA₃
fi
st
sm
iml
VB
eml
ml
ic
GP
ot
st
ac

CdP

lot

rf

SC
bic
cp
Ap
II
5n
5r

IC

icp
5t

5t
sg

9 8 7 6 5 4 3 2 1 0 −1

Sagittal 3.0

−5 −6 −7 −8 −9 −10 −11 −12 −13 −14 −15

P&W

Plate 119

AAA Anterior Amygdaloid Area
AHZ Amygdalo-Hippocampal Area
Ap Anterior Pituitary (adenohypophysis)
BMA Basomedial Nucleus of the Amygdala
CdP Caudoputamen
CeA Central Nucleus of the Amygdala
Cl Claustrum
CO$_d$ Cochlear Nucleus (dorsal)
CoA$_p$ Cortical Nucleus of the Amygdala (posterior)
De Dentate (Lateral) Cerebellar Nucleus
DG Dentate Gyrus
ENT Entorhinal Area
EP Endopiriform Nucleus
GP Globus Pallidus (external or lateral)
IA Intercalated Mass of the Amygdala
IC Inferior Colliculus
In Insular Area (Cortex)
Int Interpositus (Intermediate or Interposed) Cerebellar Nucleus
LD Lateral Dorsal Nucleus (Thalamus)
LGd Dorsal Lateral Geniculate Nucleus (Thalamus)
LL$_d$ Nucleus of Lateral Lemniscus (dorsal)
LL$_v$ Nucleus of Lateral Lemniscus (ventral)
LP Lateral Posterior Nucleus (Thalamus)

LV Lateral Ventricle
Ma Motor Agranular Cortical Area
MeA Medial Nucleus of the Amygdala
MGm Magnocellular Medial Geniculate Nucleus (Thalamus)
MGp Principal/Parvicellular Medial Geniculate Nucleus (Thalamus)
Of Orbitofrontal Area (Cortex)
pa5 Paratrigeminal Nucleus
Pam Periamygdaloid Cortex
Pbg Parabigeminal Nucleus
Pir Piriform Area (Cortex)
pp Peripeduncular Nucleus
Pr5 Principal Sensory Trigeminal Nucleus
R Reticular Nucleus (Thalamus)
rf Rhinal Fissure
S5 Supratrigeminal Nucleus
s5c Spinal Trigeminal Nucleus (caudal)
s5i Spinal Trigeminal Nucleus (interpolar)
s5o Spinal Trigeminal Nucleus (oral)
SI Substantia Innominata
STF Striatal Fundus
Sub Subiculum
VB Ventrobasal Nuclear Complex (Thalamus)
ZI Zona Incerta
5r Trigeminal Nerve Root
5s Trigeminal Nerve (sensory or major division)

Ap: Anterior Pituitary (also called Adenohypophysis and Neurohypophysis). Most of the pituitary has been torn off in the sections of this atlas, largely due to the fragility of the thin hypophysial stalk, but some remnants are visible in some sections.

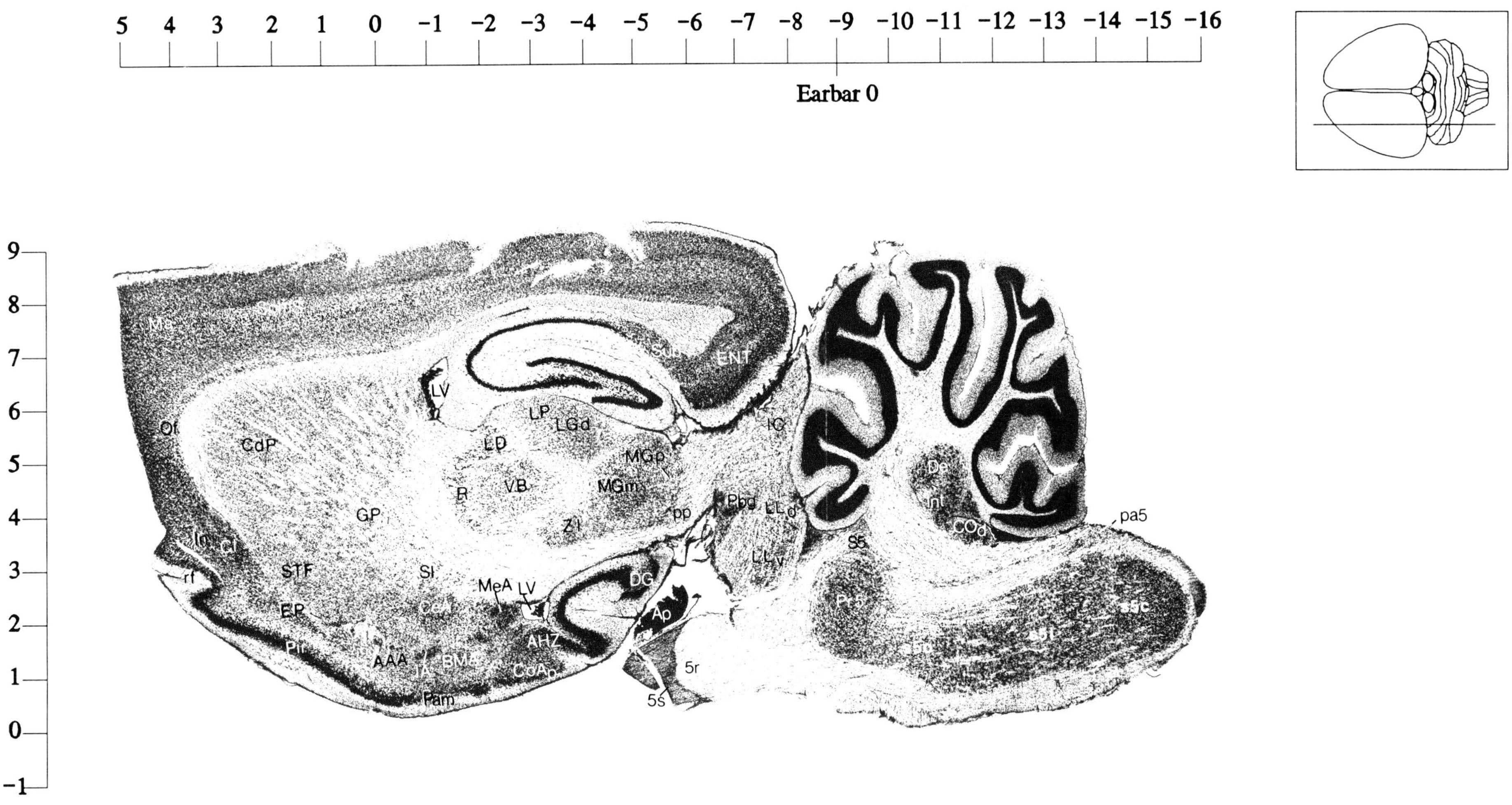

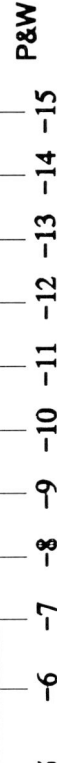

Earbar 0

5 4 3 2 1 0 -1 -2 -3 -4 -5 -6 -7 -8 -9 -10 -11 -12 -13 -14 -15 -16

pa5

S3

Crus

Sim

ENT

IC

Pbg

PP

S5

P

4v

5r

5s

MGd

Sd

MGm

Dc

Ap

LP

LGd

ZI

AHZ

LV

EP

VB

MeA

LV

CoAp

R

St

BM

Pam

GPi

AAA

STF

EP

Pir

CdP

lm ci

rf

Ot

V

9 8 7 6 5 4 3 2 1 0 -1

Sagittal 3.3

P&W

-5 -6 -7 -8 -9 -10 -11 -12 -13 -14 -15

Plate 120

alv Alveus
bic Brachium of Inferior Colliculus
CdP Caudoputamen
cp Cerebral Peduncle
fi Fimbria
ic Internal Capsule
icp Inferior Cerebellar Peduncle
LGd Dorsal Lateral Geniculate Nucleus (Thalamus)
ll Lateral Lemniscus
lot Lateral Olfactory Tract
mcp Middle Cerebellar Peduncle
ml Medial Lemniscus
SI Substantia Innominata
st Stria Terminalis
STF Striatal Fundus
5n Trigeminal Nerve
5r Trigeminal Nerve Root
5t Trigeminal Tract

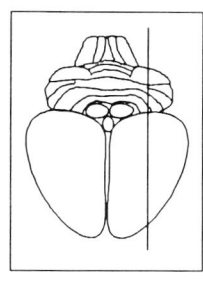
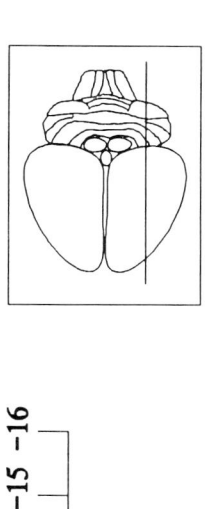

Earbar 0

−16 −15 −14 −13 −12 −11 −10 −9 −8 −7 −6 −5 −4 −3 −2 −1 0 1 2 3 4 5

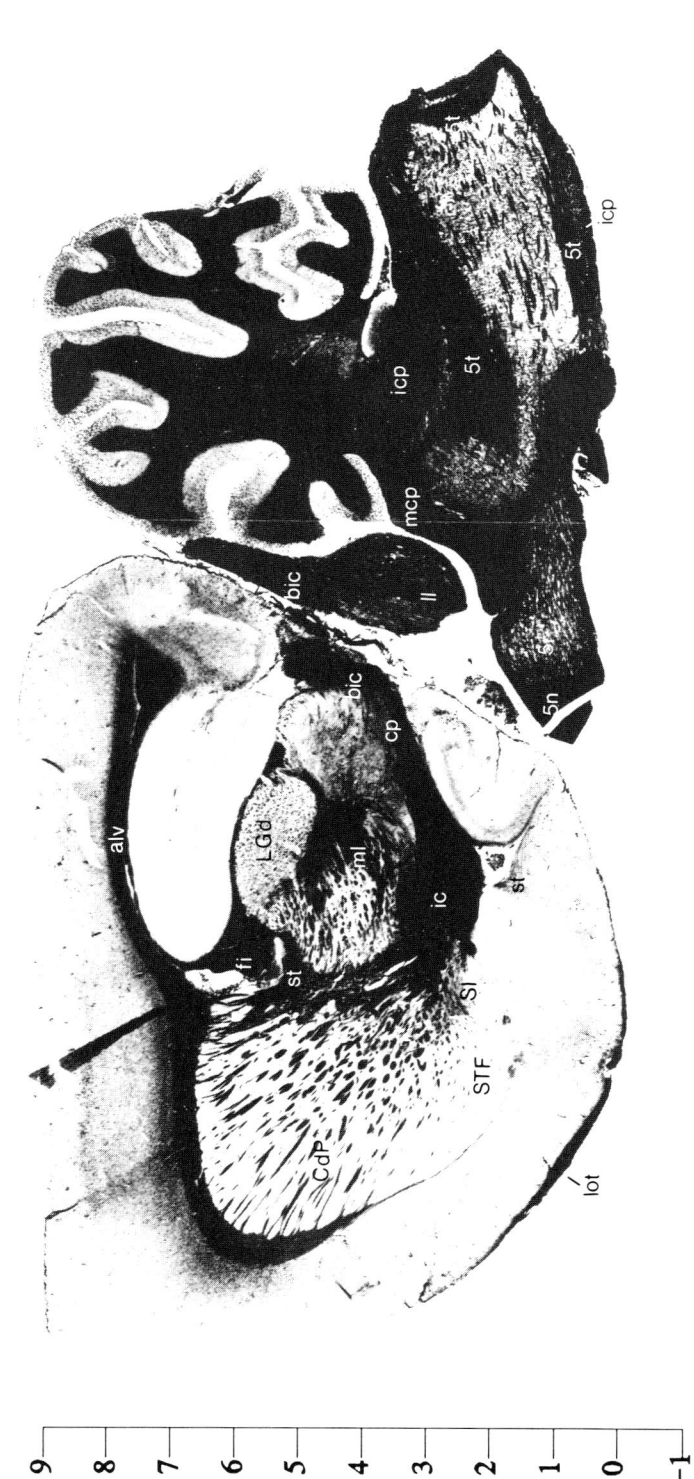

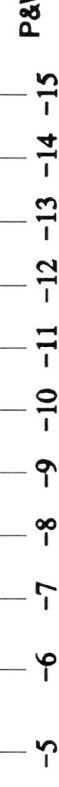

9 8 7 6 5 4 3 2 1 0 −1

P&W

−15 −14 −13 −12 −11 −10 −9 −8 −7 −6 −5

Sagittal 3.3

Plate 121

AHZ Amygdalo-Hippocampal Area
BLA Basolateral Nucleus of the Amygdala
BMA Basomedial Nucleus of the Amygdala
CA₁ Hippocampal Area CA$_1$ (Ammon's Horn)
CA₃ Hippocampal Area CA$_3$ (Ammon's Horn)
CdP Caudoputamen
CeA Central Nucleus of the Amygdala
Cl Claustrum
CO$_d$ Cochlear Nucleus (dorsal)
CoA$_p$ Cortical Nucleus of the Amygdala (posterior)
De Dentate (Lateral) Cerebellar Nucleus
DG Dentate Gyrus
EP Endopiriform Nucleus
IA Intercalated Mass of the Amygdala
In Insular Area (Cortex)
LGd Dorsal Lateral Geniculate Nucleus (Thalamus)
LGv Ventral Lateral Geniculate Nucleus (Thalamus)
LV Lateral Ventricle
IT Terminal Nuclei of the Accessory Optic Root (basal root)
MGp Principal/Parvicellular Medial Geniculate Nucleus (Thalamus)
Pam Periamygdaloid Cortex
Pir Piriform Area (Cortex)
pp Peripeduncular Nucleus
R Reticular Nucleus (Thalamus)
s5i Spinal Trigeminal Nucleus (interpolar)
SI Substantia Innominata
SomS Somatosensory Cortex
STF Striatal Fundus
Sub Subiculum
VB$_l$ Ventrobasal Nuclear Complex (lateral) (Thalamus)

MGp: Principal Medial Geniculate Nucleus (Thalamus). The principal nuclear division of the medial geniculate body constitutes the main relay of auditory fibers and is sometimes designated the parvicellular nucleus in various mammals. It has been divided into dorsal and ventral components in recent accounts of the rat (Paxinos and Watson, '86; Clerici and Coleman, '90; Swanson, '92). Winer and Larue ('87) and Clerici and Coleman ('90) also recognize a medial (ovoid) subnucleus in the ventral portion. Although difficult to delineate in transverse Nissl sections, it is strikingly evident in fiber architecture (Plates 28, 121, and 122). A marginal subnucleus, recognized by Paxinos and Watson ('86), is designated the *caudo-dorsal nucleus of the dorsal division* by Clerici and Coleman ('90); it is most easily seen in our sagittal and horizontal sections, although not separately labeled. Justification for subdivision into dorsal, ventral, and medial sectors or subnuclei can be derived from experimental studies of connections (Le Doux et al., '85; Winer and Larue, '87) from which homologies with other mammals have been derived. The concept of a "posterior intralaminar complex" is presented in a superbly illustrated account of the opossum medial geniculate body by Winer et al. ('88), who divide this entire group differently. The reader should regard the designations of this atlas as merely tentative; we have been unable to translate the superior analysis of opossum to the rat brain. The differences in fiber architecture support such subdivision, but this is difficult when employing strictly cytoarchitectural crtieria for nuclear delineation and it is likely that the nomenclature employed here is unduly conservative. The ventrocaudal margin of MGp is bounded by a distinct nucleus that is separated by a fibrous capsule easily seen with the Loyez stain (Plates 28 and 29); it is generally designated the *peripeduncular nucleus* (pp) and constitutes the lateral extension of the subparafascicular (sPf) nuclear wing, with which it may be functionally related (see Kruger et al., '88a). Some authors designate this latter the *suprapeduncular nucleus* (Winer and Larue, '87).

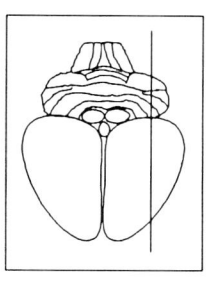
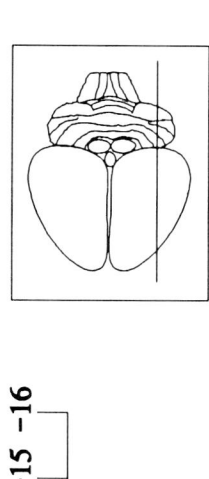

Earbar 0

-16 -15 -14 -13 -12 -11 -10 -9 -8 -7 -6 -5 -4 -3 -2 -1 0 1 2 3 4 5

9 8 7 6 5 4 3 2 1 0

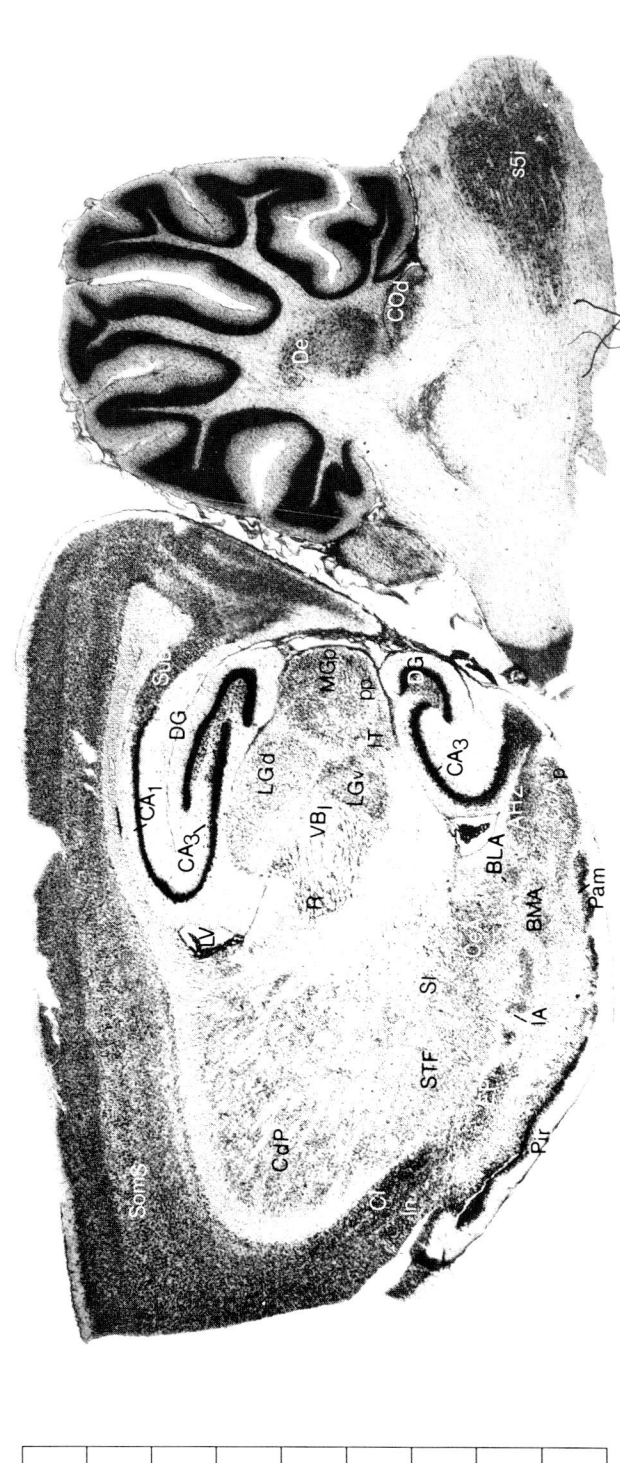
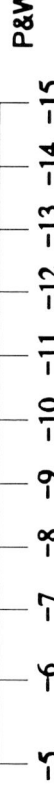

CA₁
CA₃
DG
SU
MGd
pp
DG
CA₃
LGd
LGv
HT
VBl
R
De
COd
S5l
LV
BLA
HL
p
SI
BMA
Pam
STF
IA
CdP
Pir
SomS
Ci
In

P&W

Sagittal 3.6

Plate 122

bic Brachium of Inferior Colliculus
fi Fimbria
GP Globus Pallidus (external or lateral)
ic Internal Capsule
icp Inferior Cerebellar Peduncle
LGd Dorsal Lateral Geniculate Nucleus (Thalamus)
LGv Ventral Lateral Geniculate Nucleus (Thalamus)
lot Lateral Olfactory Tract
mcp Middle Cerebellar Peduncle
MGp Principal/Parvicellular Medial Geniculate Nucleus (Thalamus)
ot Optic Tract
R Reticular Nucleus (Thalamus)
st Stria Terminalis
5t Trigeminal Tract

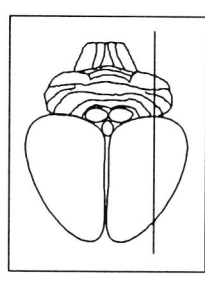

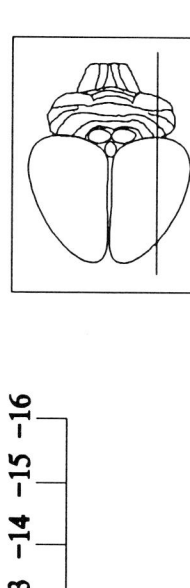

Earbar 0

-16 -15 -14 -13 -12 -11 -10 -9 -8 -7 -6 -5 -4 -3 -2 -1 0 1 2 3 4 5

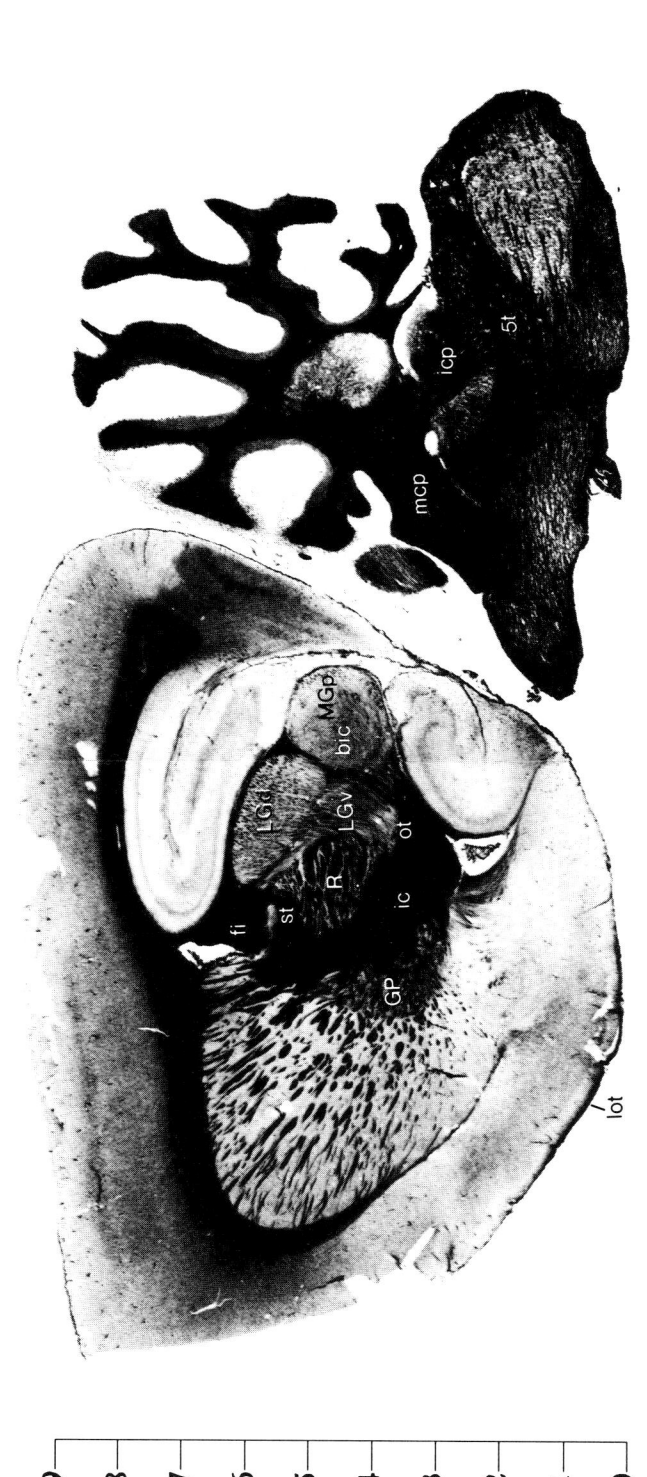

fi
st
LGd
R
LGv
MGp
bic
ot
ic
GP
lot

mcp
icp
5t

9 8 7 6 5 4 3 2 1 0

P&W

-5 -6 -7 -8 -9 -10 -11 -12 -13 -14 -15

Sagittal 3.6

Plate 123

AHZ Amygdalo-Hippocampal Area
BLA$_a$ Basolateral Nucleus of the Amygdala (anterior)
BLA$_p$ Basolateral Nucleus of the Amygdala (posterior)
BMA$_p$ Basomedial Nucleus of the Amygdala (posterior)
CA$_1$ Hippocampal Area CA$_1$ (Ammon's Horn)
CA$_3$ Hippocampal Area CA$_3$ (Ammon's Horn)
CdP Caudoputamen
CeA Central Nucleus of the Amygdala
CO$_d$ Cochlear Nucleus (dorsal)
CO$_p$ Cochlear Nucleus (ventral posterior)
CO$_v$ Cochlear Nucleus (ventral anterior)
CoA$_p$ Cortical Nucleus of the Amygdala (posterior)
De Dentate (Lateral) Cerebellar Nucleus
DG Dentate Gyrus
EP Endopiriform Nucleus
IA Intercalated Mass of the Amygdala
LGd Dorsal Lateral Geniculate Nucleus (Thalamus)
LGv Ventral Lateral Geniculate Nucleus (Thalamus)
MGp Principal/Parvicellular Medial Geniculate Nucleus (Thalamus)
Of Orbitofrontal Area (Cortex)
Pir Piriform Area (Cortex)
s5i Spinal Trigeminal Nucleus (interpolar)
SI Substantia Innominata
SomS Somatosensory Cortex
STF Striatal Fundus
Sub Subiculum
8n Vestibular-Cochlear (Acoustic) Nerve

CO: Cochlear Nuclei. The cochlear nuclei have been subdivided into three tonotopic (spatial representation of audio frequencies) subdivisions mapped in the cat: dorsal, anteroventral, and posteroventral subnuclei (CO$_{d,v,p}$), essentially following the three divisions in the widely used description of Harrison and Warr ('62). A smallcelled, densely packed superficial cap is also evident (Osen, '69) and is sometimes designated as the granular layer or nucleus, but further subdivisions noted in other species are not readily applicable to the rat.

EP: Endopiriform Nucleus. This "nucleus" appears to form the olfactory part of the claustrum deep to the piriform cortex (see Gurdjian, '28; Krettek and Price, '78). It is separated from layer 3 of the piriform cortex by a conspicuous cell-poor zone, a ventral extension of the extreme capsule.

IA: Intercalated Mass of the Amygdala (also called Massa Intercalata). Discrete dense clusters of neurons intercalated among fiber tracts below the lateral-basolateral and central amygdaloid nuclei are described collectively as the IA. The intercalated cells resemble neurons in the corpus striatum, leading Millhouse ('86), who has described this area in detail, to question whether they constitute a ventral extension of the overlying striatum.

LGv: Ventral Lateral Geniculate Nucleus (Thalamus). The ventral lateral geniculate nucleus is clearly nonhomogeneous in cell density and can be variously subdivided – for example, into dorsal and ventral components designated subnuclei by Paxinos and Watson ('86), who also recognize a lateral magnocellular and medial parvicellular division – but rigorous criteria for functional architectural separation are difficult to identify. A separate ventral nucleus, called the *subgeniculate*, is also recognized by these authors at the transition from the LGv to the zona incerta (ZI). This ventrolateral wedge caudally, at the level of the medial geniculate body, forms the lateral terminal nucleus of the optic tract.

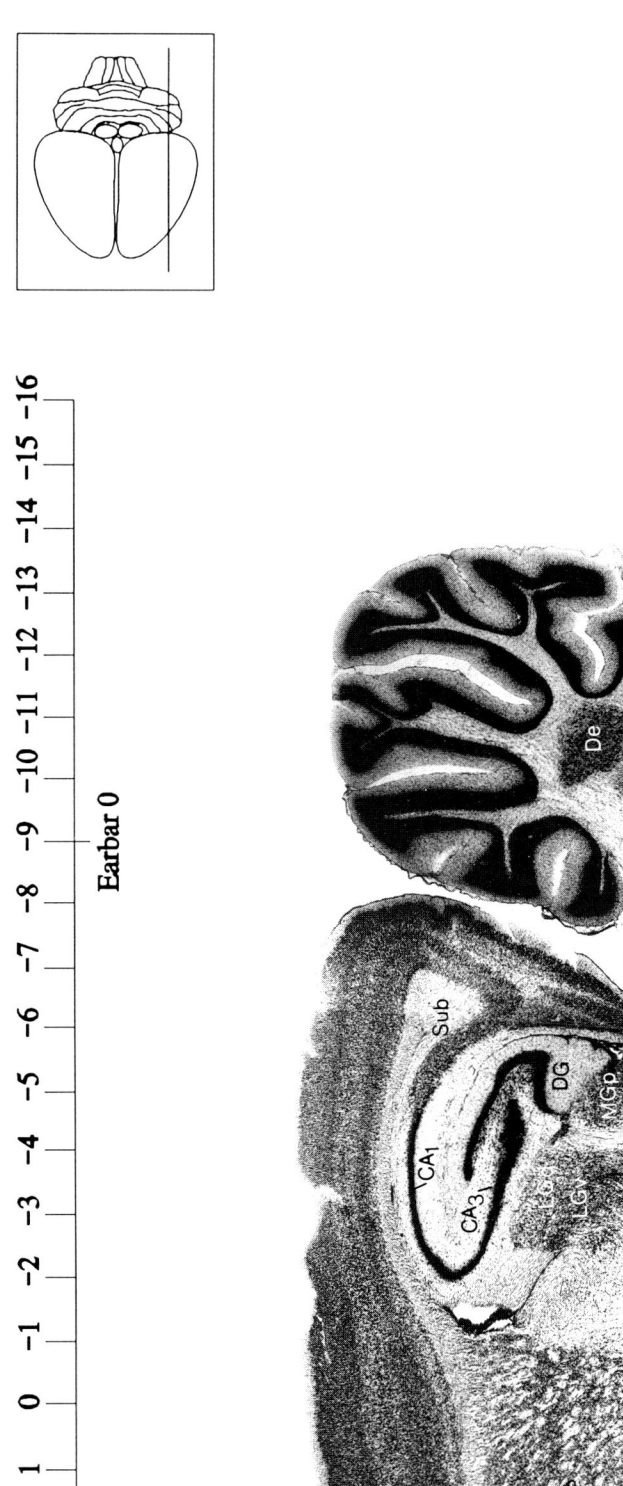
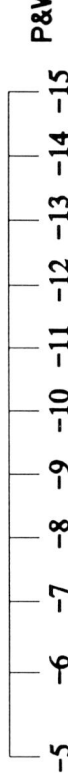
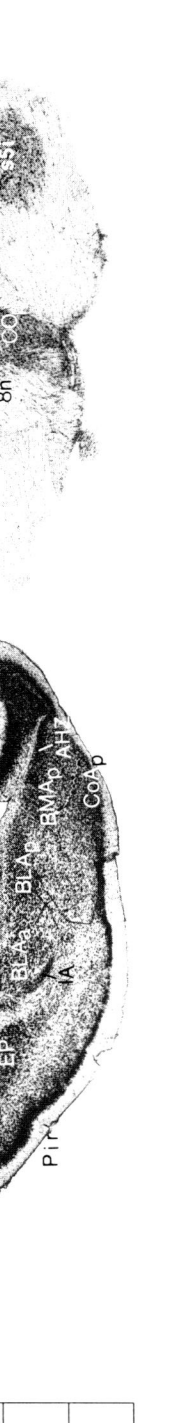

Earbar 0

5 4 3 2 1 0 −1 −2 −3 −4 −5 −6 −7 −8 −9 −10 −11 −12 −13 −14 −15 −16

P&W

−5 −6 −7 −8 −9 −10 −11 −12 −13 −14 −15

Sagittal 3.9

Plate 124

fi   Fimbria
GP   Globus Pallidus (external or lateral)
ic   Internal Capsule
LGd   Dorsal Lateral Geniculate Nucleus (Thalamus)
LGv   Ventral Lateral Geniculate Nucleus (Thalamus)
lot   Lateral Olfactory Tract
mcp   Middle Cerebellar Peduncle
MGp   Principal/Parvicellular Medial Geniculate Nucleus (Thalamus)
R   Reticular Nucleus (Thalamus)
s5i   Spinal Trigeminal Nucleus (interpolar)
5t   Trigeminal Tract
8n   Vestibular-Cochlear (Acoustic) Nerve

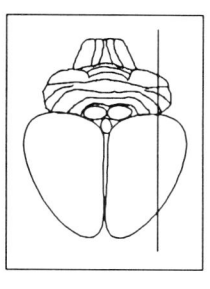

Earbar 0

-16 -15 -14 -13 -12 -11 -10 -9 -8 -7 -6 -5 -4 -3 -2 -1 0 1 2 3 4 5

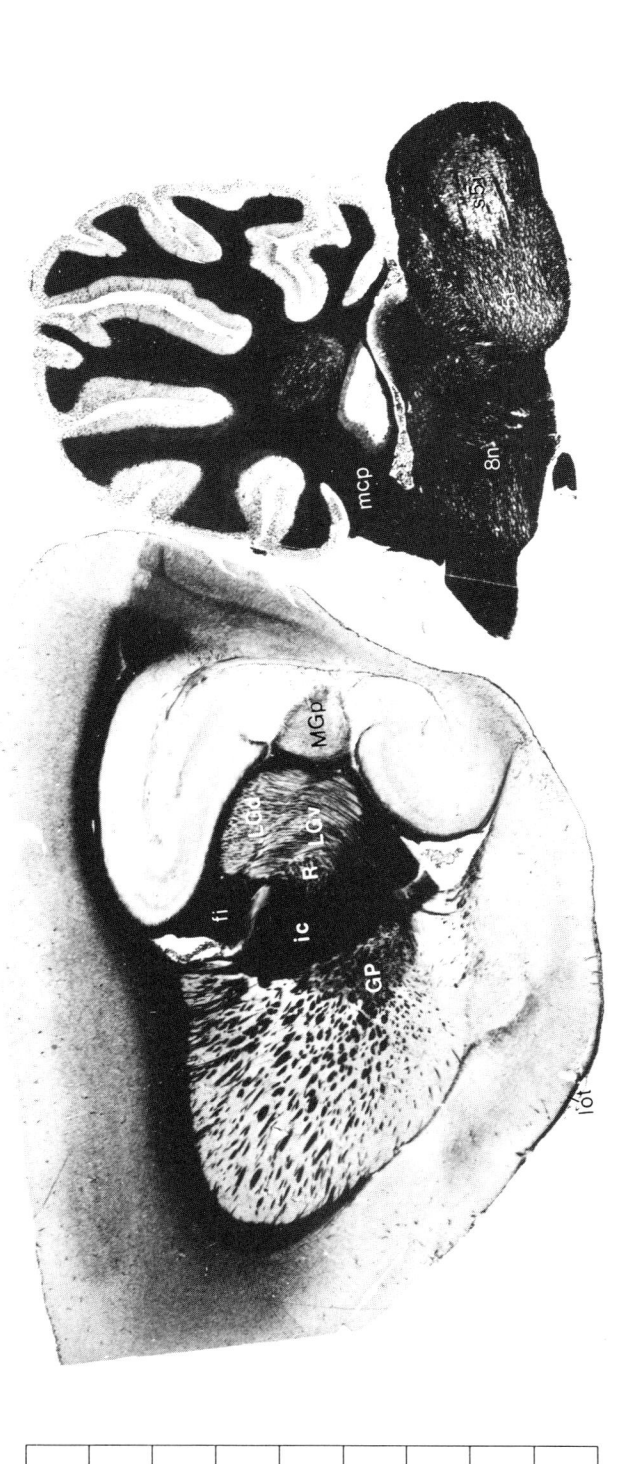

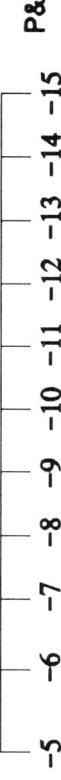

mcp

8n

SCs

SC

MGp

LGd

R

LGv

fi

ic

GP

iot

9 8 7 6 5 4 3 2 1 0

-5 -6 -7 -8 -9 -10 -11 -12 -13 -14 -15

Sagittal 3.9

Plate 125

BLA$_a$ Basolateral Nucleus of the Amygdala (anterior)
BLA$_p$ Basolateral Nucleus of the Amygdala (posterior)
CA$_1$ Hippocampal Area CA$_1$ (Ammon's Horn)
CA$_3$ Hippocampal Area CA$_3$ (Ammon's Horn)
CdP Caudoputamen
CeA Central Nucleus of the Amygdala
DG Dentate Gyrus
ENT Entorhinal Area
EP Endopiriform Nucleus
Ma Motor Agranular Cortical Area
Of Orbitofrontal Area (Cortex)
Pam Periamygdaloid Cortex
Pir Piriform Area (Cortex)
SomS Somatosensory Cortex
Sub Subiculum

CdP: Caudoputamen. The caudate nucleus and the putamen constitute the two largest components of the human basal ganglia and are distinctive and easily separated in sections of the human brain. In the rat, separation would constitute an exaggeration or distortion, although the homology is generally accepted on solid grounds. Fusing these two nuclear aggregates is common practice in contemporary usage.

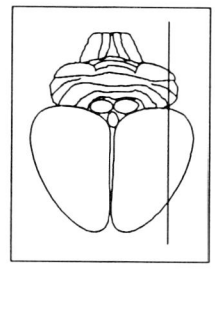
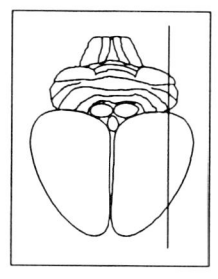

Earbar 0

-14 -13 -12 -11 -10 -9 -8 -7 -6 -5 -4 -3 -2 -1 0 1 2 3 4 5

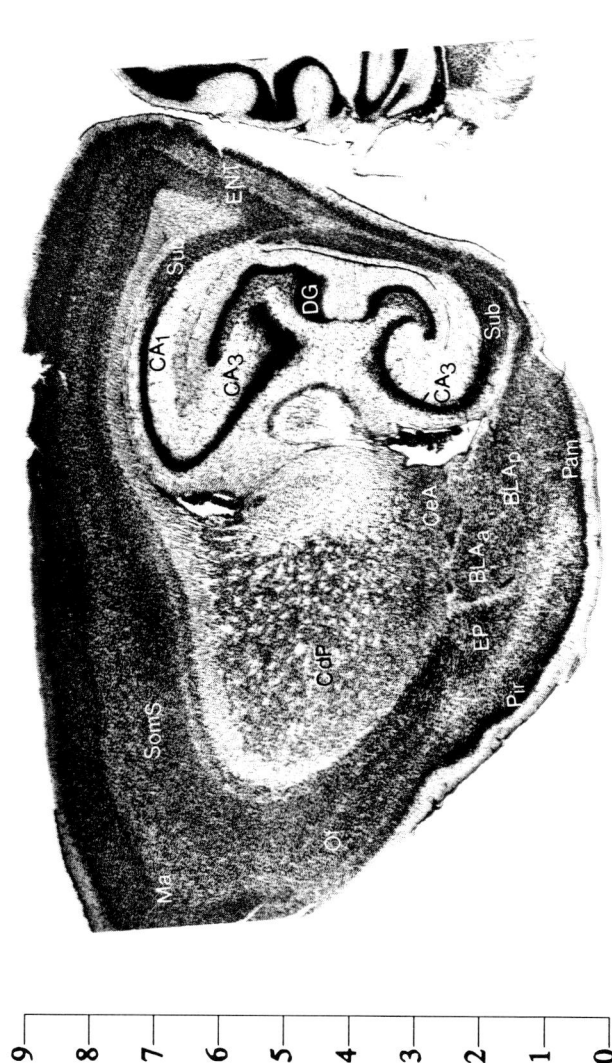

ENT
Su
CA1
DG
CA3
CA3
Sub
CeA
BLAa
BLAp
Pam
EP
Pir
CdP
SomS
Ot
Ma

9 8 7 6 5 4 3 2 1 0

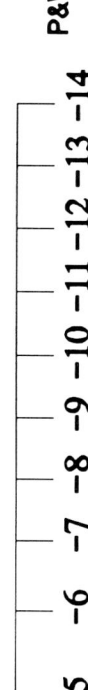

-14 -13 -12 -11 -10 -9 -8 -7 -6 -5

P&W

Sagittal 4.2

Plate 126

alv Alveus
CA₁ Hippocampal Area CA$_1$ (Ammon's Horn)
CA₂ Hippocampal Area CA$_2$ (Ammon's Horn)
CA₃ Hippocampal Area CA$_3$ (Ammon's Horn)
CdP Caudoputamen
co Cochlear Nerve
DG Dentate Gyrus
ec External Capsule
fi Fimbria
ic Internal Capsule
icp Inferior Cerebellar Peduncle
lot Lateral Olfactory Tract
st Stria Terminalis
5t Trigeminal Tract
8n Vestibular-Cochlear (Acoustic) Nerve

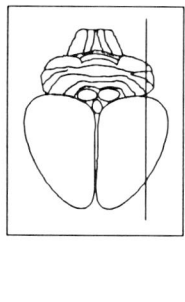
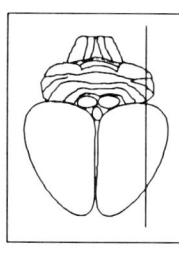

Earbar 0

−14 −13 −12 −11 −10 −9 −8 −7 −6 −5 −4 −3 −2 −1 0 1 2 3 4 5

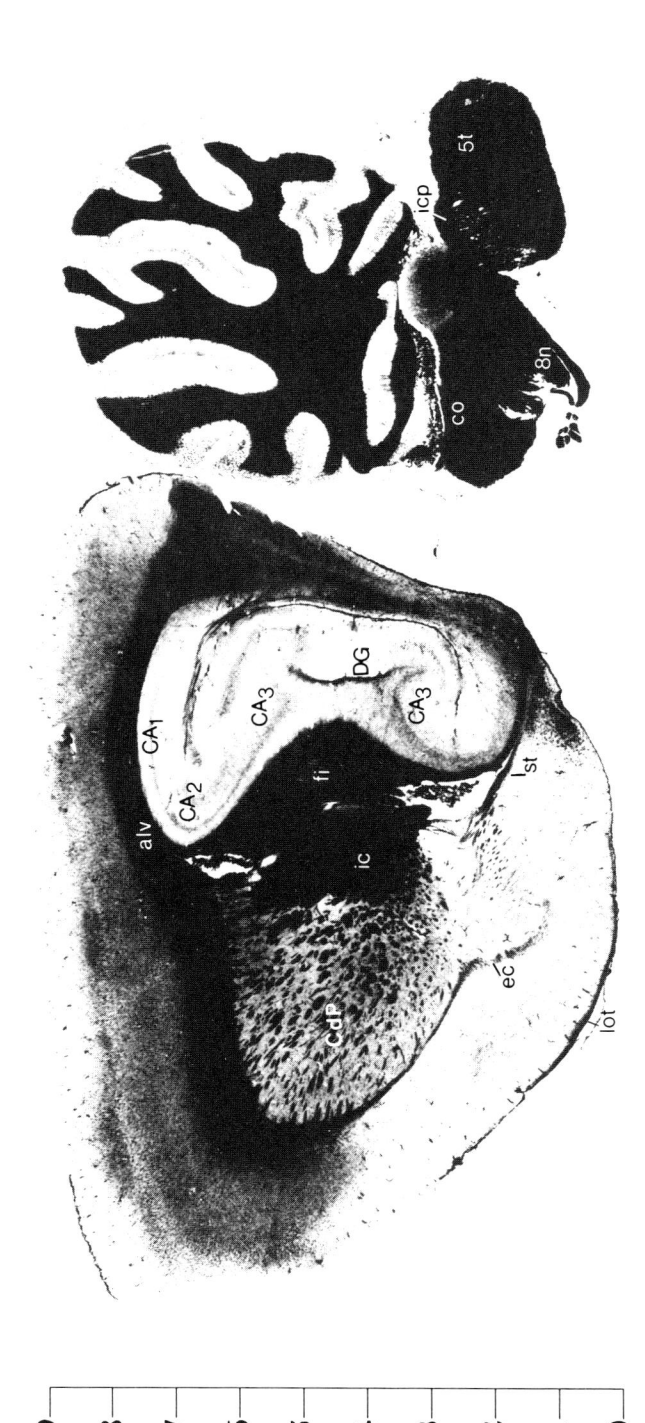
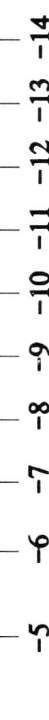

5t

icp

8n

co

CA1

CA2

alv

CA3

fi

DG

CA3

ic

lst

CdP

ec

lot

9 8 7 6 5 4 3 2 1 0

P&W

−14 −13 −12 −11 −10 −9 −8 −7 −6 −5

Sagittal 4.2

Plate 127

AHZ: Amygdalo-Hippocampal Area. This transitional zone where the caudal portion of the basolateral amygdala fuses with the ventral portion of the subiculum and hippocampal field CA$_3$, is sometimes called the *posterior amygdaloid nucleus* (e.g., Swanson, '92) and is further subdivided into subsectors by Paxinos and Watson ('86), who further recognized the transitional zone as an entity distinct from the posterior amygdala. Various atlases differ on the limits and it may be practical to recognize that there is a condensed, distinct zone forming a transition below the corticomedial amygdala bridging the basolateral amygdala and subiculum.

Sub: Subiculum. This distinctive cortical region has thick plexiform and pyramidal layers, each of which may be subdivided. It lies principally between Ammon's horn (field CA$_1$) and the entorhinal field and appears in Nissl preparations to extend ventrally around the tip of the angular bundle into the caudal end of the amygala. It is most easily and commonly subdivided on the basis of fiber architecture. The European school of the early 20th century variously recognized pre-, pro-, post-, and parasubicular regions.

AHZ   Amygdalo-Hippocampal Area
BLA   Basolateral Nucleus of the Amygdala
CA$_1$ Hippocampal Area CA$_1$ (Ammon's Horn)
CA$_3$ Hippocampal Area CA$_3$ (Ammon's Horn)
CdP   Caudoputamen
chp   Choroid Plexus
CO$_v$ Cochlear Nucleus (ventral anterior)
DG   Dentate Gyrus
ENT   Entorhinal Area
EP   Endopiriform Nucleus
LA   Lateral Nucleus of the Amygdala
Ma   Motor Agranular Cortical Area
Pam   Periamygdaloid Cortex
Pir   Piriform Area (Cortex)
SomS   Somatosensory Cortex
Sub   Subiculum

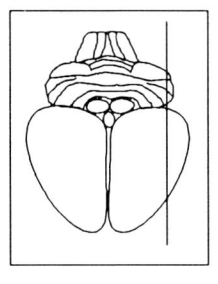

−14 −13 −12 −11 −10 −9 −8 −7 −6 −5 −4 −3 −2 −1 0 1 2 3 4 5

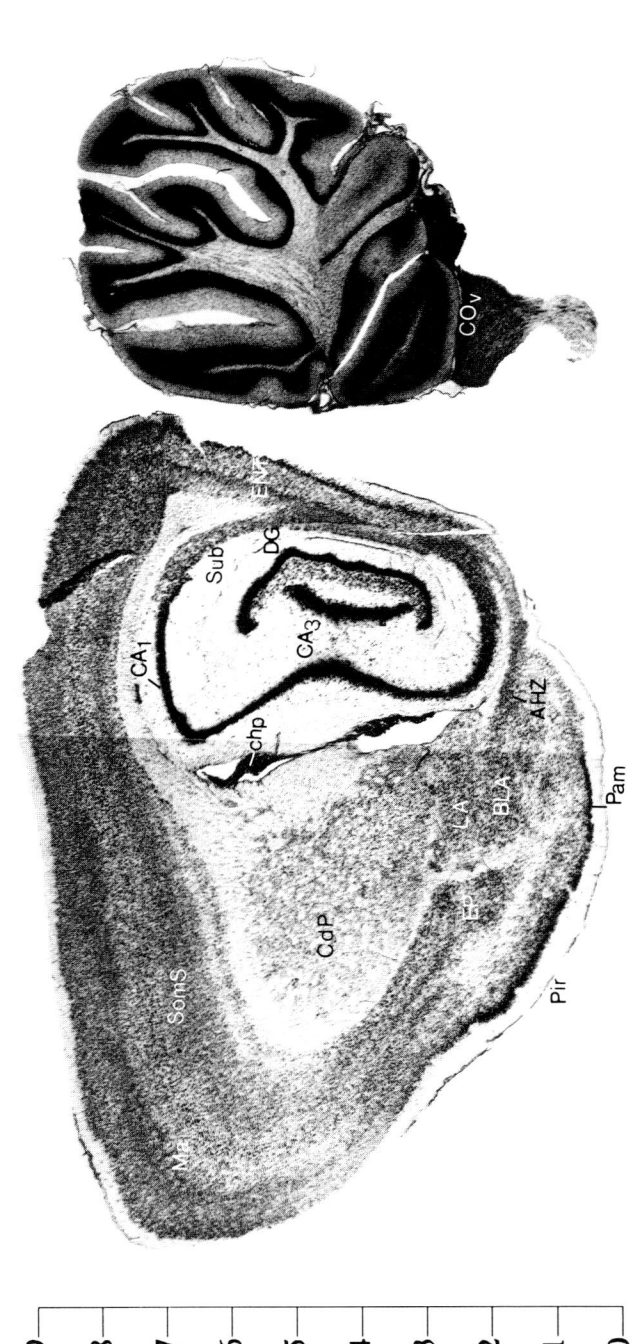

CA1

Sub

DG

CA3

chp

LA

BLA

AHZ

Pam

SomS

CdP

EP

Pir

Ma

COv

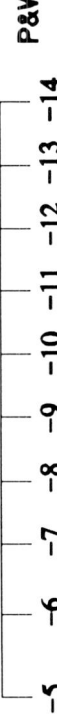

P&W

−14 −13 −12 −11 −10 −9 −8 −7 −6 −5

9 8 7 6 5 4 3 2 1 0

Sagittal 4.5

Plate 128

alv Alveus
CdP Caudoputamen
co Cochlear Nerve
ec External Capsule
fi Fimbria
floc Flocculus (Cerebellum)
ic Internal Capsule
lot Lateral Olfactory Tract
LV Lateral Ventricle
prf Perforant Path
st Stria Terminalis
8n Vestibular-Cochlear (Acoustic) Nerve

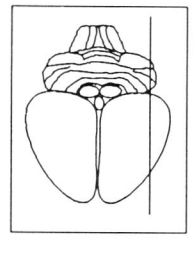

Earbar 0

-14 -13 -12 -11 -10 -9 -8 -7 -6 -5 -4 -3 -2 -1 0 1 2 3 4 5

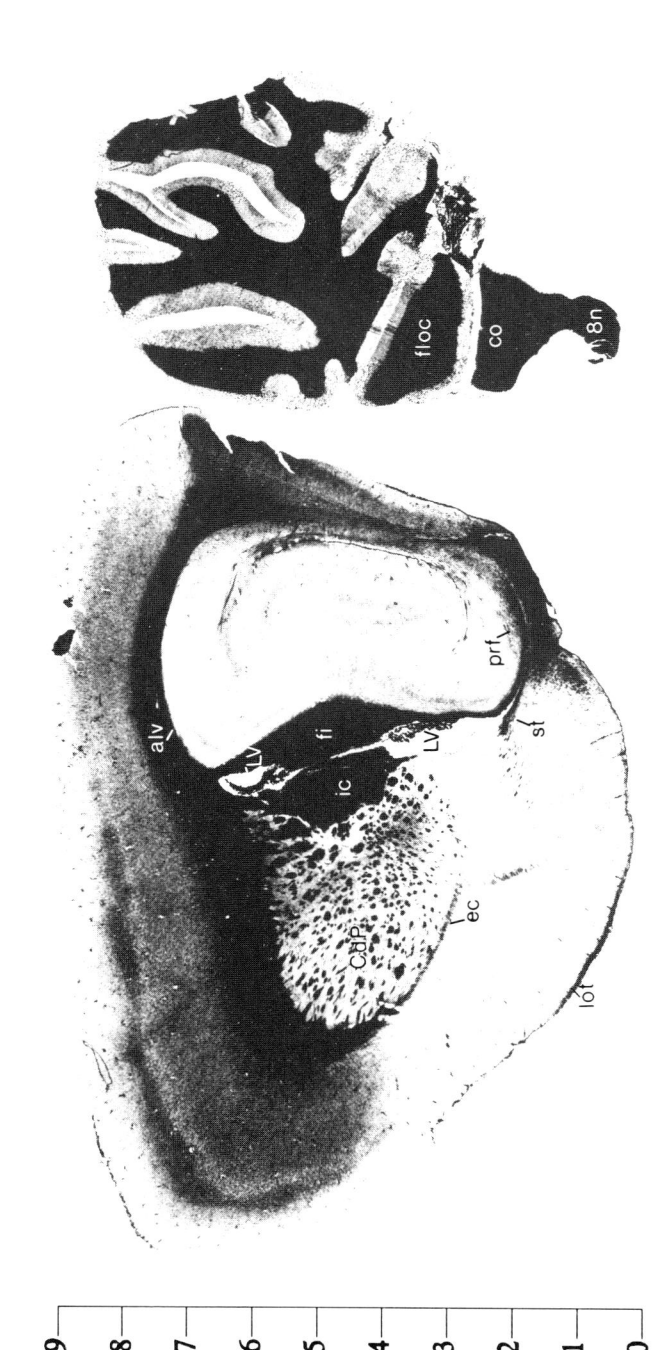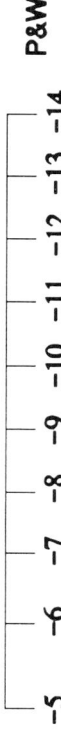

floc

co

8n

alv

fi

LV

LV

prf

ic

st

CbP

ec

lot

9 8 7 6 5 4 3 2 1 0

P&W

-5 -6 -7 -8 -9 -10 -11 -12 -13 -14

Sagittal 4.5

REFERENCES

Akert, K., H. D. Potter and J. W. Anderson (1961). The subfornical organ in mammals. I. Comparative and topographical anatomy. *J. Comp. Neurol., 116:* 1–14.

Albe-Fessard, D., F. Stutinsky, and S. Libouban (1971). *Atlas Stéréotaxique du Diencéphale du Rat Blanc,* Editions du Centre National de la Recherche Scientifique, Paris.

Allen, G. V., and D. A. Hopkins (1988). Mamillary body in the rat: A cytoarchitectonic, Golgi, and ultrastructural study. *J. Comp. Neurol., 275:* 39–64.

Allen, G. V., and D. A. Hopkins (1990). Topography and synaptology of mammillary body projections to the mesencephalon and pons in the rat. *J. Comp. Neurol., 301:* 214–231.

Altschuler, S. M., X. Bao, and R.R. Miselis (1991). Dendritic architecture of nucleus ambiguus motoneurons projecting to the upper alimentary tract in the rat. *J. Comp. Neurol., 309:* 402–414.

Anderson, J. (1929). *How to Stain the Nervous System,* E. Livingstone, Edinburgh.

Andrezik, J. A., and A. J. Beitz (1985). Reticular formation, central gray and related tegmental nuclei. In: *The Rat Nervous System, Vol. 2: Hindbrain and Spinal Cord,* G. Paxinos (ed.), Academic Press, New York, pp. 1–28.

Andrezik, J. A., V. Chan-Palay, and S. L. Palay (1981). The nucleus paragigantocellularis lateralis in the rat. *Anat. Embryol., 161:* 355–371.

Åström, K. E. (1953). The central course of afferent fibres in the trigeminal, facial, glossopharyngeal and vagal nerves and their nuclei in the mouse. *Acta Physiol. Scand. Suppl., 106:* 209–320.

Baker, M. L., and G. J. Giesler, Jr. (1984). Anatomical studies of the spinocervical tract of the rat. *Somatosens. Res., 2:* 1–18.

Bayer, S. A. (1980). Quantitative-^{3}H-thymidine radiographic analysis of neurogenesis in the rat amygdala. *J. Comp. Neurol., 194:* 845–876.

Bayer, S. A. (1985). Hippocampal region. In: *The Rat Nervous System, Vol. 2, Hindbrain and Spinal Cord,* G. Paxinos (ed.), Academic Press, New York, pp. 335–352.

Berman, A. L. (1968). *The Brain Stem of the Cat: A Cytoarchitectonic Atlas with Stereotaxic Coordinates,* University of Wisconsin Press, Madison, 175 pp.

Berman, A. L., and E. G. Jones (1982). *The Thalamus and Basal Telencephalon of the Cat: A Cytoarchitectonic Atlas with Stereotaxic Coordinates,* University of Wisconsin Press, Madison, 164 pp.

Björklund, A., and O. Lindvall (1984). Dopamine-containing systems in the CNS. In: *Handbook of Chemical Neuroanatomy, Vol. 2: Classical Transmitters in the CNS, Part I,* A. Björklund and T. Hökfelt (eds.), Elsevier, New York, pp. 55–122.

Blackstad, T. W. (1956). Commissural connections of the hippocampal region in the rat, with special reference to their mode of termination. *J. Comp. Neurol., 105:* 417–538.

Bleier, R., W. Byne, and I. Siggelkow (1982). Cytoarchitectonic sexual dimorphisms of the medial preoptic and anterior hypothalamic areas in guinea pig, rat, hamster and mouse. *J. Comp. Neurol., 212:* 118–130.

Bleier, R., P. Cohn, and I. R. Siggelkow (1979). A cytoarchitectonic atlas of the hypothalamus and hypothalamic third ventricle of the rat. In: *Handbook of the Hypothalamus, Vol. 1: Anatomy of the Hypothalamus*, P. J. Morgane and J. Panksepp (eds.), Marcel Dekker, New York, pp. 137–220.

Bloch, G. J., and R. A. Gorski (1988). Estrogen/progesterone treatment in adulthood affects the size of several components of the medial preoptic area in the male rat. *J. Comp. Neurol., 275:* 613–622.

Bodian, D. (1939). Studies on the diencephalon of the Virginia opossum. Part I. The nuclear pattern of the adult. *J. Comp. Neurol., 71:* 259–323.

Brichta, A. M., and G. Grant (1985). Cytoarchitectural organization of the spinal cord. In: *The Rat Nervous System, Vol. 2: Hindbrain and Spinal Cord*, G. Paxinos (ed.), Academic Press, New York, pp. 293–301.

Brodal, A. (1947). The amygdaloid nucleus in the rat. *J. Comp. Neurol., 87:* 1–16.

Brodal, A., and O. Pompeiano (1957). The vestibular nuclei in the cat. *J. Anat., 91:* 438–454.

Brodmann, K. (1909). *Vergleichende Lokalisationslehre der Grosshirnrinde in ihren Prinzipien dargestellt auf Grund des Zellenbaues*, Barth, Leipzig, 324 pp.

Broman, J., and A. Blomqvist (1989). Substance P-like immunoreactivity in the lateral cervical nucleus of the owl monkey (*Aotus trivirgatus*): A comparison with the cat and rat. *J. Comp. Neurol., 289:* 111–117.

Brown, J. O. (1943). The nuclear pattern of the nontectal portions of the midbrain and isthmus in the dog and cat. *J. Comp. Neurol., 78:* 365–405.

Brown, L. T. (1974). Corticorubral projections in the rat. *J. Comp. Neurol., 154:* 149–168.

Bucher, V. M., and W. J. H. Nauta (1954). A note on the pretectal cell groups in the rat's brain. *J. Comp. Neurol., 100:* 287–295.

Burton, H., and A. D. Craig, Jr. (1983). Spinothalamic projections in cat, raccoon and monkey: A study based on anterograde transport of horseradish peroxidase. In: *Somatosensory Integration in the Thalamus*, G. Macchi, A. Rustioni, and R. Spreafico (eds.), Elsevier, Amsterdam, pp. 17–41.

Cajal, S. Ramón y (1901). Estructura del septum lucidum. *Trabajos del Laboratorio de Investigaciones Biologicas de la Universidad de Madrid, 1:* 159–188.

Cajal, S. Ramón y (1909, 1911). *Histologie du Système Nerveux de l'Homme et des Vertébrés*, 2 vols., Norbert Maloine, Paris.

Campbell, S. K., T. D. Parker, and W. Welker (1974). Somatotopic organization of the external cuneate nucleus in albino rats. *Brain Res., 77:* 1–23.

Canteras, N. S., R. B. Simerly, and L. W. Swanson (1992a). The connections of the posterior nucleus of the amygdala. *J. Comp. Neurol., 324:* 143–179.

Canteras, N. S., R. B. Simerly, and L. W. Swanson (1992b). Projections of the ventral premammillary nucleus. *J. Comp. Neurol., 324:* 195–212.

Canteras, N. S., and L. W. Swanson (1992). Projections of the ventral subiculum to the amygdala, septum, and hypothalamus: A PHAL anterograde tract-tracing study in the rat. *J. Comp. Neurol., 324:* 180–194.

Castaldi, L. (1923). Studî sulla struttura e sulla sviluppo del mesencefalo: Ricerche in *Cavia cobaya I*. Parte 1. *Arch. Ital. Anat. Embriol., 20:* 23–225.

Castaldi, L. (1926). Studî sulla struttura e sulla sviluppo del mesencefalo: Ricerche in *Cavia cobaya I*. Parte 3. *Arch. Ital. Anat. Embriol., 23:* 481–609.

Cechetto, D. F., and C. B. Saper (1987). Evidence for a viscerotopic sensory representation in the cortex and thalamus in the rat. *J. Comp. Neurol., 262:* 27–45.

Chan-Palay, V. (1988). Galanin hyperinnervates surviving neurons of the human basal nucleus of Meynert in dementias of Alzheimer's and Parkinson's disease: A hypothesis for the role of galanin in accentuating cholinergic dysfunction in dementia. *J. Comp. Neurol., 273:* 543–557.

Chronister, R. B., R. W. Sikes, T. W. Trow, and J. D. De France (1981). The organization of nucleus accumbens. In: *The Neurobiology of the Accumbens*, R. B. Chronister and J. D. De France (eds.), Haer Institute for Electrophysiological Research, pp. 97–146.

Clark, W. E. Le Gross (1932). The structure and connections of the thalamus. *Brain, 55:* 406–470.

Clark, W. E. Le Gros (1938). Morphological aspects of the hypothalamus. In: *The Hypothalamus: Morphological, Functional, Clinical and Surgical Aspects*, W. E. Le Gros Clark, J. Beattie, G. Riddoch, and N. M. Dott (eds.), Oliver and Boyd, Edinburgh, pp. 2–58.

Clerici, W. J., and J. R. Coleman (1990). Anatomy of the rat medial geniculate body: I. Cytoarchitecture, myeloarchitecture, and neocortical connectivity. *J. Comp. Neurol., 297:* 14–31.

Contreras, R. J., M. M. Gomez, and R. Norgren (1980). Central origins of cranial nerve parasympathetic neurons in the rat. *J. Comp. Neurol., 190:* 373–394.

282

Cornwall, J, J. D. Cooper, and O. T. Phillipson (1990). Afferent and efferent connections of the laterodorsal tegmental nucleus in the rat. *Brain Res. Bull., 25,* 271–284.

Cowan, W. M., R. W. Guillery, and T. P. S. Powell (1964). The origin of the mammillary peduncle and other hypothalamic connexions from the midbrain. *J. Anat. (Lond.).,* 98: 345–363.

Craig, A. D., and H. Burton (1981). Spinal and medullary lamina I projection to nucleus submedius in medial thalamus: A possible pain center. *J. Neurophysiol.,* 45: 433–466.

Crosby, E. C., and T. Humphrey (1941). Studies of the vertebrate telencephalon. II. The nuclear pattern of the anterior olfactory nucleus, tuberculum olfactorium and the amygdaloid complex in adult man. *J. Comp. Neurol., 74:* 309–352.

Dahlström, A., and K. Fuxe (1964). Evidence for the existence of monoamine-containing neurons in the central nervous system. I. Demonstration of monoamines in cell bodies of brain stem neurons. *Acta Physiol. Scand., 62* (Suppl. 232): 1–55.

Davis, B. J., F. Macrides, W. H. Youngs, S. P. Schneider and D. L. Rosene (1978). Efferents and centrifugal afferents of the main and accessory olfactory bulbs in the hamster. *Brain Res. Bull., 3:* 59–72.

de Groot, J. (1959). The rat forebrain in stereotaxic coordinates. *Akademie van Wetenschappen und Natuurkunde, N. V. Noord- Hollandsche Uitgevers Mattschappy, Amsterdam, 40 pp.*

De Olmos, J., G. F. Alheid, and C. A. Beltramino (1985). Amygdala. In: *The Rat Nervous System, Vol. 1, Forebrain and Midbrain,* G. Paxinos (ed.), Academic Press, New York, pp. 223–334.

Diepen, R. (1962). Der Hypothalamus. In: *von Möllendorff's Handbuch der Mikroskopischen Anatomie des Menschen, Nervensystem, IV/7,* Springer, Berlin, 525 pp.

Divac, I., S. Marinkovic, J. Mogensen, W. Schwerdtfeger, and J. Regidor (1987). Vertical ascending connections in the isocortex. *Anat. Embryol., 175:* 443–455.

Dorner, G., and J. Staudt (1968). Structural changes in the hypothalamic ventromedial nucleus of the male rat, following neonatal castration and androgen treatment. *Neuroendocrinology, 4:* 278–281.

Dostrovsky, J. A., and G. Guilbaud (1988). Noxious stimuli excite neurons in nucleus submedius of the normal and arthritic rat. *Brain Res., 460:* 269–280.

Edwards, S. B. (1975). Autoradiographic studies of the projections of the midbrain reticular formation: Descending projections of nucleus cuneiformis. *J. Comp. Neurol., 161:* 341–358.

Eisenman, J. S., and E. C. Azmitia (1982). Physiological stimulation enhances HRP marking of salivary neurons in rats. *Brain Res. Bull., 8:* 73–78.

Everitt, B. J., B. Meister, T. Hökfelt, T. Melander, L. Terenius, Á. Rökaeus, E. Theodorsson-Norheim, G. Dockray, J. Edwardson, C. Cuello, R. Elde, M. Goldstein, H. Hemmings, C. Ouimet, I. Walaas, P. Greengard, W. Vale, E. Weber, J.-Y. Wu, and K.-J. Chang (1986). The hypothalamic arcuate nucleus-median eminence complex: Immunohistochemistry of transmitters, peptides and DARPP-32 with special reference to coexistence in dopamine neurons. *Brain Res. Rev., 11:* 97–155.

Fabri, M., and H. Burton (1991). Topography of connections between primary somatosensory cortex and posterior complex in rat: A multiple fluorescent tracer study. *Brain Res., 538:* 351–357.

Falls, W. M. (1986). Morphology and synaptic connections of myelinated primary axons in the ventrolateral region of rat trigeminal nucleus oralis. *J. Comp. Neurol., 244:* 96–110.

Falls, W. M., and M. M. Albin (1986). Morphological features of identified trigeminocerebellar projection neurons in the border zone of rat trigeminal nucleus oralis. *Somatosens. Res., 4:* 1–12.

Falls, W. M., B. J. Moore, and M. T. Schneider (1990). Fine structural characteristics and synaptic connections of trigeminocerebellar projection neurons in rat trigeminal nucleus oralis. *Somatosens. Mot. Res., 7:* 1–18.

Falls, W. M., R. E. Rice, and J. P. Van Wagner (1985). The dorsomedial portion of trigeminal nucleus oralis (Vo) in the rat: Cytology and projections to the cerebellum. *Somatosens. Res., 3:* 89–110.

Faull, R. L. M., and W. R. Mehler (1985). Thalamus. In: *The Rat Nervous System, Vol. 1: Forebrain and Midbrain,* G. Paxinos (ed.), Academic Press, New York, pp. 129–168.

Feldman, S. G., and L. Kruger (1980). An axonal transport study of the ascending projection of medial lemniscal neurons in the rat. *J. Comp. Neurol., 192:* 427–454.

Flügge, A. Jurdzinski, S. Brandt, and E. Fuchs (1990). Alpha2-adrenergic binding sites in the medulla oblongata of tree shrews demonstrated by *in vitro* autoradiography: Species related differences in comparison to the rat. *J. Comp. Neurol., 297:* 253–266.

Fox, C.A. (1940). Certain basal telencephalic centers in the cat. *J. Comp. Neurol., 72:* 1–62.

Fox, E. A., and T. L. Powley (1992). Morphology of identified preganglionic neurons in the dorsal motor nucleus of the vagus. *J. Comp. Neurol., 322:* 79–98.

Frederickson, C. J., and D. R. Trune (1986). Cytoarchitecture and saccular innervation of nucleus Y in the mouse. *J. Comp. Neurol., 252:* 302–322.

Fry, F. J., and W. M. Cowan (1972). A study of retrograde cell degeneration in the lateral mammillary nucleus of the cat, with special reference to the role of axonal branching in the preservation of the cell. *J. Comp. Neurol., 144:* 1–24.

Fulwiler, C. E., and C. B. Saper (1984). Subnuclear organization of the efferent connections of the parabrachial nucleus in the rat. *Brain Res. Rev., 7:* 229–259.

Geeraedts, L. M. G., R. Nieuwenhuys and J. G. Veening (1990). Medial forebrain bundle of the rat: IV. Cytoarchitecture of the caudal (lateral hypothalamic) part of the medial forebrain bundle bed nucleus. *J. Comp. Neurol., 294:* 537–568.

Giolli, R. A., R. H. I. Blanks, and Y. Torigoe (1984). Pretectal and brain stem projections of the medial terminal nucleus of the accessory optic system of the rabbit and rat as studied by anterograde and retrograde neuronal tracing methods. *J. Comp. Neurol., 227:* 228–251.

Giolli, R. A., R. H. I. Blanks, Y. Torigoe, and D. D. Williams (1985). Projections of the medial terminal accessory optic nucleus, ventral tegmental nuclei, and the substantia nigra of rabbit and rat as studied by retrograde neuronal transport of horseradish peroxidase. *J. Comp. Neurol., 232:* 99–116.

Giolli, R. A., R. J. Clarke, R. H. I. Blanks, Y. Torigoe, and J. H. Fallon (1989). Organization of rat medial terminal accessory optic nucleus; axon collateralization of neurons and its GABAergic neurons. *Anat. Rec., 223:* 43A.

Gobel, S., W. M. Falls, and S. Hockfield (1977). The division of the dorsal and ventral horns of the mammalian caudal medulla into eight layers using anatomical criteria. In: *Pain in the Trigeminal Region,* D. J. Anderson and B. Matthews (eds.), Elsevier/North-Holland, Amsterdam, pp. 443–453.

Goldberg, J. M., and R. Y. Moore (1967). Ascending projections of the lateral lemniscus in the cat and monkey. *J. Comp. Neurol., 129:* 143–156.

Gorski, R. A., J. H. Gordon, J. E. Shryne, and A. M. Southam (1978). Evidence for a morphological sex difference within the medial preoptic area of the rat brain. *Brain Res., 148:* 333–346.

Graybiel, A. M., and C. W. Ragsdale, Jr. (1979). Fiber connections of the basal ganglia. *Prog. Brain Res., 51:* 239–283.

Gregory, K. M. (1985). The dendritic architecture of the visual pretectal nuclei of the rat: A study with the Golgi-Cox method. *J. Comp. Neurol., 234:* 122–135.

Gritti, I., L. Mainville and B. E. Jones (1993). Codistribution of GABA- with acetylcholine-synthesizing neurons in the basal forebrain of the rat. *J. Comp. Neurol., 329:* 438–457.

Groenewegen, H. J., S. Ahlenius, S. N. Haber, N. W. Kowall, and W. J. H. Nauta (1986). Cytoarchitecture, fiber connections, and some histochemical aspects of the interpeduncular nucleus in the rat. *J. Comp. Neurol., 249:* 65–102.

Gstoettner, W., and M. Burian (1987). Vestibular nuclear complex in the guinea pig: A cytoarchitectonic study and map in three planes. *J. Comp. Neurol., 257:* 176–188.

Gurdjian, E. S. (1925). Olfactory connections in the albino rat, with special reference to the stria medullaris and the anterior commissure. *J. Comp. Neurol., 38:* 128–163.

Gurdjian, E. S. (1927). The diencephalon of the albino rat. *J. Comp. Neurol., 43:* 1–114.

Gurdjian, E. S. (1928) The corpus striatum of the rat. *J. Comp. Neurol., 45:* 249–281.

Gwyn, D. G., G. P. Nicholson, and B. A. Flumerfelt (1977). The inferior olivary nucleus of the rat. A light and electron microscopic study. *J. Comp. Neurol., 174:* 489–501.

Haberly, L. B., and J. L. Price (1978). Association and commissural fiber systems of the olfactory cortex of the rat. II. Systems originating in the olfactory peduncle. *J. Comp. Neurol., 181:* 781–808.

Hamill, G. S., J. A. Olschowka, N. J. Lenn, and D. M. Jacobowitz (1984). The subnuclear distribution of substance P, cholecystokinin, vasoactive intestinal peptide, somatostatin, leu-enkephalin, dopamine-b-hydroxylase, and serotinin in the rat interpeduncular nucleus. *J. Comp. Neurol., 226:* 580–596.

Harrison, J. M., and W. B. Warr (1962). A study of the cochlear nucleus and ascending auditory pathways of the medulla. *J. Comp. Neurol., 119:* 341–379.

Haug, F. S. (1976). Sulphide silver pattern and cyto-architectonics of parahippocampal areas in the rat. Special reference to the subdivision of area entorhinalis (area 28) and its demarcation from the pyriform cortex. *Anat. Embryol. Cell Biol., 52:* 1–73.

Haxhiu, M. A., A. S. P. Jansen, N. S. Cherniack, and A. D. Loewy (1993). CNS innervation of airway-related parasympathetic preganglionic neurons: A transneuronal labeling study using pseudorabies virus. *Brain Res., 618:* 115–134.

Hayakawa, T., and K. Zyo (1983). Comparative cytoarchitectonic study of Gudden's tegmental nuclei in some mammals. *J. Comp. Neurol., 126:* 233–244.

Hayhow, W. R., A. Sefton, and C. Webb (1962). Primary optic centers of the rat in relation to the terminal distribution of the crossed and uncrossed optic nerve fibers. *J. Comp. Neurol., 118:* 295–322.

Hayhow, W. R., C. Webb, and A. Jervie (1960). The accessory optic fiber system in the rat. *J. Comp. Neurol., 115:* 187–215.

Heimer, L. (1972). The olfactory connections of the diencephalon in the rat. *Brain Behav. Evol., 6:* 484–523.

Heimer, L., G. F. Alheid, and L. Zaborszky (1985). Basal ganglia. In: *The Rat Nervous System,* G. Paxinos (ed.), Academic Press, Sydney, Australia, pp. 37–86.

Henkel, C. K., and G. F. Martin (1977a). The vestibular complex of the American Opossum, *Didelphis virginiana.* I. Conformation, cytoarchitecture and primary vestibular input. *J. Comp. Neurol., 172:* 299–320.

Henkel, C. K., and G. F. Martin (1977b). The vestibular complex of the American Opossum, *Didelphis virginiana.* II. Afferent and efferent connections. *J. Comp. Neurol., 172:* 321–348.

Hjorth-Simonsen, A. (1972). Projection of the lateral part of the entorhinal area to the hippocampus and fascia dentata. *J. Comp. Neurol., 146:* 219–232.

Hökfelt, T., K. Fuxe, M. Goldstein, and O. Johansson (1974). Immunohistochemical evidence for the existence of adrenaline neurons in the rat brain. *Brain Res., 66:* 235–261.

Houser, C. R., G. D. Crawford, R. P. Barber, P. M. Salvaterra, and J. E. Vaughn. (1983). Organization and morphological characteristics of cholinergic neurons: An immunocytochemical study with a monoclonal antibody to choline acetyltransferase. *Brain Res., 266:* 97–119.

Imaki, T., J.-L. Nahan, C. Rivier, P. E. Sawchenko, and W. Vale (1991). Differential regulation of corticotropin-releasing factor mRNA in rat brain regions by glucocorticoids and stress. *J. Neurosci., 11:* 585–599.

Ishikawa, K., K. Katakai, S. Tanaka, S. Haga, H. Mochida, and K. Itoh (1992). Pro-opiomelanocortin-containing neurons in rat median eminence. *Neuroendocrinology, 56:* 178–184.

Jacobsohn, L. (1909). Über die Kerne des menschlichen Hirnstamms. *Königl. Preuss. Akad. Wiss.,* Berlin, 70 pp.

Jacquin, M. F., R. D. Mooney, and R. W. Rhoades (1986). Morphology, response properties, and collateral projections of trigeminothalamic neurons in brainstem subnucleus interpolaris of rat. *Exp. Brain Res., 61:* 457–468.

Jacquin, M. F., and R. W. Rhoades (1990). Cell structure and response properties in the trigeminal subnucleus oralis. *Somatosens. Mot. Res., 7:* 265–288.

Jansen, A. S. P., G. J. Ter Horst, T. C. Mettenleiter, and A. D. Loewy (1992). CNS cell groups projecting to the submandibular parasympathetic preganglionic neurons in the rat: A retrograde transneuronal viral cell body labeling study. *Brain Res., 572:* 253–260.

Johnston, J. B. (1913). The morphology of the septum, hippocampus and pallial commissures in reptiles and mammals. *J. Comp. Neurol., 23:* 371–478.

Jones, E. G. (1983). The thalamus. In: *Chemical Neuroanatomy,* P. C. Emson (ed.), Raven Press, New York, pp. 257–294.

Jones, E. G. (1985). *The Thalamus,* Plenum Press, New York, 935 pp.

Jones, E. G., and H. Burton (1974). Cytoarchitecture and somatic sensory connectivity of thalamic nuclei other than the ventrobasal complex in the cat. *J. Comp. Neurol., 154:* 395–432.

Jones, E. G., and R. Y. Leavitt (1974). Retrograde axonal transport and the demonstration of non-specific projections to the cerebral cortex and striatum from thalamic intralaminar nuclei in the rat, cat and monkey. *J. Comp. Neurol., 154:* 349–378.

Ju, G., and L. W. Swanson (1989). Studies on the cellular architecture of the bed nuclei of the stria terminalis in the rat: I. Cytoarchitecture. *J. Comp. Neurol., 280:* 587–602.

Kalia, M., and J. M. Sullivan (1982). Brainstem projections of sensory and motor components of the vagus nerve in the rat. *J. Comp. Neurol., 211:* 248–264.

Kiss, J. Z., J. Martos, and M. Palkovits (1991). Hypothalamic paraventricular nucleus – a quantitative analysis of cytoarchitectonic subdivisions in the rat. *J. Comp. Neurol.*, 313: 563–573.

Köhler, C., L. W. Swanson, L. Haglund, and Y.-Y. Wu (1985). The cytoarchitecture, histochemistry and projections of the tuberomammillary nucleus in the rat. *Neuroscience*, 16: 85–110.

Kolb, B., and R. D. Tees (eds.) (1990). *The Cerebral Cortex of the Rat*, MIT Press, Cambridge, MA, 570 pp.

König, J. F. R., and R. A. Klippel (1963). *The Rat Brain. A Stereotaxic Atlas of the Forebrain and Lower Parts of the Brain Stem*, Williams & Wilkins, Baltimore.

König, J. F. R., and R. A. Klippel (1970). *The Rat Brain. A Stereotaxis Atlas of the Forebrain and Lower Parts of the Brain Stem*, Krieger, Huntington, NY.

Korneliussen, H. K. (1968). On the morphology and subdivision of the cerebellar nuclei of the rat. *J. Hirnforsch.*, 10: 109–119.

Kosar, E., J. H. Grill, and R. Norgren (1986). Gustatory cortex in the rat. I. Physiological properties and cytoarchitecture. *Brain Res.*, 379: 329–341.

Krettek, J. E., and J. L. Price (1977a). Projections from the amygdaloid complex to the cerebral cortex and thalamus in the rat and cat. *J. Comp. Neurol.*, 172: 687–722.

Krettek, J. E., and J. L. Price (1977b). Projections from the amygdaloid complex and adjacent olfactory structures to the entorhinal cortex and to the subiculum in the rat and cat. *J. Comp. Neurol.*, 172: 723–752.

Krettek, J. E., and J. L. Price (1977c). The cortical projections of the mediodorsal nucleus and adjacent thalamic nuclei in the rat. *J. Comp. Neurol.*, 171: 157–192.

Krettek, J. E., and J. L. Price (1978). A description of the amygdaloid complex in the rat and cat with observations on intra-amygdaloid axonal connections. *J. Comp. Neurol.*, 178: 255–280.

Krieg, W. J. S. (1932). The hypothalamus of the albino rat. *J. Comp. Neurol.*, 55: 19–89.

Krieg, W. J. S. (1944). The medial region of the thalamus of the albino rat. *J. Comp. Neurol.*, 80: 381–415.

Krieg, W. J. S. (1946). Accurate placement of minute lesions in the brain of the albino rat. *Q. Bull. Northwestern Univ. Med. Sch.*, 20: 199–208.

Krieg, W. J. S. (1946a). Connections of the cerebral cortex. I. The albino rat. A. Topography of the cortical areas. *J. Comp. Neurol.*, 84: 221–276.

Krieg, W. J. S. (1946b). Connections of the cerebral cortex. I. The albino rat. B. Structure of the cortical areas. *J. Comp. Neurol.*, 84: 277–324.

Kruger, L. (1979). Functional subdivisions of the brain stem sensory trigeminal nuclear complex. In: *Advances in Pain Research and Therapy*, J. J. Bonica, J. C. Liebeskind, and D. B. Albe-Fessard (eds.), Raven Press, New York, 3: 197–209.

Kruger, L., C. Bendotti, R. Rivolta, and R. Samanin (1992). GAP-43 mRNA localization in the rat hippocampus CA3 field. *Mol. Brain Res.*, 13: 267–272.

Kruger, L., C. Bendotti, R. Rivolta, and R. Samanin (1993). The distribution of GAP-43 mRNA in the adult rat brain. *J. Comp. Neurol.*, 333: 417–434.

Kruger, L., and P. W. Mantyh (1989). Gustatory and related chemosensory systems. In: *Handbook of Chemical Neuroanatomy, Vol. 7: Integrated Systems of the CNS, Part II*, A. Björklund, T. Hökfelt, and L. W. Swanson (eds.), Elsevier, Amsterdam, pp. 323–411.

Kruger, L., S. Saporta, and S. G. Feldman (1977). Axonal transport studies of the sensory trigeminal complex. In: *Pain in the Trigeminal Region*, D. J. Anderson and B. Matthews (eds.), Elsevier Press, Amsterdam, pp. 191–201.

Kruger, L., C. Sternini, N. C. Brecha, and P. W. Mantyh (1988a). The thalamic region of calcitonin gene-related peptide (CGRP) immunoreactivity and its relation to somatosensory pathways. In: *Cellular Thalamic Mechanisms*, M. Bentivoglio and R. Spreafico (eds.), Elsevier, Amsterdam, pp. 375–386.

Kruger, L., C. Sternini, N. C. Brecha, and P. W. Mantyh (1988b). Distribution of calcitonin gene-related peptide immunoreactivity in relation to the rat central somatosensory projection. *J. Comp. Neurol.*, 273: 149–162.

Kuhlenbeck, H., and R. N. Miller (1942). The pretectal region of the rabbit's brain. *J. Comp. Neurol.*, 76: 323–365.

LaMotte, C. C., S. E. Kapadia, and C. M. Shapiro (1991). Central projections of the sciatic, saphenous, median, and ulnar nerves of the rat demonstrated by transganglionic transport of choleragenoid-HRP (B-HRP) and wheat germ agglutinin-HRP (WGA-HRP). *J. Comp. Neurol.*, 311: 546–562.

Lauterborn, J. C., P. J. Isackson, R. Montalvo, and C. M. Gall (1993). *In situ* hybridization localization of choline acetyltransferase mRNA in adult rat brain and spinal cord. *Mol. Brain Res.*, 17: 59–69.

Le Doux, J. E., D. A. Ruggiero, and D. J. Reis (1985). Projections to the subcortical forebrain from anatomically defined subregions of the medial geniculate body of the rat. *J. Comp. Neurol.*, *242*: 182–213.

Lind, R. W., G. W. Van Hoesen, and A. K. Johnson (1982). An HRP study of the connections of the subfornical organ of the rat. *J. Comp. Neurol.*, *210*: 265–277.

Loewy, A. D., and S. McKellar (1981). Serotoninergic projections from the ventral medulla to the intermediolateral cell column in the rat. *Brain Res.*, *211*: 146–152.

Loo, Y. T. (1931). The forebrain of the opossum, *Didelphis virginiana*. *J. Comp. Neurol.*, *52*: 1–148.

Lorente de Nó (1922). Contribución al conocimiento del nervio trigémino. In: *Libro en Honor de D.S. Ramón y Cajal: Trabajos Originales de sus Admiradores y Discipulos, Extranjeros y Nacionales, Tomo 2*, Madrid, pp. 13–30.

Lorente do Nó (1934). Studies on the structure of the cerebral cortex. II. Continuation of the study of the ammonic system. *J. Psychol. Neurol.*, *46*: 113–177.

Low, J. S. T., L. A. Mantle-St. John, and D. J. Tracey (1986). Nucleus Z in the rat: Spinal afferents from collaterals of dorsal spinocerebellar tract neurons. *J. Comp. Neurol.*, *243*: 510–526.

Luo, P. F., B. R. Wang, Z. Z. Peng, and J. S. Li (1991). Morphological characteristics and terminating patterns of masseteric neurons of the mesencephalic trigeminal nucleus in the rat: An intracellular horseradish peroxidase labeling study. *J. Comp. Neurol.*, *303*: 286–299.

Ma, P. M., and T. A. Woolsey (1984). Cytoarchitectonic correlates of the vibrissae in the medullary trigeminal complex of the mouse. *Brain Res.*, *306*: 374–379.

Malone, E. (1910). Über die Kerne des menschlichen Diencephalon. *Königl. Preuss. Akad. Wiss.*, Berlin, 92 pp.

Malone, E. (1914). The nuclei tuberis laterales and the so-called ganglion opticum basale. *Bull. Johns Hopk. Hosp.*, *17*: 441–480.

Mantle-St. John, L. A., and D. J. Tracey (1987). Somatosensory nuclei in the brainstem of the rat: Independent projections to the thalamus and cerebellum. *J. Comp. Neurol.*, *255*: 259–271.

Marburg, O. (1904). *Mikroskopisch-topographischer Altas des menschlichen Zentralnervensystems*, F. Deuticke, Leipzig., 125 pp.

Maslany, S., D. P. Crockett and M. D. Egger (1991). Somatotropic organization of the dorsal column nuclei in the rat: Transganglionic labelling with B-HRP and WGA-HRP. *Brain Res.*, *564*: 56–65.

Matesz, C., and G. Székely (1983). The motor nuclei of the glossopharyngeal-vagal and the accessorius nerves in the rat. *Acta Biol. Hung.*, *34*: 215–230.

Maxwell, D. S., L. Kruger, and A. Pineda (1969). The trigeminal root with special reference to the central-peripheral transition zone: An electron microscopic study in the macaque. *Anat. Rec.*, *164*: 113–125.

Meessen, H., and J. Olszewski (1949). *A Cytoarchitectonic Atlas of the Rhombencephalon of the Rabbit*, S. Karger, Basel, 52 pp.

Mehler, W. R. (1980). Subcortical afferent connections of the amygdala in the monkey. *J. Comp. Neurol.*, *190*: 733–762.

Mehler, W. R., and Rubertone, J. A. (1985). Anatomy of the vestibular nucleus complex. In: *The Rat Nervous System, Vol. 2, Hindbrain and Spinal Cord*, G. Paxinos (ed.), Academic Press, New York, pp. 185–219.

Miletic, V., and J. E. Coffield (1989). Responses of neurons in the rat nucleus submedius to noxious and innocuous mechanical cutaneous stimulation. *Somatosens. Mot. Res.*, *6*: 567–587.

Millhouse, O. E. (1986). The intercalated cells of the amygdala. *J. Comp. Neurol.*, *247*: 246–271.

Millhouse, O. E., and L. Heimer (1984). Cell configurations in the olfactory tubercle of the rat. *J. Comp. Neurol.*, *228*: 571–597.

Mizuno, N. (1970). Projection fibers from the main sensory trigeminal nucleus and the supratrigeminal region. *J. Comp. Neurol.*, *139*: 457–471.

Morest, K. D. (1961). Connexions of the dorsal tegmental nucleus in rat and rabbit. *J. Anat.*, *95*: 229–246.

Narkiewicz, (1965). Degenerations in the claustrum after regional neocortical ablations in the cat. *J. Comp. Neurol.*, *123*:335–356.

Nauta, W. J. H., and W. Haymaker (1969). Hypothalamic nuclei and fiber connections. In: *The Hypothalamus*, W. Haymaker, E. Anderson and W. J. H. Nauta (eds.), C. C. Thomas, Springfield, pp. 136–209.

Nauta, W. J. H., and J. J. van Straaten (1947). The primary optic centres of the rat. An experimental study by the "bouton" method. *J. Anat.*, *81*: 127–134.

Newman, D. B. (1985). Distinguishing rat brainstem reticulospinal nuclei by their neuronal morphology. I. Medullary Nuclei. *J. Hirnforsch.*, 26: 187–226.

Nicholson, J. E., and C. M. Severin (1981). The superior and inferior salivatory nuclei in the rat. *Neurosci. Lett.*, 21: 149–154.

Nieuwenhuys, R., L. M. G. Geeraedts, and J. G. Veening (1982). The medial forebrain bundle of the rat. *J. Comp. Neurol.*, 206: 49–81.

Noback, C. R., and L. Gross (1959). Brain of a gorilla. I. Surface anatomy and cranial nerve nuclei. *J. Comp. Neurol.*, 111: 321–343.

Nord, S. (1967). Somatotropic organization in the spinal trigeminal nucleus, the dorsal column nuclei and related structures in the rat. *J. Comp. Neurol.*, 130: 343–356.

Nunezabades, P. A., R. Pasaro, and A. L. Bianchi (1992). Study of the topographical distribution of different populations of motoneurons within rat's nucleus ambiguus, by means of four different fluorochromes. *Neurosci. Lett.*, 135: 103–105.

Olszewski, J., and D. Baxter (1954). *Cytoarchitecture of the Human Brain Stem*, Karger, New York, 199 pp.

Onstott, D., B. Mayer, and A. J. Beitz (1993). Nitric oxide synthase immunoreactive neurons anatomically define a longitudinal dorsolateral column within the midbrain periaqueductal gray of the rat: analysis using laser confocal microscopy. *Brain Res.*, 610: 317–324.

Osen, K. K. (1969). Cytoarchitecture of the cochlear nuclei in the cat. *J. Comp. Neurol.*, 136: 453–484.

Paxinos, G., and L. L. Butcher (1985). Organization principles of the brain as revealed by choline acetyltransferase and acetylcholinesterase distribution and projections. In: *The Rat Nervous System, Vol. 2: Hindbrain and Spinal Cord*, G. Paxinos (ed.), Academic Press, New York, pp. 487–521.

Paxinos, G., and C. Watson (1982). *The Rat Brain in Stereotaxic Coordinates*, Academic Press, New York.

Paxinos, G., and C. Watson (1986). *The Rat Brain in Stereotaxic Coordinates*, 2d ed., Academic Press, New York.

Paxinos, G., C. Watson, M. Pennisi, and A. Topple (1985). Bregma, lambda and the interaural midpoint in stereotaxic surgery with rats of different sex, strain and weight. *J. Neurosci. Meth.*, 13: 139–143.

Pellegrino, L. J., A. S. Pellegrino, and A. J. Cushman (1979). *A Stereotaxic Atlas of the Rat Brain*. Plenum Press, New York.

Petrovicky, P. (1985). Gudden's tegmental nuclei and their connections to the hypothalamus and the reticular formation. I. An experimental study using retrograde labelling with HRP or iron-dextran in the rat. *J. Hirnforsch.*, 26: 531–537.

Petrovicky, P. (1990). Thalamic afferents from the brain stem. An experimental study using retrograde single and double labelling with HRP and iron-dextran in the rat. I. Medial and lateral reticular formation. *J. Hirnforsch.*, 31: 359–374.

Petrovicky, P., D. Kolesárová, and V. Slavinská (1990). Thalamic afferents from the brain stem. An experimental study using retrograde single and double labelling with HRP and iron-dextran in the rat. II. Nucleus laterodorsalis and subnucleus compactus nuclei pedunculo-pontini. *J. Hirnforsch.*, 31: 375–383.

Phillipson, O. T. (1979). A golgi study of the ventral tegmental area of Tsai and interfascicular nucleus in the rat. *J. Comp. Neurol.*, 187: 99–116.

Powell, T. P. S., and W. M. Cowan (1955). An experimental study of the efferent connexions of the hippocampus. *Brain, 78:* 115–135.

Pribram, K. H., and L. Kruger (1954). Functions of the "olfactory brain." *Ann. N.Y. Acad. Sci., 58:* 109–138.

Price, J. L. (1973). An autoradiographic study of complementary laminar patterns of termination of afferent fibers to the olfactory cortex. *J. Comp. Neurol., 150:* 87–108.

Price, J. L. (1981). Toward a consistent terminology for the amygdaloid complex. In: *The Amygdaloid Complex*, Y. Ben-Ari (ed.), Elsevier/North Holland Biomedical Press, Amsterdam, pp. 13–18.

Price, J. L. (1987). The central olfactory and accessory olfactory systems. In: *Neurobiology of Taste and Smell*, T. E. Finger and W. L. Silver (eds.), Wiley & Sons, New York, pp. 179–203.

Price, J. L., and Powell, T. P. S. (1970). An experimental study of the origin and the course of the centrifugal fibers to the olfactory bulb in the rat. *J. Anat., 107:* 215–237.

Price, J. L., F. T. Russchen, and D. G. Amaral (1987). The limbic region. II. The amygdaloid complex. In: *Handbook of Chemical Neuroanatomy, Vol. 5: Integrated Systems of the CNS, Part I*, A. Björklund, T.

Hökfelt and L. W. Swanson (eds.), Elsevier, Amsterdam, pp. 279–388.

Price, J. L., and B. M. Slotnik (1983). Dual olfactory representation in the rat thalamus: An anatomical and electrophysiological study. *J. Comp. Neurol., 215:* 63–77.

Raisman, G. (1966). The connexions of the septum. *Brain, 89:* 317–348.

Reid, J. M., D. G. Gwyn, and B. A. Flumerfelt (1975). A cytoarchitectonic and golgi study of the red nucleus in the rat. *J. Comp. Neurol., 162:* 337–362.

Renehan, W. E., M. F. Jacquin, R. D. Mooneh, and R. W. Rhoades (1986). Structure-function relationships in rat medullary and cervical dorsal horns. II. medullary dorsal horn cells. *J. Neurophysiol., 55:* 1187–1201.

Rexed, B., and A. Brodal (1951). The nucleus cervicalis lateralis – a spinocerebellar relay nucleus. *J. Neurophysiol., 14:* 399–407.

Rioch, D. M. (1929). Studies on the diencephalon of *Carnivora.* Part II: Certain nuclear configurations and fiber connections of the subthalamus and the midbrain of the dog and cat. *J. Comp. Neurol., 49:* 121–153.

Rokx, J. T. M., P. J. W. Juch, and J. D. van Willigen (1986). Arrangement and connections of mesencephalic trigeminal neurons in the rat. *Acta Anat., 127:* 7–15.

Rose, J. E. (1939). The cell structure of the mammillary body in the mammals and in man. *J. Anat., 74:* 91–115.

Rose, J. E. (1942a). The ontogenetic development of the rabbit's diencephalon. *J. Comp. Neurol., 77:* 61–129.

Rose, J. E. (1942b). The thalamus of the sheep: Cellular and fibrous structure and comparison with pig, rabbit and cat. *J. Comp. Neurol., 77:* 469–523.

Rose, J. E., and V. B. Mountcastle (1952). The thalamic tactile region in rabbit and cat. *J. Comp. Neurol., 97:* 441–490.

Rose, J. E., and C. N. Woolsey (1943). A study of thalamocortical relations in the rabbit. *Bull. Johns Hopkins Hosp., 73:* 65–128.

Rose, J. E., and C. N. Woolsey (1948). Structure and relations of limbic cortex and anterior thalamic nuclei in rabbit and cat. *J. Comp. Neurol., 89:* 279–347.

Rose, J. E., and C. N. Woolsey (1958). Cortical connections and functional organization of the thalamic auditory system of the cat. In: *Biological and Biochemical Bases of Behavior,* H. F. Harlow and C. N. Woolsey (eds.), University of Wisconsin Press, Madison, pp. 127–150.

Rose, M. (1935). Das Zwischenhirn des Kaninchens. *Mem. Acad. Pol. Sci. (Ser. B) No. 8,* pp. 1–108 and 25 plates.

Ross, C. A., D. A. Ruggiero, T. H. Joh, D. H. Park, and D. J. Reis (1984). Rostral ventrolateral medulla: Selective projections to the thoracic autonomic cell column from the region containing C1 neurons. *J. Comp. Neurol., 228:* 168–185.

Roste, G. K. (1989). Observations on the projection from the perihypoglossal nuclei to the cerebellar cortex and nuclei in the cat. *Anat. Embryol., 180:* 521–533.

Ruggiero, D. A., C. A. Ross, M. Anwar, D. H. Park, T. H. Joh, and D. J. Reis (1985). Distribution of neurons containing phenylethanolamine N-methyltransferase in medulla and hypothalamus of rat. *J. Comp. Neurol., 239:* 127–154.

Rutherford, J. G., and D. G. Gwyn (1982). A light and electron microscopic study of the interstitial nucleus of Cajal in rat. *J. Comp. Neurol., 205:* 327–340.

Rye, D. B., C. B. Saper, J. H. Lee, and B. H. Wainer (1987). Pedunculopontine tegmental nucleus of the rat: Cytoarchitecture, cytochemistry, and some extrapyramidal connections of the mesopontine tegmentum. *J. Comp. Neurol., 259:* 483–528.

Sadjadpour, K., and A. Brodal (1968). The vestibular nuclei in man. A morphological study in the light of experimental findings in the cat. *J. Hirnforsch., 10:* 299–323.

Saper, C. B. (1984). Organization of cerebral cortical afferent systems in the rat. II. Magnocellular basal nucleus. *J. Comp. Neurol., 222:* 313–342.

Saper, C. B., and A. D. Loewy (1980). Efferent connections of the parabrachial nucleus in the rat. *Brain Res., 197:* 291–317.

Saper, C. B., L. W. Swanson, and W. M. Cowan (1978). The efferent connections of the anterior hypothalamic area of the rat, cat and monkey. *J. Comp. Neurol., 182:* 575–600.

Scalia, F. (1972). The termination of retinal axons in the pretectal region of mammals. *J. Comp. Neurol., 145:* 223–258.

Scalia, F., and S. S. Winans (1975). The differential projections of the olfactory bulb and accessory olfactory bulb in mammals. *J. Comp. Neurol., 161:* 31–56.

289

Schober, W. (1986). The rat cortex in stereotaxic coordinates. *J. Hirnforsch., 27:* 121–143.

Segovia, S., and A. Guillamon (1993). Sexual dimorphism in the vomeronasal pathway and sex differences in reproductive behaviors. *Brain Res. Rev., 18:* 51–74.

Senba, E., P. E. Daddona, and J. I. Nagy (1987). A subpopulation of preganglionic parasympathetic neurons in the rat contain adenosine deaminase. *Neuroscience, 20:* 487–502.

Shibata, H. (1987). Ascending projections to the mammillary nuclei in the rat: A study using retrograde and anterograde transport of wheat germ agglutinin conjugated to horseradish peroxidase. *J. Comp. Neurol., 264:* 205–215.

Shintani, Y. K. (1959). The nuclei of the pretectal region of the mouse brain. *J. Comp. Neurol., 113:* 43–60.

Simerly, R. B., C. Chang, M. Muramatsu, and L. W. Swanson (1990). Distribution of androgen and estrogen receptor mRNA-containing cells in the rat brain: An *in situ* hybridization study. *J. Comp. Neurol., 294:* 76–95.

Simerly, R. B., and L. W. Swanson (1987). The distribution of neurotransmitter-specific cells and fibers in the anteroventral periventricular nucleus: Implications for the control of gonadotropin secretion in the rat. *Brain Res., 400:* 11–34.

Simerly, R. B., L. W. Swanson, and R. A. Gorski (1984). Demonstration of a sexual dimorphism in the distribution of serotonin-immunoreactive fibers in the medial preoptic nucleus of the rat. *J. Comp. Neurol., 225:* 151–166.

Siminoff, R., H. O. Schwassmann, and L. Kruger (1967). Unit analysis of the pretectal nuclear group in the rat. *J. Comp. Neurol., 130:* 329–342.

Slotnick, B. M., and D. L. Brown (1980). Variability in the stereotaxic position of cerebral points in the albino rat. *Brain Res. Bull., 5:* 135–139.

Slotnick, B. M., and S. Hersch (1980). A stereotaxic atlas of the rat olfactory system. *Brain Res. Bull., 5* (Suppl. 5): 1–55.

Slotnick, B. M., and C. M. Leonard (1975). *A Stereotaxic Atlas of the Albino Mouse Forebrain*, U.S. Department of Health, Education and Welfare, Rockville, MD, 174 pp.

Sofroniew, M. V., F. Eckenstein, H. Thoenen, and A. C. Cuello (1982). Topography of choline acetyltransferase-containing neurons in the forebrain of the rat. *Neurosci. Lett., 33:* 7–12.

Spencer, S. E., W. B. Sawyer, H. Wada, K. B. Platt, and A. D. Loewy (1990). CNS projections to the pterygopalatine parasympathetic preganglionic neurons in the rat: A retrograde transneuronal viral cell body labeling study. *Brain Res., 534:* 149–169.

Steinbusch, H. W. M., and R. Nieuwenhuys (1983). The raphé nuclei of the rat brainstem: A cytoarchitectonic and immunohistochemical study. In: *Chemical Neuroanatomy*, P.C. Emson (ed.), Raven Press, New York, pp. 131–207.

Sun, M.-K., B. S. Young, J. T. Hackett, and P. G. Guyenet (1988). Rostral ventrolateral medullary neurons with intrinsic pacemaker properties are not catecholaminergic. *Brain Res., 451:* 345–349.

Sutin, J. (1966). The periventricular stratum of the hypothalamus. *Int. Rev. Neurobiol., 9:* 263–300.

Swanson, L. W. (1976). An autoradiographic study of the efferent connections of the preoptic region in the rat. *J. Comp. Neurol., 167:* 227–256.

Swanson, L. W. (1982). The projections of the ventral tegmental area and adjacent regions: A combined fluorescent retrograde tracer and immunofluorescence study in the rat. *Brain Res. Bull., 9:* 321–353.

Swanson, L. W. (1987). The hypothalamus. In: *Handbook of Chemical Neuroanatomy, Vol. 5, Integrated Systems of the CNS, Part I*, A. Björklund, T. Hökfelt, and L. W. Swanson (eds.), Elsevier, Amsterdam, pp. 1–124.

Swanson, L. W. (1992). *Brain Maps: Structure of the Rat Brain*, Elsevier, Amsterdam, 240 pp.

Swanson, L. W., and W. M. Cowan (1979). The connections of the septal region in the rat. *J. Comp. Neurol., 186:* 621–656.

Swanson, L. W., G. J. Mogenson, C. R. Gerfen, and P. Robinson (1984). Evidence for a projection from the lateral preoptic area and substantia innominata to the 'mesencephalic locomotor region' in the rat. *Brain Res., 295:* 161–178.

Székely, G., and C. Matesz (1982). The accessory motor nuclei of the trigeminal, facial, and abducens nerves in the rat. *J. Comp. Neurol., 210:* 258–264.

Taber, E., A. Brodal, and F. Walberg (1960). The raphé nuclei of the brain stem in the cat. I. Normal topography and cytoarchitecture and general discussion. *J. Comp. Neurol., 114:* 161–187.

Toga, A. W., M. Samaie, and B. A. Payne (1989). Digital rat brain: A computerized atlas. *Brain Res. Bull., 22:* 323–333.

Tohyama, M., K. Satoh, T. Sakumoto, Y. Kimoto, Y. Takahashi, K. Yamamato, and T. Itakura (1978). Organization and projections of the neurons in the dorsal tegmental area of the rat. *J. Hirnforsch.* 19: 165–176.

Torigoe, Y., R. H. I. Blanks and W. Precht (1986). Anatomical studies on the nucleus reticularis tegmenti pontis in the pigmented rat. I. Cytoarchitecture, topography, and cerebral cortical afferents. *J. Comp. Neurol., 243*: 71–87.

Törk, I. (1985). Raphé nuclei and serotonin containing systems. In: *The Rat Nervous System, Vol. 2: Hindbrain and Spinal Cord,* G. Paxinos (ed.), Academic Press, New York, pp. 43–78.

Torvik, A. (1956). Afferent connections to the sensory trigeminal nuclei, the nucleus of the solitary tract and adjacent structures: An experimental study in the rat. *J. Comp. Neurol., 106*: 51–142.

Tsang, Y. C. (1940). Supra- and post-optic commissures in the brain of the rat. *J. Comp. Neurol., 72*: 535–567.

Tucker, D. C., C. B. Saper, D. A. Ruggiero, and D. J. Reis (1987). Organization of central adrenergic pathways: I. Relationships of ventrolateral medullary projections to the hypothalamus and spinal cord. *J. Comp. Neurol., 259*: 591–603.

Valverde, F. (1962). Reticular formation of the albino rat's brain stem. Cytoarchitecture and corticofugal connections. *J. Comp. Neurol., 119*: 25–53.

Valverde, F., M. V. Facal-Valverde, M. Santacana, and M. Heredia (1989). Development and differentiation of early generated cells of sublayer VIb in the somatosensory cortex of the rat: A correlated Golgi and autoradiographic study. *J. Comp. Neurol., 290*: 118–140.

Van Houten, M., and J. R. Brawer (1978). Cytology of neurons in the hypothalamic ventromedial nucleus in the adult male rat. *J. Comp. Neurol., 178*: 89–116.

Veazey, R. B., and C. M. Severin (1980). Efferent projections of the deep mesencephalic nucleus (*Pars Lateralis*) in the rat. *J. Comp. Neurol., 190*: 231–244.

Vogt, B. A., and A. Peters (1981). Form and distribution of neurons in rat cingulate cortex: Areas 32, 24, and 29. *J. Comp. Neurol., 195*: 605–625.

Voogd, J., N. M. Gerritts, and E. Marani (1985). Cerebellum. In: *The Rat Nervous System, Vol. 2: Hindbrain and Spinal Cord,* G. Paxinos (ed.), Academic Press, New York, pp. 251–291.

Watanabe, K., and E. Kawana (1982). The accessory nucleus of Luys in the rat. *Okajimas Folia Anat. Jpn., 58*: 859–873.

Whitehead, M. C. (1988). Neuronal architecture of the nucleus of the solitary tract in the hamster. *J. Comp. Neurol., 220*: 378–395.

Winer, J. A., and D. T. Larue (1987). Patterns of reciprocity in auditory thalamocortical and corticothalamic connections: Study with horseradish peroxidase and autoradiographic methods in the rat medial geniculate body. *J. Comp. Neurol., 257*: 282–315.

Winer, J. A., D. K. Morest, and I. T. Diamond (1988). A cytoarchitectonic atlas of the medial geniculate body of the opossum, *Didelphys virginiana,* with a comment on the posterior intralaminar nuclei of the thalamus. *J. Comp. Neurol., 274*: 422–448.

Wünscher, W., W. Schober, and L. Werner (1965). *Arkitektonischer Atlas vom Hirnstamm der Ratte,* S. Hirzel, Leipzig, 61 pp.

Wyss, J. M., and K. Sripanidkulchai (1983). The indusium griseum and anterior hippocampal continuation in the rat. *J. Comp. Neurol., 219*: 251–272.

Yamaguchi, K., A. Morimoto, and N. Murakami (1993). Organum vasculosum laminae terminalis (OVLT) in rabbit and rat: Topographic studies. *J. Comp. Neurol., 330*: 352–362.

Zilles, K., and A. Wree (1985). Cortex: Areal and laminar structure. In: *The Rat Nervous System, Vol. 1: Forebrain and Midbrain,* G. Paxinos (ed.), Academic Press, Orlando, FL, pp. 375–415.

INDEX

Commentaries concerning various structures located on the page opposite plate numbers are indicated in **bold**.

293